D1748717

Global Change – The IGBP Series

Springer
Berlin
Heidelberg
New York
Hong Kong
London
Milan
Paris
Tokyo

Will Steffen · Jill Jäger · David J. Carson · Clare Bradshaw (Eds.)

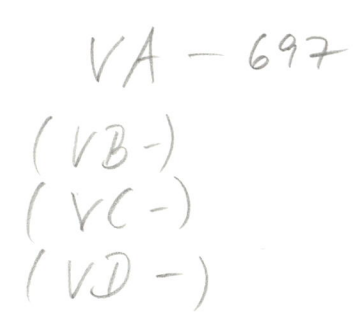

Challenges of a Changing Earth

Proceedings of the Global Change
Open Science Conference,
Amsterdam, The Netherlands, 10–13 July 2001

With 101 Figures and 7 Tables

Springer

Editors

Dr. Will Steffen

Executive Director, International Geosphere-Biosphere Programme
The Royal Swedish Academy of Sciences
Box 50005, SE 10405 Stockholm, Sweden
will@igbp.kva.se

Dr. Jill Jäger

Executive Director, International Human Dimensions Programme on Global Environmental Change
Walter-Flex-Strasse 3, D-53113 Bonn, Germany
fuj.jaeger@nextra.at

Dr. David J. Carson

Director, World Climate Research Programme, World Meteorological Organization
Case Postale No. 2300, 7 Bis Avenue de la Paix, 1211 Geneva 2, Switzerland
carson_d@gateway.wmo.ch

Dr. Clare Bradshaw

International Geosphere-Biosphere Programme, The Royal Swedish Academy of Sciences
Box 50005, SE 10405 Stockholm, Sweden
clare.bradshaw@igbp.kva.se

ISSN 1619-2435
ISBN 3-540-43308-2   Springer-Verlag Berlin Heidelberg New York

Die Deutsche Biliothek - CIP-Einheitsaufnahme

Challenges of a changing earth : proceedings of the Global Change Open
Science Conference, Amsterdam, The Netherlands, 10 - 13 July 2001 ;
with 7 tables / Will Steffen ... (ed.). - Berlin ; Heidelberg ; New York ;
Barcelona ; Hong Kong ; London ; Milan ; Paris ; Tokyo : Springer, 2002
  (Global change - the IGBP series)
   ISBN 3-540-43308-2

This work is subject to copyright. All rights are reserved, whether the whole or part of the material is concerned, specifically the rights of translation, reprinting, reuse of illustrations, recitation, broadcasting, reproduction on microfilms or in any other way, and storage in data banks. Duplication of this publication or parts thereof is permitted only under the provisions of the German Copyright Law of September 9, 1965, in its current version, and permission for use must always be obtained from Springer-Verlag. Violations are liable for prosecution under the German Copyright Law.

Springer-Verlag Berlin Heidelberg New York
a member of BertelsmannSpringer Science+Business Media GmbH
http://www.springer.de
© Springer-Verlag Berlin Heidelberg 2002
Printed in Germany

The use of general descriptive names, registered names, trademarks, etc. in this publication does not imply, even in the absence of a specific statement, that such names are exempt from the relevant protective laws and regulations and therefore free for general use.

Cover Design: Erich Kirchner, Heidelberg
Cover Photographs: *Cars* by IGBP, *Clouds* by CLIVAR, *Coral* by Terence Done, *Fire* by Michael F. Ryan
Dataconversion: Büro Stasch, Bayreuth

SPIN: 10867941   3130 – 5 4 3 2 1 0 –   Printed on acid-free paper

Preface

This volume is based on plenary presentations from Challenges of a Changing Earth, a Global Change Open Science Conference held in Amsterdam, The Netherlands, in July 2001. The meeting brought together about 1 400 scientists from 105 countries around the world to describe, discuss and debate the latest scientific understanding of natural and human-driven changes to our planet. It examined the effects of these changes on our societies and our lives, and explored what the future might hold.

The presentations drew upon global change science from an exceptionally wide range of disciplines and approaches. Issues of societal importance – the food system, air quality, the carbon cycle, and water resources – were highlighted from both policy and science perspectives. Many of the talks presented the exciting scientific advances of the past decade of international research on global change. Several challenged the scientific community in the future. What are the visionary and creative new approaches needed for studying a complex planetary system in which human activities are intimately interwoven with natural processes?

This volume aims to capture the timeliness and excitement of the science presented in Amsterdam. The plenary speakers were given a daunting task: to reproduce their presentations in a way that delivers their scientific messages accurately and in sufficient detail but at the same time reaches a very broad audience well beyond their own disciplines. Furthermore, they were required to do this in just a few pages. The result, contained in the following pages, is impressive.

The science presented in this volume relies, in large part, on research carried out under the auspices of the three international global change programmes that sponsored the Global Change Open Science Conference – IGBP (International Geosphere-Biosphere Programme), IHDP (International Human Dimensions Programme on Global Environmental Change) and WCRP (World Climate Research Programme) – and on DIVERSITAS, an international programme of biodiversity science. We thank the thousands of scientists around the world who contribute to these programmes and thus provide the rich array of research that forms the basis for our improved understanding of global change.

Several people provided invaluable assistance in the preparation of this volume. Susannah Eliott inspired us with the original idea for the book and encouraged us to proceed, John Bellamy transformed many of the figures into publication quality, Paola Fastmark carried out some of the initial editing of the manuscripts, and Luisa Tonarelli provided much help and advice on the technical production of the volume.

Without the financial support to the Global Change Open Science Conference itself, this book would not have been possible. We are most grateful to the following agencies and organisations for their support: The Netherlands Organization for Scientific Research (NWO), The Dutch Ministry of Education, Culture and Science, The Netherlands National Programme on Global Air Pollution and Climate Change, The Royal Netherlands Academy of Arts and Sciences, the US Global Change Research Program, the European Commission, the UK Natural Environment Research Council, the Natural Sciences and Engineering Research Council of Canada, the Climate

Change Action Fund – Canada, the Office for Scientific, Technical and Cultural Affairs – Belgium, the International Council for Science (ICSU), and the Centre National d' Etudes Spatiales (CNES) – France.

Will Steffen   *Jill Jäger*   *David J. Carson*   *Clare Bradshaw*
Stockholm   Bonn   Geneva   Stockholm

Contents

Part I Opening 1

1 Opening Address 3

2 Challenges of a Changing Earth 7
    References 17

Part II Achievements and Challenges 19

Part IIa Food, Land, Water, and Oceans 20

3 Toward Integrated Land-Change Science: Advances in 1.5 Decades of Sustained International Research on Land-Use and Land-Cover Change 21
  3.1 Trends 21
  3.2 Causes 23
  3.3 Model-Methods 24
  3.4 Summary and Observations 25
      References 25

4 Climate Variability and Ocean Ecosystem Dynamics: Implications for Sustainability 27
  4.1 Introduction 27
  4.2 Climate Effects on Marine Ecosystems 28
  4.3 Implications for Sustainability 29
      References 29

5 Food in the 21st Century: Global Climate of Disparities 31
  5.1 Food in the 21st Century: Global Climate of Disparities 31
  5.2 The Critical Role of Knowledge 31
  5.3 Environment and Sustainable Development 32
  5.4 Global Environmental Change 32
  5.5 Global Agro-Ecological Assessment 33
  5.6 Global AEZ Findings 34
  5.7 Global Warming and Climate Change 34
  5.8 Impact of Climate Change on Worldwide Cereal Production 35
  5.9 Food Security and Climate Change 37
  5.10 Climate Change Impact: Fairness and Equity? 38
  5.11 Concluding Remark 38
      References 38

6 Equity Dimensions of Dam-Based Water Resources Development: Winners and Losers 39
  6.1 Introduction 39
  6.2 Why is Equity Relevant in River Basin Development Contexts? An Illustration from the Senegal River 39

6.3	Role of Dams in Water Resources Allocation	40
6.4	Addressing Equity Dimensions of Dam-Based Water Resources Development	41
	6.4.1 The Concept of Equity	41
	6.4.2 Ideas for Improving the Equity Performance of Dam Projects	41
6.5	Conclusions	42
	References	43

Part IIb Out of Breath: Air Quality in the 21st Century ... 44

7 Atmospheric Chemistry in the "Anthropocene" ... 45

8 Fires, Haze and Acid Rain:
The Social and Political Framework of Air Pollution in ASEAN and Asia ... 49

8.0	Introduction: Air Pollution and Asia in Context	49
8.1	Part 1: The Problem in Perspective	49
	8.1.1 The Haze: Summary of a Recurring Disaster	49
	8.1.2 Northeast Asian Acid Rain: The Smog of Growth	50
8.2	Part 2: International Principles and Practice	51
	8.2.1 International Law and Practice in Other Regions	51
	8.2.2 The ASEAN Way and Environmental Cooperation	51
	8.2.3 Cooperation in Northeast Asia	52
8.3	Part 3: Assessing and Improving Regional Cooperation on the Environment	52
	8.3.1 Assessment and Prospects for Improvement	52
	8.3.2 How to Improve ASEAN?	53
	8.3.3 Northeast Asian Cooperation	53
8.4	Conclusion: Strengthening the Social and Political Frameworks	54
	General References	55
	Specific References	55

Part IIc Managing Planetary Metabolism? The Global Carbon Cycle ... 56

9 Carbon and the Science-Policy Nexus: The Kyoto Challenge ... 57

9.1	The Global Carbon Cycle	58
9.2	Human Perturbation of the Carbon Cycle	58
9.3	The Terrestrial Sink	59
9.4	Location of the Terrestrial Sink	59
9.5	Variability of the Terrestrial Sink	60
9.6	Origin of the Terrestrial Sink	60
9.7	The Future of the Terrestrial Carbon Cycle	60
	References	64

10 Industry Response to the CO_2 Challenge ... 65

10.1	Introduction	65
10.2	The Energy Context	65
10.3	The Industry Response	66
	10.3.1 Reducing GHG Emissions	66
	10.3.2 Future Innovations	67
	10.3.3 Renewable Energy	68
	10.3.4 Transport Fuels	69
10.4	Relations between Business and the Scientific Community	69
10.5	Summary of the Key Points	71

Part IId Summary: Global Change and the Challenge for the Future ... 72

11 Global Change and the Challenge for the Future ... 73

Part III Advances in Understanding 75

Part IIIa Global Biogeochemistry:
Understanding the Metabolic System of the Planet 76

12   Ocean Biogeochemistry: A Sea of Change 77
12.1   Introduction 77
12.2   The Oceanic Carbon Cycle 77
12.3   Ocean Time-Series Programmes 78
12.4   Summary 80
       References 80

13   The Past, Present and Future of Carbon on Land 81
13.1   What Controls the Behaviour of Biospheric Carbon? 81
13.2   The Carbon Cycle is Constrained by Other Elemental Cycles .. 82
13.3   The Nitrogen-Carbon Link 82
13.4   Do Land Ecosystems Retain Nitrogen Deposited From the Air? .. 83
13.5   The Phosphorus-Nitrogen Link 83
13.6   The Carbon Cycle During the Past 420 Thousand Years 83
       References 85

**14   Can New Institutions Solve Atmospheric Problems?
       Confronting Acid Rain, Ozone Depletion and Climate Change** 87
14.1   Does the Regime Have Appropriate Behavioural Mechanisms? .. 88
14.2   Has the Regime Given Rise to a Robust Social Practice? 89
14.3   Does the Regime Have a Sensitive Steering System? 90
14.4   Implications for the Climate Regime 91
       References 91

Part IIIb Land-Ocean Interactions: Regional-Global Linkages 92

**15   Emissions from the Oceans to the Atmosphere, Deposition from
       the Atmosphere to the Oceans and the Interactions Between Them** .. 93
15.1   Atmospheric Inputs to the Oceans 93
       15.1.1 Nitrogen 93
       15.1.2 Iron 94
15.2   Emissions from the Ocean to the Atmosphere 94
15.3   Interacting Cycles – the Challenge for the Future 95
       References 96
       Bibliography 96

16   The Impact of Dams on Fisheries: Case of the Three Gorges Dam .. 97
16.1   The Effect of Dams on Deltas and Estuaries 97
16.2   The Case of the Three Gorges Dam 98
16.3   Threat to Other Shelves 98
       References 99

17   Global Change in the Coastal Zone: The Case of Southeast Asia .. 101
17.1   Where is the Coastal Zone and Why Southeast Asia? 101
17.2   Pressures on the Coastal Zone:
       Population and a Resource-Dependent Economy 101
       References 104

Part IIIc The Climate System: Prediction, Change and Variability 106

18   Climate Change Fore and Aft: Where on Earth Are We Going? .. 107

19 Climate Change – Past, Present and Future: A Personal Perspective 109
19.1 Lessons from the Past 110
19.2 Global Temperature Change 110
19.3 The Future in Perspective of the Past 111
       References 112

20 The Changing Cryosphere: Impacts of Global Warming in the High Latitudes 113
20.1 What Is the Cryosphere and How Does It Contribute to Global Climate Change? 113
20.2 What Is the Empirical Evidence of Climate Change in the High Latitudes? 113
20.3 Impacts of Global Warming in the High Latitudes 114

21 The Coupled Climate System: Variability and Predictability 117
21.1 Introduction 117
21.2 Seasonal to Interannual Climate Variability 117
21.3 Decadal Climate Variability 118
21.4 Detection and Attribution of Climate Change 120

Part IIId Hot Spots of Land-Use Change and the Climate System: A Regional or Global Concern 122

22 Hot Spots of Land-Use Change and the Climate System: A Regional or Global Concern? 123
       References 124

23 Africa: Greening of the Sahara 125
23.1 Africa: A Hot Spot of Nonlinear Atmosphere-Vegetation Interaction 125
23.2 The African Wet Period 126
23.3 Abrupt Changes in North Africa 126
23.4 Will North Africa Become Green Again in the Near Future? 127
23.5 Outlook 128
       Acknowledgements 128
       References 128

24 The Role of Large-Scale Vegetation and Land Use in the Water Cycle and Climate in Monsoon Asia 129
24.1 The Asian Monsoon As a Huge Water Cycling System 129
24.2 Atmospheric Water Cycle over Monsoon Asia and the Eurasian Continent 129
24.3 Is Monsoon Rainfall Decreasing? The Impact of Deforestation on the Water Cycle in Thailand 129
24.4 Do Water-Fed Rice Paddy Fields Increase Rainfall in Monsoon Asia? 130
24.5 A Hydro-Climate Memory Effect of the Taiga-Permafrost System in Siberia 131
       References 132

25 Can Human-Induced Land-Cover Change Modify the Monsoon System? 133
25.1 History of Land-Cover and Land-Use Changes over East Asia ... 133
25.2 Design of the Numerical Experiments 133
25.3 Changes of Surface Dynamic Parameters under Two Types of Vegetation Cover 135
25.4 Changes of East Asian Monsoon by Human-Induced Land-Cover Changes ... 135
25.5 Conclusions 136
       References 136

26	**The Amazon Basin and Land-Cover Change: A Future in the Balance?**	**137**
26.1	Land-Use Change	137
26.2	Impacts on the Carbon Cycle	138
26.3	Impacts on the Atmosphere	139
26.4	Impacts on Water Chemistry	140
26.5	The Future	140
	Acknowledgements	140
	References	140

Part IV Looking to the Future 143

Part IVa Simulating and Observing the Earth System 144

27	**Virtual Realities of the Past, Present and Future**	**145**
27.1	What Are General Circulation Models (GCMs)?	145
27.2	How Do We Use Climate Models?	145
27.3	How Well Do Models Simulate Present Climate?	146
27.4	How Well Do Models Predict Past Climate Change?	146
27.5	What Are the Predictions for the Future?	147
27.6	What Developments Are Likely in the Future?	148
27.7	How Can We Deal with Uncertainty in Predictions?	148
27.8	Concluding Remarks	149
	References	149

28	**Coping with Earth System Complexity and Irregularity**	**151**
28.1	The Challenge	151
28.2	Great Cognitive Barriers	151
28.3	Breaches and Bypasses to Understanding	152
28.4	Adaptive Planetary Stewardship	156
	References	156

29	**Simulating and Observing the Earth System: Summary**	**157**
29.1	Earth System	157
29.2	Simulators	158
29.3	Observations	158
29.4	The Way Forward	158

Part IVb Does the Earth System Need Biodiversity? 160

30	**Marine Biodiversity: Why We Need It in Earth System Science**	**161**
	Reference	163

31	**Does Biodiversity Matter to Terrestrial Ecosystem Processes and Services?**	**165**
31.1	Is Biodiversity Important to the Functioning of the Earth System?	165
31.2	What is Biodiversity?	165
31.3	The Most Abundant Plants Are Important for Ecosystem Functioning	165
31.4	The Number of Functionally Similar Species Is Important in Facing Environmental Change	166
31.5	Implications for Conservation and Sustainable Management	166

32	**Biodiversity Loss and the Maintenance of Our Life-Support System**	**169**
32.1	How Does Biodiversity Affect Ecosystem Functioning at Small Scales?	169
32.2	Scaling Up in Time: Biodiversity As Insurance Against Environmental Changes	171
32.3	Scaling Up in Space: Biodiversity Effects at Landscape and Regional Scales	172
32.4	Conclusions	172
	References	172

Part IVc Can Technology Spare the Planet? 174

**33 Maglevs and the Vision of St. Hubert –
Or the Great Restoration of Nature: Why and How** 175
33.1 Introduction 175
33.2 The Vision of St. Hubert 175
33.3 Our Triune Brain 175
33.4 Sparing Sea Life 177
33.5 Sparing Farmland 177
33.6 Sparing Forests 179
33.7 Sparing Pavement 180
33.8 Cardinal Resolutions 181
References 182

**34 Industrial Transformation:
Exploring System Change in Production and Consumption** 183
34.1 Human Choice on Issues Involving a Time Scale of Decades to Centuries 183
34.2 Transformation: Why, How Does It Work and What Are the Options 184
34.3 From Green Products to System Transformation 185
34.4 Five Major Foci for Transformation Research 186
34.5 International Cooperation 186
34.6 In Summary 187
References 188

35 Will Technology Spare the Planet? 189
References 191

Part IVd Towards Global Sustainability 192

36 Challenges and Road Blocks for Local and Global Sustainability 193
36.1 Advances 193
36.2 International Instruments: Urgency of Goals and Synergies 193
36.3 The Biodiversity Convention 193
36.4 The Climate Change Convention 194
36.5 Synergies 195
36.6 New Inputs 195

37 Research Systems for a Transition Toward Sustainability 197
Acknowledgements 199
References 199

38 Summary: Towards Global Sustainability 201

Part IVe Closing Session 202

39 Closing Address 203
39.1 Science and Policy 203
39.2 Prospects for the Climate Change Negotiations 205

40 The Amsterdam Declaration on Global Change 207

Index 209

Contributors

Oleg Anisimov

Head of Climatological Department, State Hydrological Institute
St.Petersburg, Russia
oleg@OA7661.spb.edu

Jesse H. Ausubel

Director, Program for the Human Environment, The Rockefeller University
1230 York Avenue, New York, NY 10021-6399, USA
http://phe.rockefeller.edu and ausubel@mail.rockefeller.edu

Raymond S. Bradley

Climate System Research Center, Department of Geosciences, University of Massachusetts
Amherst, MA 01003, USA
rbradley@geo.umass.edu

Victor Brovkin

Potsdam Institute for Climate Impact Research
PO Box 60 12 03, D-14412 Potsdam, Germany
brovkin@pik-potsdam.de

Antonio J. Busalacchi

Director and Professor, Earth System Science Interdisciplinary Center, University of Maryland
College Park, MD 20742-2425, USA
tonyb@essic.umd.edu

Julia Carabias Lillo

SEMARNAP
Blvd. Adolfo Ruiz Cortines #4209, Jardines en la Montaña
Tlalpan, C.P. 14210, México D.F.
jcarabias@miranda.ecologia.unam.mx

Chen-Tung Arthur Chen

Institute of Marine Geology and Chemistry, National Sun Yat-Sen University
Kaohsiung, Taiwan, Republic of China
ctchen@mail.nsysu.edu.tw

William C. Clark

Belfer Center for Science and International Affairs, John F. Kennedy School of Government, Harvard University
79 Kennedy Street, Cambridge, MA 02138, USA
William_Clark@harvard.edu

Martin Claussen

Potsdam Institute for Climate Impact Research
PO Box 60 12 03, D-14412 Potsdam, Germany
claussen@pik-potsdam.de

Contributors

Paul Crutzen

Max-Planck-Institute for Chemistry
PO Box 3060, D-55020 Mainz, Germany
air@mpch-mainz.mpg.de

Sandra Díaz

Multidisciplinary Institute of Plant Biology (IMBIV)
and Department of Biodiversity and Ecology
National University of Córdoba and Argentine National Research Council
Córdoba, Argentina
sdiaz@com.uncor.edu

Michael J. Fogarty

National Oceanic and Atmospheric Administration, National Marine Fisheries Service
Woods Hole, MA 02543, USA
michael.fogarty@noaa.gov

Congbin Fu

START Regional Center for Temperate East Asia, Institute of Atmospheric Physics
Chinese Academy of Sciences
Beijing, 100029, China
fcb@ast590.tea.ac.cn

Andrey Ganopolski

Potsdam Institute for Climate Impact Research
PO Box 60 12 03, D-14412 Potsdam, Germany
ganopolski@pik-potsdam.de

Jill Jäger

Executive Director, International Human Dimensions Programme on Global Environmental Change
Walter-Flex-Strasse 3, D-53113 Bonn, Germany
fuj.jaeger@nextra.at

Timothy Jickells

School of Environmental Sciences, University of East Anglia
Norwich NR4 7TJ, United Kingdom
T.Jickells@uea.ac.uk

Pavel Kabat

Dept. of Water & the Environment, ALTERRA Green World Research
Wageningen University & Research Centre
P.O. Box 49, NL-6700 AA Wageningen, The Netherlands
p.kabat@alterra.wag-ur.nl

David M. Karl

University of Hawaii, School of Ocean and Earth Science and Technology
University of Hawaii
Honolulu, HI 96822, USA
dkarl@soest.hawaii.edu

Michel Loreau

Laboratoire d'Ecologie, UMR 7625, Ecole Normale Supérieure
46 rue d'Ulm, F-75230 Paris Cedex 05, France
Loreau@ens.fr

Liana Talaue-McManus

Marine Science Institute, University of the Philippines
Quezon City, Philippines
lmcmanus@epic.net
and Rosenstiel School of Marine and Atmospheric Science, University of Miami
Florida 33149, USA

John Mitchell

Hadley Centre for Climate Prediction and Research, UK Meteorological Office
Bracknell, United Kingdom
john.f.mitchell@metoffice.com

Berrien Moore III

Director, Institute for the Study of Earth, Oceans and Space (EOS), University of New Hampshire
39 College Road, 305 Morse Hall, Durham NH 03824-3524, USA
b.moore@unh.edu

Madiodio Niasse

Independent Consultant, B.P. 16911, Dakar-Fann, Senegal
maniasse@hotmail.com

Charles C. Nicholson

Group Senior Advisor, BP plc, International Headquarters, Britannic House
1 Finsbury Circus, London, EC2M 7BA, UK
nicholcc@bp.com

Ian R. Noble

CEO, CRC for Greenhouse Accounting, Research School of Biological Sciences
Australian National University
Canberra ACT 0200, Australia
Ian.Noble@greenhouse.crc.org.au

Carlos Nobre

Centro de Previsao de Tempo e Estudos Climaticos (CPTEC),
Instituto Nacional de Pesquisas Espaciais (INPE)
Rod. Presidente Dutra (km 40), 12630-000 Cachoeiro Paulista, SP, Brazil
nobre@cptec.inpe.br

Thomas F. Pedersen

Earth and Ocean Sciences, University of British Columbia
6270 University Boulevard, Vancouver, B.C., V6T 1Z4, Canada
pedersen@eos.ubc.ca

Jan Pronk

President of the sixth Conference of the Parties to the United Nations Framework Convention
on Climate Change, Minister, Ministry of Housing, Spatial Planning and Environment
The Hague, Netherlands

Chris Rapley

British Antarctic Survey, High Cross
Madingley Road, Cambridge CB3 0ET, United Kingdom
C.Rapley@bas.ac.uk

Katherine Richardson

Department of Marine Ecology, University of Aarhus
Finlandsgade 14, DK-8200 Aarhus N, Denmark
richardson@biology.au.dk

Hans Joachim Schellnhuber

Potsdam Institute for Climate Impact Research
PO Box 60 12 03, D-14412 Potsdam, Germany
john@pik-potsdam.de

Robert J. Scholes

CSIR Division of Water, Environment and Forest Technology
PO Box 395, Pretoria, 0001 South Africa
bscholes@csir.co.za

Mahendra Shah

International Institute for Applied Systems Analysis (IIASA)
A-2361 Laxenburg, Austria
shah@iiasa.ac.at

Will Steffen

International Geosphere-Biosphere Programme, The Royal Swedish Academy of Sciences
Box 50005, SE 10405 Stockholm, Sweden
will@igbp.kva.se

Simon Tay

Faculty of Law, National University of Singapore
10 Kent Ridge Crescent, Singapore 119260
lawtaysc@nus.edu.sg

Billie L. Turner II

Graduate School of Geography and George Perkins Marsh Institute, Clark University
950 Main Street, Worcester, MA 01610-1477, USA
BTurner@clarku.edu

Peter D. Tyson

Climatology Research Group, University of the Witwatersrand, Johannesburg
PO Wits, 2050, South Africa
141pdt@cosmos.wits.ac.za

Pier Vellinga

Institute for Environmental Studies (IVM)
De Boelelaan 1115, 1081 HV Amsterdam, The Netherlands
pier.vellinga@ivm.vu.nl

Robert T. Watson

Chair, Intergovernmental Panel on Climate Change, Chief Scientist, World Bank
Environmental Department, World Bank
1818 H. Street, NW, Washington DC 20433, USA
rwatson@worldbank.org

Tetsuzo Yasunari

Frontier Research System for Global Change (FRSGC), and
Hydrospheric-Atmospheric Research Center (HyARC),
Nagoya University, Nagoya 464-86-1, Japan, and
Adjunct Professor, Institute of Geoscinece, University of Tsukuba,
Ibaraki 305-8571, Japan
yasunari@ihas.nagoya-u.ac.jp

Oran R. Young

Professor of Environmental Studies, Institute of Arctic Studies, Dartmouth College
6214 Fairchild, Hanover, NH 03755, USA
oran.young@dartmouth.edu

Michael Zammit Cutajar

Executive Secretary, UNFCCC, (United Nations Framework Convention on Climate Change)
P.O. Box 260124, D-53153 Bonn, Germany
secretariat@unfccc.int

Part I
Opening

Chapter 1

Opening Address

Jan Pronk

Ladies and gentlemen:

It gives me great pleasure to be able to speak to you at this Global Change Open Science Conference on the Challenges of a Changing Earth.

Scientific Research and Climate Change Policy

Over the past decade, scientific research has taught us a lot about the causes and effects of climate change and its policy implications. Three things struck me in particular in the IPCC's (Intergovernmental Panel on Climate Change) Third Assessment Report. Firstly, we now know that in all human history, the climate has never changed as fast as it is changing today.

Secondly, what was merely a suspicion a decade ago is now a practical certainty: the greenhouse gases we produce are already having a visible impact on the environment. The effects of climate change are irreversible – for ecosystems, agriculture, water supply and health. The less we do and the longer we wait to tackle the roots of this problem, the more serious the effects and the greater the strain on the resilience of people, plants and animals. A recent study has shown that sea level rise resulting from climate change threatens the Wadden Sea, the ecosystem of the Frisian Islands. The Wadden Sea is a Dutch wetland of international importance, for migratory birds among others. Even more alarming is the fact that developing countries, which are least of all to blame for this predicament, will suffer the most devastating consequences. And on top of that, their economic resilience is lowest. The damage caused by climate change aggravates the socio-economic inequalities that already exist. The poorest people often live in the toughest locations in the world: the driest, least productive, and the most vulnerable. They are the first hit and the least able to defend themselves.

The third point that struck me in the Third Assessment Report has to do with solutions. We *can* tackle climate change, and we can afford to do so. Innovative instruments like emissions trading can bring down the costs considerably.

Over the past year, scientists have thus helped substantiate the urgency of climate change policy. Strong scientific arguments are essential, because the relationship between cause and effect is not as clear as it is with other environmental problems.

Scientists must find a clear way to convince the world what the causes are, and what effects we can expect to see. This is difficult when it comes to climate change, for three reasons:

1. The causes are difficult to explain: it is not clear what proportion of the damage is due to human activity.
2. The effects are not immediately visible: it will be thirty years before we can clearly identify significant changes to our climate.
3. The causes are difficult to prove: we can't compare the existing situation with how the world would have been without climate change. The only information we have are observations and analyses made over long periods in the past and projections for the future. Climate models are our only way of making such projections, and they are not one hundred per cent reliable.

It is therefore critical that we assess the results of scientific research – including all the uncertainties and controversies – as objectively and thoroughly as possible, and that we use the most up-to-date information available.

Political Developments

How politicians will respond to the scientific facts we now have at our disposal is not yet clear. You will no doubt have closely followed the recent developments in climate change politics. In late 2000, the sixth Conference of the Parties to the UN Framework Convention on Climate Change (UNFCCC) – otherwise known as COP6 – was held at The Hague. Decisions were to be taken on issues related to the Climate Change Convention and Kyoto Protocol. These decisions would enable ratification of the Kyoto Protocol by all Parties to the Climate Change Convention. After three years of technical negotiations since Kyoto, it was time to round off the remaining political issues.

On the basis of the negotiations and informal consultations with ministers, I drafted a compromise document during COP6. The document aimed at a balanced package covering the major issues at stake. Unfortunately, it proved impossible to reach a compromise. Ministers were greatly disappointed with the outcome, and decided to resume the talks in late July. The basis for further negotiations would be my document and the negotiating texts that were on the table.

Then the new US administration made headlines all over the world with the official announcement of its position against the Protocol. The US did however reaffirm that it remained committed to negotiations in the framework of the UNFCCC, which provides the basis for voluntary actions. If the governments were to follow the US, there would be little hope for the Kyoto Protocol.

To gain an idea of countries' willing to negotiate at COP6, two weeks ago I held an informal round of consultations in Scheveningen. To speed up progress, I issued a new negotiating text for that meeting. I asked the one hundred and twenty-seven countries attending to focus on the texts on the table, talk to each other and advise me on possible ways forward.

The main outcomes of Scheveningen are that:

- The Consolidated Negotiating Text will be used as a tool in the negotiations at the resumed session of COP6 in Bonn in July, and the agenda for the meeting has been set;
- My proposals were discussed, although this mainly involved reiterating old arguments. The major challenge in Bonn will be to find a solution to the problems of sinks, the financial issues and the compliance regime;
- The European Union, eastern Europe and Switzerland adopted a generally positive stance; the developing countries have a constructive attitude;
- The US declared it would make statements only on matters that affect the Climate Convention and on issues which might serve as a precedent for other UN negotiations that in itself can have far reaching implications; Canada and Australia took a rather negative position, thereby not reflecting recent political statements by their heads of government;
- Japan announced that it would be aiming to ratify the Kyoto Protocol in 2002, but insisted on US participation.

All in all, in Scheveningen we created a basis for the negotiations in Bonn this July. Most other countries want to reach a consensus at COP6, even if the United States does not participate in the Kyoto Protocol. But it is clearly not going to be easy.

Political Principles and Flexibility in Climate Policy

Climate experts are supplying politicians with more facts about the causes and effects of climate change. Needless to say, there are still doubts and uncertainties about the forecasts. After all, no one can look into the future. But the scientific community agrees that the risks are present and that they threaten the entire world. In the Climate Change Convention politicians also referred to the scientific doubts and uncertainties. They felt it would be irresponsible to wait for hard scientific proof before taking measures, and therefore undertook to adhere to the precautionary principle. The Climate Change Convention states that:

> The parties should take precautionary measures to anticipate, prevent or minimise the causes of climate change and mitigate its adverse effects. Where there are threats of serious and irreversible damage, lack of full scientific certainty should not be used as a reason for postponing such measures.

The Climate Change Convention also enshrines a number of other important principles. The principle of 'common but differentiated responsibilities', is an example. We cannot expect developing countries to do as much to fight climate change as industrialised countries. Therefore another important principle is laid down in the Convention: developed countries should lead the way in the battle against climate change. They, after all, are responsible for the lion's share of emissions. All Parties involved in the climate change negotiations accept these basic elements of the convention, although contrary to the principles which underpins the Kyoto Protocol there is international pressure on developing countries to take on commitments now.

Of course we can be flexible in the way we implement climate policy. The Kyoto Protocol already contains important flexibility features such as the Kyoto mechanisms, the five-year commitment period, six greenhouse gases, and the use of sinks. There is also a great deal of room for negotiation on how we put the protocol into practice and fulfil our commitments under the Convention. But the principles of the Climate Change Convention are not up for negotiation.

The United States' decision and the unclear position of Japan have put climate policy in jeopardy. This is the time for politicians to hold firm and agree to start reducing greenhouse gas emissions. We can no longer postpone this crucial decision, because if we do, the Kyoto Protocol really will become nothing more than a dead letter. Then governments will no longer be able to turn around their policies, and the targets will be beyond our reach. Parliaments and the public expect politicians to reach an agreement in Bonn. Politicians can rely not only on their support at those negotiations, but also on the Climate Change Convention and the latest scientific knowledge.

Future Research

But scientists also have responsibilities to live up to. You have to ensure that you present your results as objectively as possible, including all the uncertainties and controversial issues.

There will also have to be major investments in further climate research. Policy makers and politicians want to hear where and how often extreme weather conditions – floods, storms, and drought – are likely to occur. They also want to know what effect the changes will have on a regional level on basic necessities like health, safety, water and food supplies. A good international monitoring system will be essential. It is also important that we detect the warning signals of climate change in good time. What's happening to the Gulf Stream, the carbon cycle and the radiation budget? We must invest more in the Global Climate Observing Systems, which are currently under performing.

Mitigation is another area that requires more research. The international community has agreed that it must strive to prevent dangerous human interference with our climate. Ultimately, we will probably have to cut greenhouse gas emissions by sixty percent in order to reach this goal. We will all have to put our shoulders to the wheel and use every instrument at our disposal. To improve these instruments, we need to conduct much more research into new techniques for reducing emissions of greenhouse gases or storing them.

These are major challenges for science. Your future research must continue to provide a sound basis for future climate policy.

Thank you.

Chapter 2

Challenges of a Changing Earth

Berrien Moore III

> To see the Earth as we now see it, small and beautiful
> in that eternal silence where it floats,
> is to see ourselves as riders on the Earth together,
> brothers on that bright loveliness in the unending night.
>
> *Archibald MacLeish, commenting on the first pictures of Earth from the Moon*

In the Prologue to the "The Family of Man", the extraordinary 1955 photographic exhibition at the Museum of Modern Art in New York, Carl Sandburg writes

> The first cry of a newborn baby in Chicago or Zamboango, in Amsterdam or Rangoon, has the same pitch and key, each is saying, "I am! I have come through! I belong! I am of the Family" …
>
> People! Flung wide and far, born into toil, struggle, blood and dreams, among lovers, eaters, drinkers, workers, loafers, fighters, players, gamblers. Here are ironworkers, bridgemen, musicians, miners, builders of huts and skyscrapers, jungle hunters, landlords and landless, the loved and the unloved, the lonely and abandoned, the brutal and the compassionate-one big family hugging close to the ball of Earth for its life and being …
>
> Everywhere the sun, moon and stars, the climates and weathers, have meanings for people. Though meanings vary, we are alike in all countries and tribes in trying to read what the sky, land, and sea say to us. Alike and ever alike we are on all continents in need of love, food, clothing, work, speech, worship, sleep, games, dancing, fun. From tropics to arctics humanity lives with these needs so alike, so inexorably alike.

These remarkable words speak clearly to us across the near half century, and they now quietly strike raw nerves and sensitive areas: Are there perhaps too many of us in this Family? And what now are the sky, land, and sea saying to us? And are we listening?

There may well be too many of us, and therefore there likely will continue to be too many of us. There is inherent circularity in this complex, and it will be broken only through persistent changes in the behaviour of this and subsequent generations. There is a natural, demographic dynamic that must be acknowledged. But for the moment, let us take all of the Family and focus on what the Family is doing and what the sky, land, and sea are saying.

Unfortunately, as a global society, are not listening. And we should (Fig. 2.1).

- In a few generations humankind is in the process of exhausting fossil fuel reserves that were generated over several hundred million years;
- The carbon dioxide concentration in the atmosphere has increased by more than 30% since the beginning of the Industrial Revolution; methane has increased by 100%;
- Human activity has increased the present total global annual flux of sulphur to the atmosphere by more than 50% from a preindustrial level: 228 Tg S yr^{-1} to a current flux of more than 340 Tg S;

Fig. 2.1.
Global change comprises a wide range of changes in the global environment caused by human activities. Many of the changes are accelerating and interact with each other and with other environmental changes at local and regional scales (sources: NOAA; Reid and Miller 1989, with permission from the World Resources Institute, Washington DC; and Vitousek 1994 with permission from the Ecological Society of America)

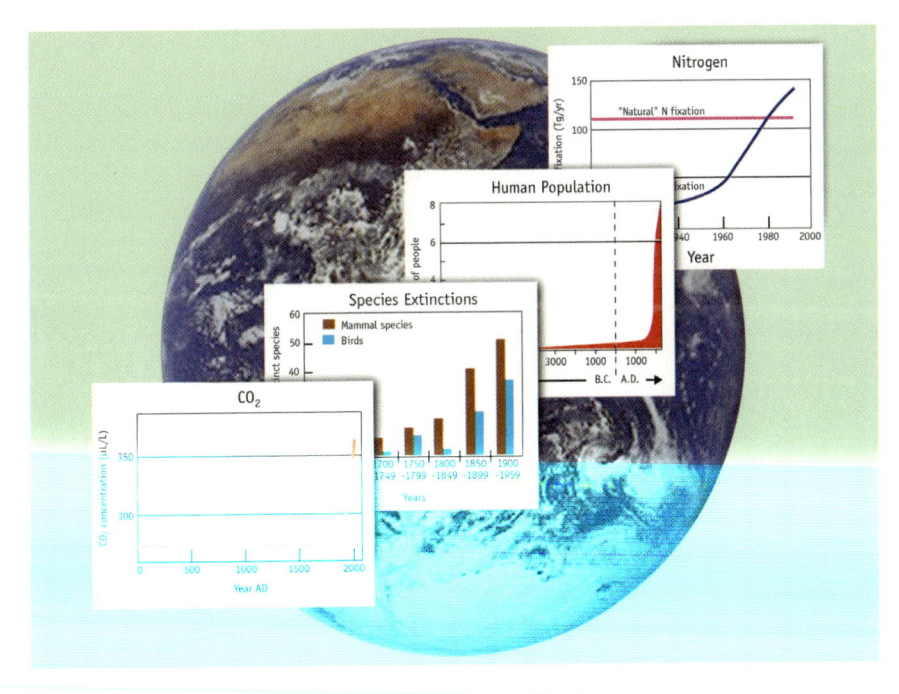

- Nearly 50% of the land surface has been transformed by direct human action, with significant consequences for biodiversity, nutrient cycling, soil structure and biology, and climate. More than one-fifth of land ecosystems have been converted into permanent croplands; more than a quarter of the world's forests have been cleared;
- More nitrogen is now fixed synthetically and applied as fertilisers in agriculture than is fixed naturally in all terrestrial ecosystems;
- Rainforests cover 7 percent of Earth's land surface and 2 percent of the total surface, but they are home to more than 50 percent of the world's plants and animals. At the current rate of clearing, the rainforests will be essentially gone within 100 years, causing unknown effects to the global climate and the extinction of most of those plant and animal species;
- More than half of all accessible fresh water is used directly or indirectly by humankind, and underground water resources are being depleted rapidly in many areas;
- Dam construction has increased the volume of water held in river courses by 700% over the last 50 years;
- Coastal and marine habitats are being dramatically altered; 50% of mangroves have been removed and wetlands have shrunk by one-half;
- About 22% of recognised marine fisheries are overexploited or already depleted, and 44% more are at their limit of exploitation;
- Extinction rates are increasing sharply in marine and terrestrial ecosystems around the world; the Earth is now in the midst of its first great extinction event caused by the activities of a single biological species (humankind);

But still we, as a global society, are not listening.

But there are moments when even the deaf must listen. Much as when one hears the music of Bach, even those who cannot play an instrument, who can not sing a note, even those who do not want to listen – when Bach's music flows through the room it flows through ourselves and makes us sing. And we listen to the music.

Much as when one sees the paintings of Vermeer, Rembrandt, Van Gogh, Leonardo Da Vinci or Heroshige or the work of an unknown master of sculpture, then even those who cannot paint a chair, who cannot draw a circle, even those who do not want to see – when Rembrandt's painting flows through the room, it flows through ourselves and make us see afresh.

And we listen.

When David Keeling began making measurements at Mauna Loa, two years after the opening of the "Family of Man" photographic exhibit, David Keeling was asking what the sky, land, and sea was saying. Perhaps because Keeling is also a wonderful pianist, he proved to have acute hearing. And this is what he, and now we, have heard (Keeling et al. 1976).

The fundamental cycle of the planet, the carbon cycle, is changing. It is changing naturally throughout the year in its seasonal cycle, which is but another expression of "dust to dust" with inorganic carbon dioxide being incorporated into organic material through photosynthesis and then returned to the atmosphere through respiration and decay. This magical capturing in the Keeling record (Fig. 2.2) of the metabolic activity of the planet is often almost overlooked as we stare at the unmistakable imprint of human activity on the global atmosphere – the rising trend line. This record has now

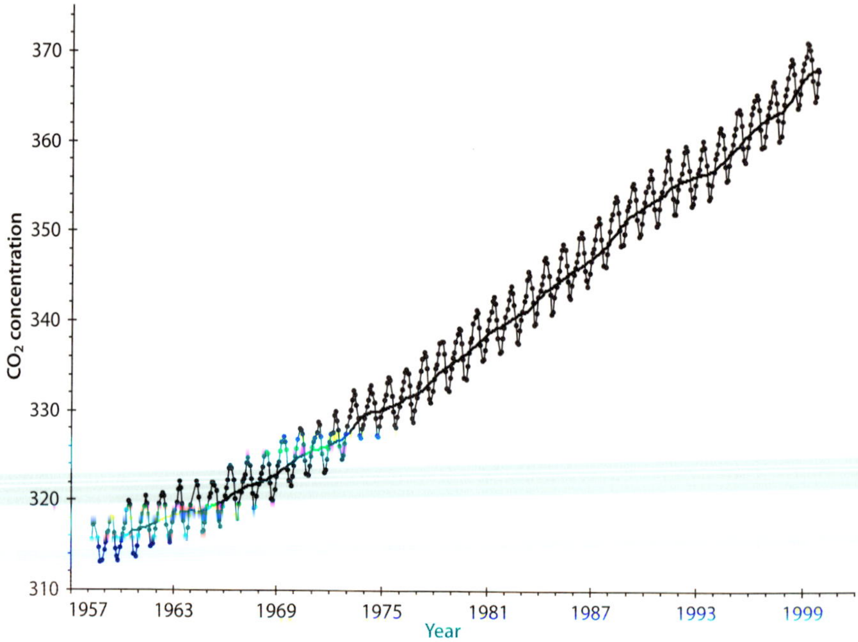

Fig. 2.2.
The "Keeling curve", showing the steady increase in atmospheric CO_2 concentration recorded monthly at Mauna Loa in Hawaii, 1958–1999 (adapted from Keeling and Whorf 2000)

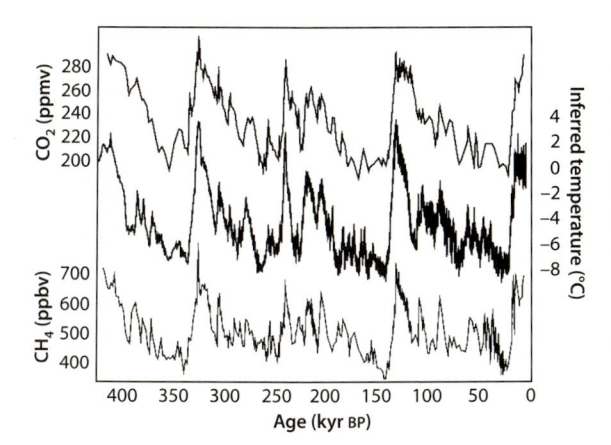

Fig. 2.3. The 420 000 year Vostok ice core record, showing the regular pattern of atmospheric CO_2 and CH_4 concentrations and inferred temperature through four glacial-interglacial cycles (reproduced from Petit et al. 1999, with permission from Nature, A9 1999, McMillan Magazines Ltd.)

become almost a symbol of global change; it would be recognised by many without labels on the axis – it is like the Mona Lisa.

The increasing concentration of carbon dioxide in the atmosphere that is so clearly evident in the Keeling record is a reflection of the industrial metabolic system interacting with the planet's metabolic system. This human-induced global change is even more apparent when we look back over the last 1 000 years.

On longer time scales, there is a remarkably regular pattern of change: an ebb and flow within and between different glacial cycles. When Petit and their colleagues "listened" to the Vostok ice core, this is what they heard: there is a remarkable and intriguing and regular signal of change (Petit et al. 1999; Fig. 2.3). It is almost as if we are hearing the rhythm of the planet's heart. The periodicity of interglacial and glacial climate periods are in a dance step with the beat of the carbon cycle as significant pools of carbon are slowly transferred from the land through the atmosphere to the ocean as the planet enters glaciation, and then there is the rapid recovery of carbon from the ocean back through the atmosphere and onto the landscape as the planet exits glaciation. This resembles a slow Viennese waltz shifting to the modern jitterbug.

The repeated pattern of a 100 ppmv decline in atmospheric CO_2 from an interglacial value of 280 to 300 ppmv to a 180 ppmv floor and then the rapid recovery as the planet exits glaciation suggests a tightly governed control system with firm stops at 280–300 and 180 ppmv. There is a similar CH_4 cycle between 320–350 ppbv (parts per billion by volume) and 650–770 ppbv in step with temperature. What begs explanation is not just the linked periodicity of carbon and glaciation (reflected in the temperature record), but also the apparent hard stops in the carbon system. What were the controls, and why are there apparently "hard stops"?

Today's atmosphere, imprinted with the fossil fuel CO_2 signal, stands at nearly 100 ppmv above the previous "hard stop" of 280–300 ppmv. The current CH_4 value is even further (percentage-wise) from its previous interglacial high values. In essence, carbon has been moved from a relatively immobile pool (in fossil fuel reserves) in the slow carbon cycle to the relatively mobile pool (the atmosphere) in the fast carbon cycle, and the ocean, terrestrial vegetation and soils have yet to equilibrate with this "rapidly" changing concentration of carbon dioxide in the atmosphere.

And when we look to the future through the lens of the past, then the IPCC's 92a industrial-carbon scenario takes on an expanded meaning. One *cannot help but wonder* about the characteristics of the carbon cycle in the future regardless of possible changes in climate.

These two records: the Vostok/Petit record and the Mauna Loa/Keeling record are Bach and Rembrandt, Mozart and Da Vinci, Brahms and Van Beethoven, and Heroshige all rolled into one grand sound-painting of what the Earth is saying. One cannot help but listen; even the deaf must hear.

The primary human activities contributing to the current change in the global carbon cycle are fossil fuel combustion and modifications of global vegetation through land-use (e.g., biomass burning and conversion to agriculture). For the decade of the 1990s, an average of about 6 Gt C per year as CO_2 was released to the atmosphere from the burning of fossil fuels, and it is estimated that an average of about 0.5–1.5 Gt C per year was emitted due to deforestation and land-use change during the same interval (IPCC 2001). For fossil fuel it is roughly a tonne of carbon (as CO_2) per person per year – though the country to country differences are significant and important.

Only 50–60% of the CO_2 from fossil fuel combustion has remained in the atmosphere. Analysis using isotopes and oxygen-nitrogen ratios in the atmosphere has shown that the land and oceans have sequestered the non-airborne fraction, in approximately equal proportions; however, the proportional balance varies in time and space. The current state of the science cannot completely account for the growth rate and interannual variations of atmospheric CO_2 with confidence. The variability of the year-to-year growth in the concentration of carbon dioxide in the atmosphere cannot be explained by the variability in fossil fuel use; it appears to be primarily changes in terrestrial ecosystems and connected with large-scale weather and climate patterns. But we do not understand this pattern – and the pattern is important.

The geographic distribution of the land and ocean sinks has likewise remained elusive. This uncertainty is likewise important. As nations seek to develop strategies to manage their carbon emissions and sequestration, the capability to quantify the present-day *regional* carbon sources/sinks and to understand the underlying mechanisms are prerequisites to prediction and informed policy decisions. Knowledge of today's regional CO_2 sources and sinks, including their mechanisms and their sensitivity to climate perturbations is central to theoretical predictions of future levels. We need to understand the fundamental biogeochemical cycle of the planet; we need to listen.

The increase in the atmospheric CO_2 concentration (as well as other radiatively active trace gases) due to human activity has produced serious concern regarding the heat balance of the global atmosphere. Specifically, the increasing concentrations of these gases will lead to an intensification of the Earth's natural greenhouse effect. Shifting the heat balance will force the global climate system in ways which are not well understood, given the complex interactions and feedbacks involved, but there is a general consensus that global patterns of temperature and precipitation will change, though the magnitude, distribution and timing of these changes are far from certain.

There is, however, an emerging consensus that is reflected in the scientific literature and again in the recent Third Assessment Report (TAR) of the IPCC (IPCC 2001; Fig. 2.4):

- global mean surface temperature has increased more than 0.5 °C since the beginning of the 20th century, with this warming likely being the largest during any century over the past 1 000 years for the Northern hemisphere;
- an increasing body of observations of climatic and other changes in physical and ecological systems gives a collective picture of a warming world;
- global temperature will rise from 1.4–5.8 °C over this century unless greenhouse gas emissions are greatly reduced; and
- There is new and stronger evidence that most of the warming observed over the last 50 years is attributable to human activities.

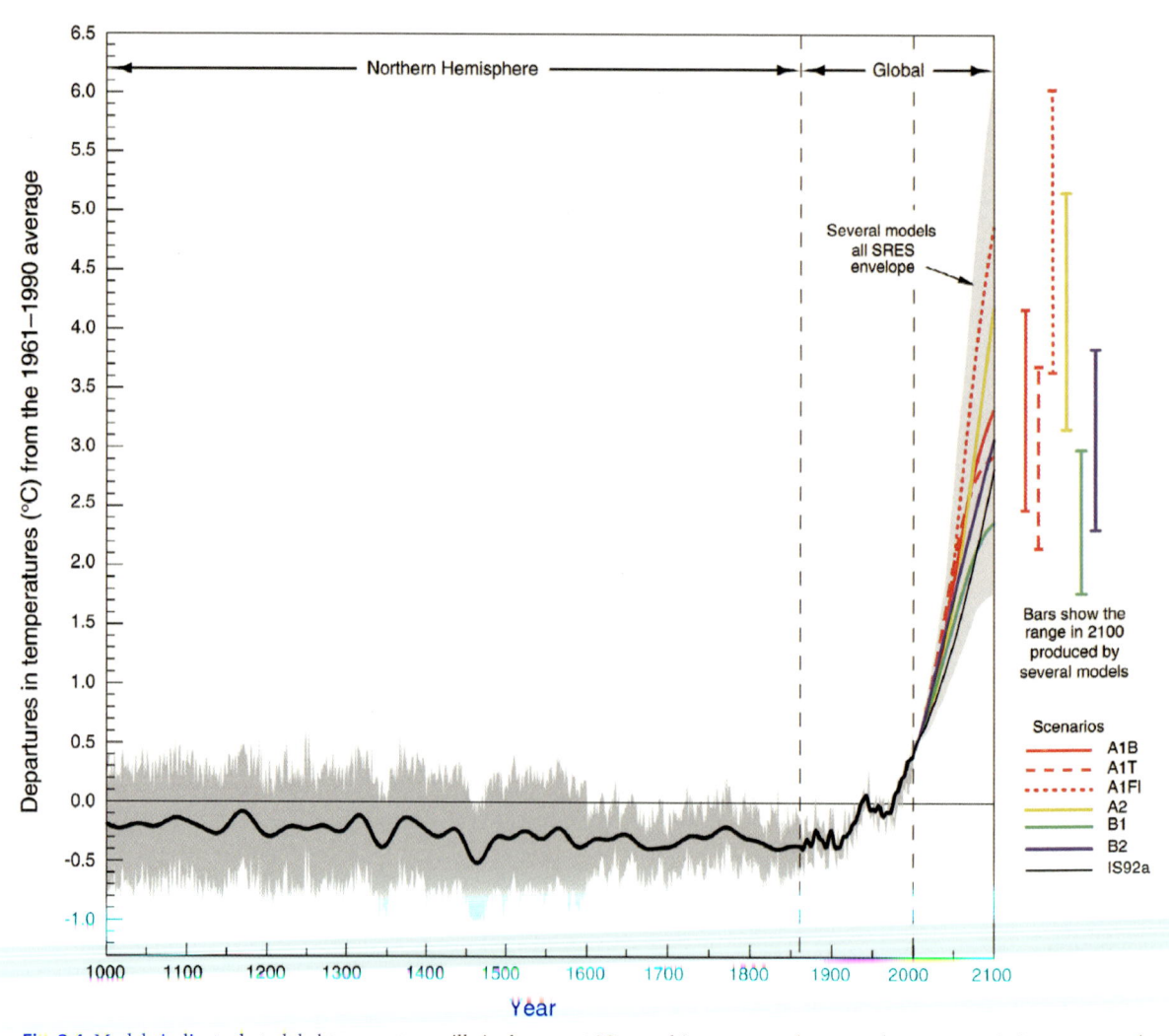

Fig. 2.4. Models indicate that global temperature will rise by 1.4–5.8 °C over this century unless greenhouse gas emissions are greatly reduced (reproduced with permission from Cubasch et al. 2001, © IPCC 2001)

Are we listening to what the sky, land, and ocean are saying?

There obviously remain many important scientific challenges to improving our understanding of the climate system.

An overriding scientific challenge is prediction – what are the likely futures of our climate system and how will the future of the planet unfold? This challenge is particularly acute when predictive capability is sought for a system, such as the climate system or more generally the Earth System, that is chaotic, that has significant nonlinearities, that has widely varying time constants, that interacts with the biogeochemical system with its complexities, and that directly involves humans and their institutions.

There is evidence in the past of rapid shifts in regional climates that involve subtle nonlinearities and interactions. The couplings and linkages may at times be quite sensitive and contain surprises.

The climate of 6 000 BP was different from today's climate. The ecosystem distribution included a more poleward extent of northern forests and the African and Asian monsoon regions were greatly expanded. The climate in the Sahel-Sahara region was much more humid than today, with vegetation cover resembling that of a modern-day African savanna in areas that are now desert. But, about 5 500 years ago, an abrupt change in the regional climate occurred, triggering a rapid conversion of the Sahara into its present desert condition (Claussen et al. 1999).

The ultimate cause was a small, subtle change in Earth's orbit, leading to a small change in the distribution of solar radiation on Earth's surface. Model simulations suggest that this small change nudged the Earth System across a threshold that triggered a series of biophysical feedbacks that led, in turn, to a drying climate. Vegetation changed sharply in response to the decreases in rainfall, and the region shifted into the present-day desert.

Model simulations suggest that it was an interplay of atmosphere, ocean, vegetation and sea ice changes in widely separated parts of the planet that amplified the original orbital forcing with, as one consequence, the abrupt change from savannah to desert in North Africa. Thus we know that abrupt changes in the Earth System can occur when thresholds are crossed, that the dynamical behaviour is the consequence of the interplay of the components of the system, and that aspects of that interplay may involve rather distant and unsuspected teleconnections.

Making predictions about the climate of the planet is *clearly not simple*.

Consequently, we must have means of testing our understanding given the complexity of the system. There need to be ways of evaluating the prognostic skill of any model and understanding the characteristics of this skill. In the case of weather prediction, one can test the skill – we do this daily. One can even question the skill and perhaps fire the forecaster. But for climate, the problem is fundamentally different. The question of predictability of climate is wrapped up with understanding the physics behind the low frequency variability of climate and distinguishing the signal of climate change. In other words, there are the paired challenges of capturing (predicting) "natural" variability of climate as well as the emerging anthropogenically forced climate signal. This dual challenge is distinctively climatic in nature; moreover, the longer-term character of climate projections is unavoidable and problematic.

Fortunately, there appear to be coherent modes of behaviour that not only support a sense of optimism in attacking the prediction problem, but also possibly offer measurable prediction targets that can be used as benchmarks for evaluating our understanding of the climate system. In addition, predicting these modes represents a valuable contribution in its own right.

The intraseasonal to interdecadal modes of climate variability (e.g., El Niño-Southern Oscillation, Pacific Decadal Oscillation, and North Atlantic Oscillation) offer opportunities to test prognostic climate skill (Busalacchi, this volume). Here, some predictive skill for the climate system appears to exist on longer time scales. One example is the ocean-atmosphere phenomenon of El Niño-Southern Oscillation (ENSO; Fig. 2.5). This skill has been advanced and more clearly demonstrated with the prediction of the 1998 El Niño, and this progress and demonstration are important. Such demonstrations and the insights gained by developing and making prognostic statements on climate modes frame an important area for further work. This success also clearly demonstrated the importance of data, both space-based such as Topex-Poisidon topography data as well as the *in situ* buoy data. In the case of El Niño, the TOGA (Tropical Ocean and Global Atmosphere) array was particularly valuable.

The success in El Niño forecasting and in unravelling the large-scale climate modes is first order. And yet, the range in climate forecasting for CO_2 remains a large problem. Much of the uncertainty about climate change and the degree of warming centres on clouds. It is generally accepted that the net effect of clouds on the radiative balance of the planet is currently negative and has an average magnitude of about 10–20 W m^{-2}; in other words, clouds have a cooling effect overall. But how will the cloud system respond to changes induced by the increase in greenhouse gases?

The effect of clouds on the radiative balance consists of a short-wave cooling (the albedo effect) of about 40–50 W m^{-2} and a long-wave warming of about 30 W m^{-2}. Unfortunately, the size of the uncertainties in this budget is large when compared to the expected anthropogenic

Fig. 2.5.
Temperature and height anomalies in the Pacific during an El Niño–La Niña cycle (source: NASA/Goddard Space Flight Center Scientific Visualization Studio)

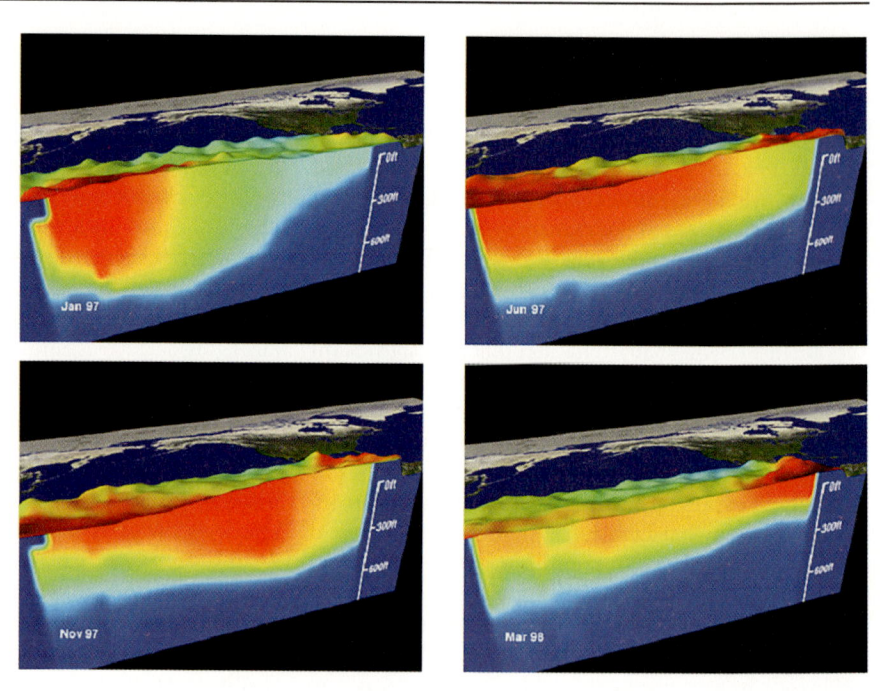

greenhouse forcing. Although we know that the overall net effect of clouds on the radiative balance is slightly negative, we do not know the sign of cloud feedback with respect to the increase of greenhouse gases, and how it may vary with the region. In fact, the basic issue of the nature of the future cloud feedback is not clear.

Understanding the cloud-climate connection is a particularly challenging (I could have said cloudy) scientific problem, because it involves processes covering a very wide range of space and time scales. For example, cloud systems extending over thousands of kilometres to cloud droplets and aerosols of microscopic size are all-important components of the climate system. The time scales of interest can range from hundreds of years (e.g., future equilibrium climates) to fractions of a second (e.g., droplet collisions). This is not to say that all cloud microphysics must be included, for instance, in modelling cloud formation and cloud properties, but the demarcation between what must be included and what can be parameterised remains unclear. Clarifying this demarcation and improving both the resulting phenomenological characterisations and parameterisations will depend critically on improved global observations of clouds.

There are hopeful signs. The CERES instruments on the NASA Terra spacecraft are providing new insights on clouds and the heat balance of the planet. The recently approved CloudSat and PICCASO-CENA missions, which will fly in formation with the NASA EOS PM (the Aqua Mission), will provide valuable profiles of cloud ice and liquid content, optical depth, cloud type, and aerosol properties. These observations combined with contributions from ESA's ENVISAT and NASDA's ADEOS-II will provide a rich new source of information about the properties of clouds.

What we must keep in mind is that the basic issue of the nature of the *future* cloud feedback is not clear. Will it remain *negative*? If the planet warms, then it is likely that evaporation will increase, which probably implies that liquid water content will increase, but the volume of clouds may not. What will be the effect, and how will the effects be distributed in time and space? What will be the effect of changes in the distributions of aerosols? *We must get at the underlying processes.*

John Mitchell of the Hadley Centre has stated the issue well: "Reducing the uncertainty in cloud-climate feedbacks is one of the toughest challenges facing atmospheric physicists."

We return to the issue of unfolding the possible futures of the planet. This challenging hurdle becomes challenging in the extreme when confronted with the need to predict extreme events.

Extreme events are, almost by definition, of particular importance to human society. Consequently, the importance of understanding potential extreme events is first order. There appear to be some consistent patterns with increased CO_2 with respect to *changes in variability*: the Pacific climate base state could be a more El Niño-like state and enhanced variability in the daily precipitation in the Asian summer monsoon with increased precipitation intensity. More generally, there is a general intensification of the hydrological cycle with increased CO_2. For possible *changes in extreme weather and climate events*, the most robust conclusions appear to be first an increased probability of extreme warm days and decreased probability of extreme cold days and second an increased chance of drought for midcontinental areas during summer with increasing CO_2.

Returning to the two issues of climate variability and of possible rapid shifts in the climate system, there is increasing evidence that there is a decline in extent and thickness of Arctic sea ice in the summer that appears to be connected with the observed recent Arctic warming (Anisimov, this volume). It is not known whether these changes reflect anthropogenic warming transmitted either from the atmosphere or the ocean, or whether they mostly reflect a major mode of multi-decadal variability. Some of this pattern of warming has been attributed to recent trends in the Arctic Oscillation; however, how the anthropogenic signal is imprinted on the natural patterns of climate variability remains a central question. What does seem clear is that the changes in Arctic sea ice are significant, and there is a positive feedback that could be triggered by declines in sea ice extent through changes in the planetary albedo (Fig. 2.6). If the Arctic shifted from being a bright summer object to a less bright summer object, then this would be an important positive feedback on a warming pattern. The shift could be rapid and dramatic.

There is another feedback loop through increased evaporation leading to increases in cloud and enhanced albedo. Which will win? Are we to have an open Arctic sea in summer? The issue is disturbing; are we listening?

There are similar trends of decreasing land glaciers; the snows of Kilimanjaro may soon be remembered only in literature.

These and other scientific questions about the functioning of the Earth's climate system remain open, and it is not clear how they will be answered. Without question, the strategy for attacking the feedback question will involve comparison of model simulations with appropriate observations on global or local scales. The interplay of observation and models, again, will be the key for progress. But remember, they will be answered by human beings, by our colleagues, and we should remember to give them our support and our thanks.

We also owe the World Climate Research Programme, the International Geosphere-Biosphere Programme, and the International Human Dimensions Programme our thanks for the remarkable progress to date on this and other challenging issues, in the complex web of climate science. But for the future, the science of climate and global change deserves more than thanks. We need to listen carefully to what it is telling us about what the planet is saying, and we need to listen even more carefully.

In thinking about climate, we may be paying too much attention to temperature – the term global warming may be actually focusing attention on a too restrictive topic. After all, most people on the planet live in areas far warmer than Amsterdam or New Hampshire. It may well be that changes in the water cycle are more significant – particularly when these changes are viewed against the backdrop of other changes including population growth.

The global water cycle is truly a complex "problematique" that intersects profoundly with the full Earth System, including particularly the human system. This full coupling profoundly complicates attempts to foresee clearly the potential changes in the global water cycle. Let us just think about the first order connections.

First, water is obviously tied up with the cloud feedback problem via atmospheric water vapour, precipitation, and other state variables, including temperature and processes in the climate system even narrowly defined. And that is just the beginning.

Precipitation is an essential element in determining the level of soil moisture. Patterns of precipitation set the stage for and are partially determined by evapotranspiration and the resulting distribution of soil moisture. In turn, terrestrial ecosystems recycle water vapour at the land-surface/atmosphere boundary and exchange numerous important trace gases with the atmosphere. Soil moisture is an important determinant of ecosystem structure and hence a principal means by which climate regulates (and is partially regulated by) ecosystem distribution. Moreover, soil moisture is a primary ingredient in evapotranspiration and thus to the apportioning of sensible and latent heat fluxes.

Soil moisture is pivotal in the formation of runoff and hence river-flow. River systems are linked to regional and continental-scale hydrology through interactions among precipitation, evapotranspiration, soil water, and runoff in terrestrial ecosystems. River systems, and more generally the entire global water cycle, control the movement of constituents over vast distances, from the continental landmasses to the world's oceans and to the atmosphere.

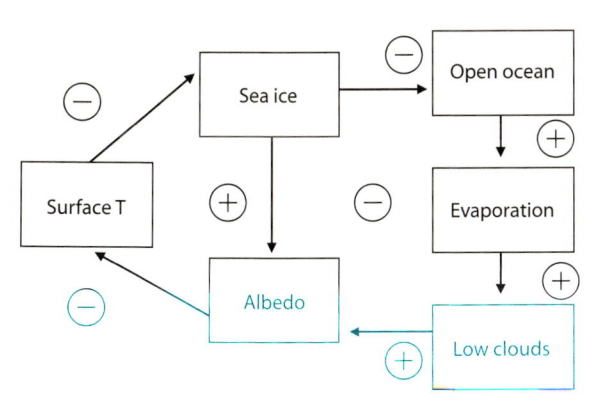

Fig. 2.6. The feedbacks involving Arctic sea ice and warming trends (reproduced with permission from Stocker et al. 2001, © IPCC 2001)

Rivers begin with rainfall to soil moisture and runoff; each of these is linked directly to land-use and land-cover. Rivers end in and are key features of the coastal zone. Finally, rivers deliver more than water to the ocean; rivers process and contribute significant quantities of nutrients to coastal oceans and hence are linked to coastal fisheries.

And did I mention: precipitation is also critical for humans, rivers are central features of human settlement and development, and the climate-fishery-human dynamical system is real and important? Any global perspective on surface hydrology must *explicitly* recognise the impact of human intervention on the water cycle, not only through climate and land-use change, but also through the operation of reservoirs, interbasin transfers, irrigation, industrial consumption, and other human uses.

Understanding the hydrological cycle and water resources is intimately bound up with understanding the human system as well as understanding the biogeochemical and climate systems.

This should not really be a surprise to us. But, unfortunately, even possible changes in precipitation are difficult to predict. And this is just the tip of the iceberg.

Fortunately, over the past decade, the International Human Dimensions Programme, the World Climate Research Programme and the International Geosphere Biosphere Programme have realised a remarkable record of scientific achievement. These accomplishments allow us to tackle, at least figuratively, the entire 300 000 000-ton iceberg and more. We are ready to put the pieces of the hydrological cycle together, to make coherent both the vertical and horizontal flows of water, to treat the changes in biogeochemistry and feedbacks from climate on water quality as well as quantity, address fully the coupling of the land with the sea, and include deeply the role of humans! The very fact that we, as a community, are ready to address the hydrological system is a very positive reflection upon the successes of the core activities within the Global Change Programmes.

We need now to get on with the job of understanding well the water cycle of our planet. And in doing so, we will also be tackling central issues facing the global change community and the planet including potential changes in the food system, in the biogeochemistry, and in biodiversity of the planet. In this latter regard, I am delighted that Diversitas is with us today and will be with us in the future.

In my opening remarks, I noted that the atmospheric CO_2 concentration is now nearly 100 ppmv higher, and has risen to that concentration at a rate at least 10 and possibly 100 times faster, than during any other time in the past 420 000 years. The palaeo records clearly show that we have driven the Earth System from the tightly bounded domain of glacial-interglacial dynamics and have now entered into carbon territory that has not been visited in the last 25 million years (Falkowski et al. 2000).

Are we in a transition period to a new, stable domain? If so, what are the main forcing factors and feedbacks of this transition? What will be the climatological features of a new domain? What will be the responses, and feedbacks, of Earth's ecosystems? If only someone had a friendly dinosaur (non-political that is) with whom we might discuss life in a high CO_2 world.

Atmospheric CO_2 exchanges rapidly with oceans and terrestrial ecosystems. The ratio between the rate at which these two reservoirs absorb atmospheric CO_2 and the rate of emissions determines the overall rate of change of atmospheric CO_2. As we saw earlier, during glacial-interglacial transitions, the atmosphere acts to transfer carbon between terrestrial ecosystems and the oceans. The remarkable consistency of the upper and lower limits of the glacial-interglacial atmospheric CO_2 concentrations and the apparent fine control over periods of many thousands of years around those limits suggest strong feedbacks that constrain the dynamics of the land-atmosphere-ocean system.

Dissolved inorganic carbon in the oceans is 50 times that of the atmosphere, and on time scales of millennia, the oceans determine atmospheric CO_2 concentrations, not vice versa. Atmospheric CO_2 continuously exchanges with oceanic CO_2 at the surface. This exchange, which amounts to roughly 90 Gt C per year in each direction, leads to rapid equilibration of the atmosphere with the surface water (Fig. 2.7). Upon dissolution in water, CO_2 forms a weak acid that reacts with carbonate anions and water to form bicarbonate. The capacity of the oceanic carbonate system to buffer changes in CO_2 is finite and depends on the addition of cations from the relatively slow weathering of rocks. Because the rate of anthropogenic CO_2 emissions is several orders of magnitude greater than the supply of mineral cations, on time scales of millennia, the ability of the surface oceans to absorb CO_2 will inevitably decrease as the atmospheric concentration of the gas increases.

The concentration of total dissolved inorganic carbon in the ocean increases markedly below the upper 300 m, where it remains significantly above the surface ocean-atmosphere equilibrium value in all ocean basins. The higher concentration of inorganic carbon in the ocean interior results from a combination of two fundamental processes: the "solubility pump" and "biological pump".

The efficiency of the solubility pump depends on the thermohaline circulation and latitudinal and seasonal changes in ocean ventilation. CO_2 is more soluble in cold waters, and sequestration of atmospheric CO_2 in the ocean interior is therefore controlled by the formation of cold, dense water masses at high latitudes, especially in the North Atlantic and in the Southern Ocean confluence and on continental margins. As these

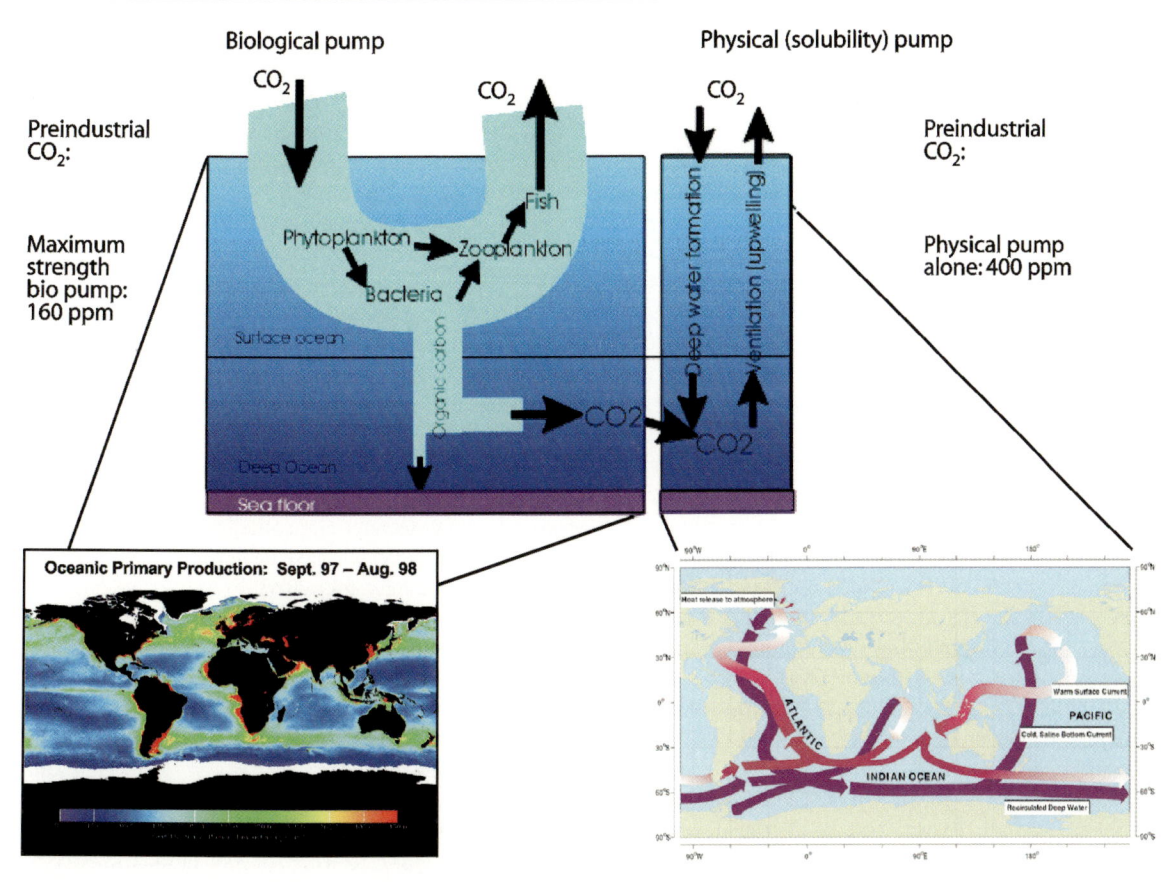

Fig. 2.7. The importance of the biological and physical pumps in the ocean carbon cycle

water masses sink into the ocean interior and are transported laterally, CO_2 is effectively prevented from re-equilibrating with the atmosphere by a cap of lighter overlying waters. Re-equilibration occurs only when waters from the ocean interior are brought back to the surface, decades to several hundreds of years later. The water upwelling today was last at the surface when atmospheric CO_2 concentrations were significantly lower, and hence the outgassing is less than it would otherwise be.

Coupled climate-ocean simulations suggest that CO_2-induced global warming will lead to increased stratification of the water column. If this occurs, the transport of carbon from the upper ocean to the deep ocean will be reduced, with a resulting decrease in the rate of sequestration of anthropogenic carbon in the ocean. The combined effects of progressive saturation of the buffering capacity, increased stratification, and warmer sea surface temperatures individually and collectively will weaken the oceanic uptake rate of anthropogenic CO_2.

Phytoplankton photosynthesis, part of the biological pump, lowers the partial pressure of CO_2 in the upper ocean and thereby promotes the absorption of CO_2 from the atmosphere. Approximately 25% of the carbon fixed in the upper ocean sinks into the interior, where it is oxidised, raising the concentration of dissolved inorganic carbon (DIC). This process keeps atmospheric CO_2 concentrations 150 to 200 ppmv lower than they would be if all the phytoplankton in the ocean were to die. In addition to the organic biological pump, several phytoplankton and zooplankton species form calcium carbonate shells that sink into the interior of the ocean, where some fraction dissolves. This inorganic carbon cycle leads to a reduction in upper ocean DIC relative to the deep ocean, and is therefore sometimes called the "carbonate pump".

There are interesting feedback possibilities as the relative strengths between the various interlocked pumps play together in different climatic regimes and in different patterns of marine ecosystem distributions. For instance, the stratification of the ocean that weakened the solubility pump may, in fact, strengthen the biological pump through greater nutrient use efficiency; however, not to the extent necessary to offset other negative feedbacks. This is an area where accomplishments have been great over the past decade, but where much more needs to be done.

A similar case of coupling occurs on land. The metabolic processes are responsible for plant growth and maintenance and the microbial turnover is associated with dead organic matter decomposition cycle carbon, nutrients, and water through plants and soil on both rapid and intermediate time scales. As mentioned earlier, these cycles affect the energy balance and provide key controls over biogenic trace gas production.

Looking at the carbon fixation-organic material decomposition as a linked process, one sees that some of the carbon fixed by photosynthesis and incorporated into plant tissue is perhaps significantly delayed from returning to the atmosphere until it is oxidised by decomposition or fire. This slower carbon loop through the terrestrial component of the carbon cycle affects the rate of growth of atmospheric CO_2 concentration, and in its shorter term expression, imposes a seasonal cycle on that trend.

The structure of terrestrial ecosystems, which respond on even longer time scales, is the integrated response to changes in climate and to the intermediate time scale dynamics of the carbon-nutrient-water machinery. As noted in the discussion on water, this loop is closed back to the climate system, since it is the structure of ecosystems, including species composition that largely sets the terrestrial boundary condition on the climate system in terms of surface roughness, albedo, and latent heat exchange.

In addition to ecophysiological considerations, land-use change plays a major role in the carbon source/sink dynamics. The increased pressure in the developing world to increase food and fibre production by converting forests to agricultural use effectively increases the flux of carbon to the atmosphere while simultaneously reducing the land area available for active sinks. Abandonment of agricultural land and regrowth of forests, largely in the temperate northern hemisphere, may be a significant terrestrial CO_2 sink at present, but cannot be sustained indefinitely. We must consider how terrestrial systems respond to multiple concurrent changes: to climate, to human land use, to changes in water availability, and to changes in the chemistry of the atmosphere. Sinks may buy some time, but unless CO_2 emissions are reduced, terrestrial sinks simply cannot mitigate against continued accumulation of the gas in Earth's atmosphere. Thus in the end, we simply must find ways of understanding better human actions and institutional structures; humans actions and institutions will need to change (Watson and Noble, this volume).

As we move further away from the domain that characterised the preindustrial Earth System, we severely test the limits of our understanding of how the Earth System will respond. Our present, imperfect models suggest that the feedbacks between carbon and other biogeochemical and climatological processes will lead to weakened sink strengths in the foreseeable future, and the prospects of retrieving anthropogenic CO_2 from the atmosphere by enhancing natural sinks are small. This condition cannot persist indefinitely. Our present state of uncertainty arises largely from lack of integration of information. Nevertheless, scientists' abilities to predict the future will always have a component of uncertainty. This uncertainty should not be confused with lack of knowledge, nor should it be used as an excuse to postpone prudent policy decisions based on the best information available at the time.

As the climate system evolves in a transient response to increasing greenhouse gases, the terrestrial and perhaps marine biotic systems will also evolve. This evolution must be considered concurrently, since the couplings are two-way. This is an important scientific challenge. It will require scientific insight from a variety of disciplines and advanced observational and computational resources. More importantly, if the challenge is to be met, then it must be addressed in a more coherent and more aggressive manner in the next decade, than it was in the past decade. The contributions of the IGBP, the IHDP, and the WCRP have been remarkable; the demands, however, have increased.

There is much change afoot, and this change is rapid, perhaps more rapid than at any time in human history and perhaps at any time in Earth's history. This change is occurring in all of the Earth's systems, including the human system. And the accumulation and interaction of these changes may well threaten our well-being and the planet's well-being.

I began expressing my view that we, as a global society, are not listening to what the sky, land, and sea are saying. We are, however, growing uneasy. There is, as when the nations gathered in 1972 in Stockholm, a sense of urgency regarding the environmental state of the planet. This sense of urgency is not misplaced.

There are shortages of clean and accessible fresh water, degradation of terrestrial and aquatic ecosystems, increases in soil erosion, loss of biodiversity, changes in the chemistry of the atmosphere, declines in fisheries, and the possibility of significant changes in climate. These changes are occurring because of human activity. The concentration of carbon dioxide in the atmosphere is higher than at any time in the last 25 million years, and it is so because of the human use of fossil fuels and because of the human alteration of the landscape.

These human-induced changes are over and above the stresses imposed by the natural variability of a dynamic planet and are intersecting with the effects of past and existing patterns of conflict, poverty, disease, and malnutrition.

The changes are changes in the relationship between humans and nature . They are recent, they are profound, and many are accelerating. They are cascading through the Earth's environment in ways that are difficult to understand and often impossible to predict. Surprises abound. At the least, these human-driven changes in the global environment will require societies to develop a multitude of creative responses and adaptation strategies. Some are adapting already; most are not. At worst, these changes may drive the Earth itself into a different state that may be much less hospitable to humans and other forms of life.

As global environmental change assumes a more central place in human affairs, science is being thrust into the unfamiliar and uncomfortable role of a primary participant in a heated and potentially divisive international debate about the nature and severity of global change and its implications for ways of life. Much is at stake, and the game is being played hard. Despite the risks, science must accept the responsibility of developing and communicating the essential knowledge base societies can use to consider and decide ultimately on how to respond to global change.

The linked challenges of confronting and coping with global environmental changes and addressing and securing a sustainable future are daunting and immediate, but they are not insurmountable. The challenges can be met, but only with a new and even more vigorous approach to understanding our changing planet and ourselves and with a concomitant commitment by all to alter our actions. Those that consume the most must take the greatest action. We simply must take some of the pressure off the Earth.

There is no time to lose; humanity must confront the future now.

References

Claussen M, Kubatzki C, Brovkin V, Ganopolski A, Hoelzmann P, Pachur HJ (1999) Simulation of an abrupt change in Saharan vegetation at the end of the mid-Holocene. Geophys Res Lett 24(14):2037–2040

Cubasch U, Meehl GA, Boer GJ, Stouffer RJ, Dix M, Noda A, Raper S, Senior CA, Yap KS (2001) Projections of future climate change. In: Houghton JT, Ding Yihui, Griggs DJ, Noguer M, van der Linden PJ, Dai X, Maskell K, Johnson CA (eds) Climate change 2001: The scientific basis. Contribution of Working Group I to the Third Assessment Report of the Intergovernmental Panel on Climate Change. Cambridge University Press, Cambridge UK

Falkowski P, Scholes RJ, Boyle E, Canadell J, Canfield D, Elser J, Gruber N, Hibbard K, Högberg P, Linder S, Mackenzie FT, Moore III B, Pedersen T, Rosenthal Y, Seitzinger S, Smetacek V; Steffen W (2000) The global carbon cycle: A test of our knowledge of Earth as a system. Science 290:291–296

IPCC (2001) Climate Change 2001:The scientific basis. Contribution of Working Group I to the Third Assessment Report of the Intergovernmental Panel on Climate Change (Houghton JT, Ding Y, Griggs DJ, Noguer M, van der Linden PJ, Dai X, Maskell K, Johnson CA (eds) Cambridge University Press, Cambridge UK

Keeling CD, Whorf TP (2000) Atmospheric CO_2 records from sites in the SIO air sampling network. In Trends: A Compendium of Data on Global Change. Carbon Dioxide Information Analysis Center, Oak Ridge National Laboratory, US Department of Energy, Oak Ridge, TN, USA

Keeling CD, Bacastow RB, Bainbridge AE, Ekdahl CA Jr., Guenther PR, Waterman LS, Chin JFS (1976) Atmospheric carbon dioxide variations at Mauna Loa Observatory, Hawaii. Tellus 28(6): 538–551

Petit JR, Jouzel J, Raynaud D, Barkov NI, Barnola J-M, Basile I, Bender M, Chappellaz J, Davis M, Delaygue G, Delmotte M, Kotlyakov VM, Legrand M, Lipenkov VY, Lorius C, Pepin L, Ritz C, Saltzman E, Stievenard M (1999) Climate and atmospheric history of the past 420 000 years from the Vostok ice core, Antarctica. Nature 399:429–436

Stocker TF, Clarke GKC, Le Treut H, Lindzen RS, Meleshko VP, Mugara RK, Palmer TN, Pierrehumbert RT, Sellers PJ, Trenberth KE, Willebrand J (2001) Physical climate processes and feedbacks. In: Houghton JT, Ding Yihui, Griggs DJ, Noguer M, van der Linden PJ, Dai X, Maskell K, Johnson CA (eds) Climate change 2001: The scientific basis. Contribution of Working Group I to the Third Assessment Report of the Intergovernmental Panel on Climate Change. Cambridge University Press, Cambridge UK

Vitousek PM (1994) Beyond global warming – ecology and global change. Ecology 75 (7):1861–1876

Part II

Achievements and Challenges

Part IIa

Food, Land, Water, and Oceans

Chapter 3

Toward Integrated Land-Change Science: Advances in 1.5 Decades of Sustained International Research on Land-Use and Land-Cover Change

B. L. Turner II

Land change is likely the most ancient of all human-induced environmental impacts on the biosphere and the first to obtain a magnitude to warrant the title "global". Evidence mounts that *Homo sapiens* was instrumental in the worldwide destruction of megafauna before the last glacial maximum, and by the seventeenth century, humankind had restructured global biota by the transcontinental movement of domesticates and ornamental flora, complete with the unintentional transport of vermin, pests, weeds, and diseases that affected ecosystems globally. Today, virtually no land surface remains untouched in some way by humankind, and ~50% of the ice-free surface of Earth is considered significantly modified by human action. Land use commands as much as 40% of the net primary productivity of the Earth (Vitousek et al. 1997), although the uncertainties are large (Rojstaczer et al 2001). About 35 300 dams have been constructed since 1950, fragmenting habitats within 60% of the major river basins worldwide. Water diversion for irrigation consuming about 70% of all water withdrawals is sufficiently significant to stop the flow of such large rivers as the Colorado, Huang Ho, and Amu Darya from reaching the sea during the dry season (Johnson et al. 2001). Land-cover changes, largely deforestation, have accounted for 33% of the increase in atmospheric CO_2 since 1850 (176 ±55 Gt C) (Watson et al. 2000), and recent assessments suggest that heretofore unaccounted parts of forests including large areas of regrowth constitute much of the missing carbon sink (Pacala et al. 2001). These and other land changes alter ecosystem services locally and globally (Daily et al. 2000) and account for most of the current "mass extinction" of biota. That the human impress on the terrestrial Earth is vast is no surprise; that land change was one of the last of the subjects to be formally entertained by global environmental change science is surprising.

This relatively late start notwithstanding, major headway has been made internationally toward the development of systematic and integrated land science that is consistent with the emerging research agendas variously labelled "sustainability science" (Kates et al. 2001), including activities falling within the auspices of the IGBP, IHDP, and IPCC, especially the joint IGBP-IHDP LUCC (Land Use/Cover Change) project (IGBP-IHDP 1995, 1999) and various parts of the IGBP's GCTE (Global Change and the Terrestrial Ecosystem) project. The IGBP-IHDP LUCC project seeks to assist the various research communities globally (*i*) in documenting the magnitude and pace of land changes and identifying the critical locations of these changes, (*ii*) in understanding and explaining these changes sufficiently (*iii*) to produce robust near-term models of change at various spatial scales, and (*iv*) to apply this knowledge to various linked issues, such as the vulnerability of coupled human-environment systems to climate change and other perturbations and stresses. Oddly, given its significance, systematically generated and readily comparable documentation of land-use and land-cover change and its causes has been sparse compared to the amount of work focused on methodological and technological issues including software and hardware development that may dramatically improve the documentation and monitoring of land changes and of the means of modelling and projecting the observed changes. The state of the art in "integrated land science" is rapidly improving, however, permitting assessments of major trends, causes, and model-method developments (Moran n.d.).

3.1 Trends

Forests. The search for the missing carbon sink and the concern for tropical deforestation has drawn considerable attention to forest-cover change, the trends of which are better known than for any other land cover, save agriculture. Since the advent of domestication, about 47% of the world's forests have been lost, yet they continue to occupy some 3.5 billion ha, 55% of which are located in the tropics. It is precisely this in geographical realm, however, where the pace of deforestation is rapid. About 10% of the world's tropical forests (or 200 million ha) were lost between 1980 and 1995 alone (FAO 1999), with hot spots of deforestation occurring throughout the tropics and some, such as those of Sunda Shelf of Indonesia, at alarmingly high rates. Most of these assessments involve "clear cutting" on the assumption that once cut, the lands will remain "open". Careful observations, however, demonstrate that much tropical reforestation is permitted to regrow (for undetermined periods), reducing the estimates noted above (Skole and Tucker 1993). These reductions, however, appear to be significantly offset by "cryptic deforestation", that was generated by selective logging and linked burning of logged lands. Thus, Nepstad and colleagues (1999) estimate that annual amounts of deforestation in Amazonia are more than doubled, when this kind of deforestation is taken into account. While others find the actual damage, minus fires, to constitute a smaller order of damage (D. Skole, pers. comm.), the scale of forest modification, broadly interpreted, is sufficiently large that this kind of change requires attention in land-change assessments.

Temperate and boreal forests, with one prominent exception, display trends contrary to those in the tropics. Tree cover in much of the developed economies of the world has increased over the past one-half decade. About 20 million ha were gained between 1980 and 1995 alone (FAO 1999), be it transitioning of farmlands to suburban and recreational uses in service sector economies or deliberate tree planting in command economies (Seto et al. 2000). Indeed, the Northern Hemisphere constitutes a terrestrial carbon sink on the order of 1 to 2 $Pg\ C\ yr^{-1}$ (Pacala et al. 2001). Theoretically, this forestation is related to the movement of the former land-uses elsewhere (deforestation displacement), although efficiencies in modern production may lead to a reduced hectarage in the replacement activities compared to the hectarage of regrowth. The major exception to temperate and boreal forestation trends is found in Siberia, the looming deforestation giant. Here, the demise of the Soviet command economy has given way to socioeconomic conditions akin to those in the tropics, and international logging has taken advantage of the vast stocks of Siberian forests.

Forest Trends
- Sustained tropical deforestation well into the new century;
- Tropical forest modification (cryptic change) increasing in importance;
- Forestation in most temperate and boreal forests well into the new century,
- Potentially large-scale, timber-led deforestation in Siberia.

Cultivated Lands. Various projections (e.g., Ausubel 1996) foresee a future in which cultivation intensifies world-wide, focused on prime agricultural lands and with marginal lands increasingly taken out of production (see Urban, below). This process has already begun in developed economies, in some cases related to the depletion of critical resources, usually involving water drawn down in aquifers, which renders land marginal for cultivation. Where marginal agriculture continues in these econo-

mies, it is supported by policies designed to protect the enterprise for political and other reasons. Intensified production, on the other hand, relies on significant chemical and fossil fuel inputs as well as irrigation, in some cases stressing catchment hydrology and releasing significant amounts of N_2O to the atmosphere (Matson et al. 1998). Such problems promise to be substantially reduced by satellite-led "smart farming" and other advanced technologies.

The overall reduction and intensification in cultivated lands is not necessarily replicated in the less developed economies, especially within tropics. Here economically marginal but life-sustaining cultivation continues to expand into the forest frontier, commonly following roads constructed for timber and other extractive activities, or corporate or large-holder investments seek to profit from inexpensive land (Geist and Lambin 2001). Few, if any indicators suggest that this expansion will cease in the near future, although it will surely ebb and flow by region. Where cropping strategies remain low-input but cultivation frequencies increase, invasive species promise to become a significant problem with this expansion (Mooney and Hobbs 2000; Turner et al. 2001).

Agricultural Trends
- Intensified production focused on prime croplands in first- and second-tier economies predicated on high input and increasing technology, such as smart farming;
- Marginal croplands moved to other uses in these economies;
- Sustained expansion of croplands and pasture lands in less developed economies, especially the tropics, where invasive species promise to pose significant problems.

Urban. Notwithstanding the impression from satellite imagery of the night lights of Earth, urban or built-up settlement areas pale in area covered compared to other land covers. Despite rapid urban expansion, the major impact of urban areas is through the urban "footprint" on resource uses and environmental impacts on other land-covers (Folke et al. 1997; Wackernagel and Yount 2000). These footprints are many, but two noted here raise special concerns for global change and sustainability science. Urban and, by definition, industrial production contribute significantly to tropospheric ozone, which undertakes various reactions to become a toxic pollutant, not only with implications for human health but for plant growth. Chameides and colleagues (1994) demonstrate the spatial correlation between urban-industrial complexes worldwide and the build up of ozone that spreads across vast tracts of prime agricultural lands. Some of the more endangered regions include the major croplands of the American Midwest, Eastern Europe, eastern China, and the Ganges Basin,

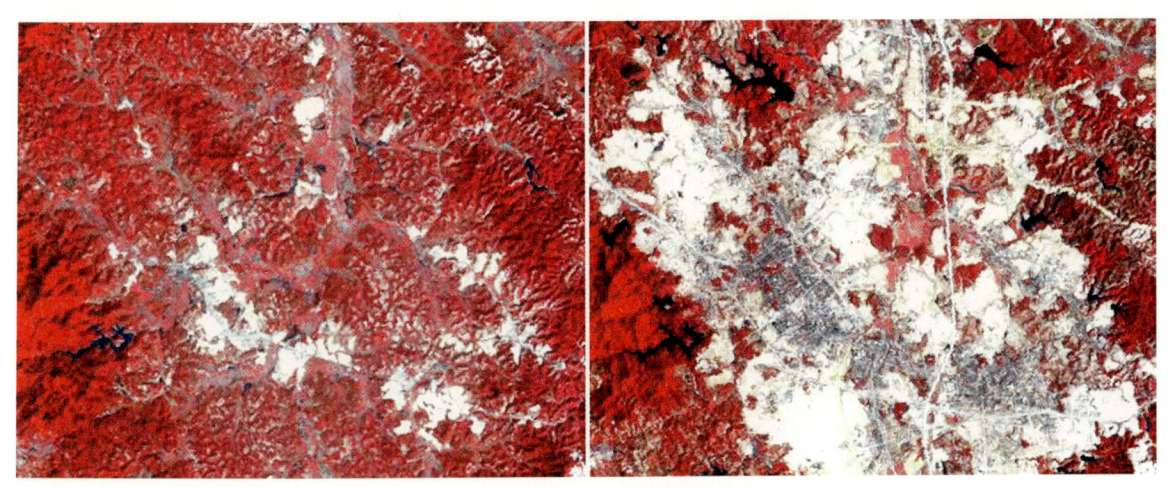

Fig. 3.1. Urban expansion into surrounding lands that were once in irrigated rice, mixed agriculture, and forest-shrub cover, Baoan (Shenzhen, Pearl River delta), China. Landsat TM, 432 RGB, March 3, 1996 (provided by Karen Seto, Center for Environmental Science and Policy, Stanford University)

which is the most ozone polluted watershed in the world. Beyond this indirect impact, urbanisation has begun to consume significant amounts of prime croplands in China, India, western Africa, and almost everywhere, where the rapid development of megacities is underway (Fig. 3.1). Urban-industrial land uses simply outbid agriculture for these lands, shifting the need to increase production on lands elsewhere. For example, in the Pearl River delta of China, 1 376 km² of prime rice land was consumed by urban growth from 1988 to 1996 (Seto et al. 2000).

Urban-related Trends

- Expanded "ecological footprint" on land use and cover, including biogeochemical linkages through pollution of troposphere;
- Loss of prime croplands to urban-industrial expansion.

Grasslands-Pastures. The land-cover conversion trends of grasslands and pastures are poorly understood. Part of the problem lies in definitions of what is to be counted in this land-cover category and how, as attested through comparisons of "systematic" assessments of the long-term trends in grasslands and pastures, which range from losses of only 1% to more than 45%. Agreement exists, however, that grasslands and by extension, arid lands have been extensively modified worldwide, largely degraded in terms of standing biomass and that many of these ecosystem operate under conditions of non-equilibria (Behnke et al. 1993). This recognition, however, does not lend support to various estimates by UNCOD (UN Convention on Desertification) indicating dramatic desertification worldwide; various land-change and environment-development communities have seriously challenged these estimates, suggesting that they are highly exaggerated (Thomas 1997).

Grassland-Pasture Trends

- No agreement on global conversion trends;
- Significant degradation of grasslands worldwide but excessive claims about desertification.

3.2 Causes

Explaining land change beyond several simple and broad lessons has proven difficult, and the problems are heightened by the need for spatially or geographically explicit explanations. It is not sufficient to know the human-environment dynamics that will expand, shrink, or intensify a particular land use; the global change community needs to know where these changes will take place (IGBP-IHDP 1995, 1999). Beyond agriculture, however, the attention paid to general land-change explanations has been sparse, and that to spatially explicit explanations, even more sparse. These circumstances are changing, in part driven by various modelling needs (Veldkamp and Lambin 2001) and by several comparative assessments of explanations that have been undertaken recently (e.g., Lambin et al. 2001).

The most detailed work of this last kind has been directed to tropical deforestation by way of comparison of modelling results and case studies. These comparisons break down the causes of tropical deforestation (as with any land-cover change) into proximate (immediate action leading to change) and distal (forces precipitating the action) kinds (Fig. 3.2). They demonstrate recurrent factors generating deforestation, but these factors come together in highly diverse ways and play out on the landscape similarly. In short, precise explanations – those on which modelling projections can be based – are highly place specific and change by scale of aggregation addressed. Thus, Angelsen and Kaimowitz (1999) find that various models of tropical deforestation recurrently

Fig. 3.2.
Proximate and distal causes of tropical deforestation (adapted from Geist and Lambin 2001)

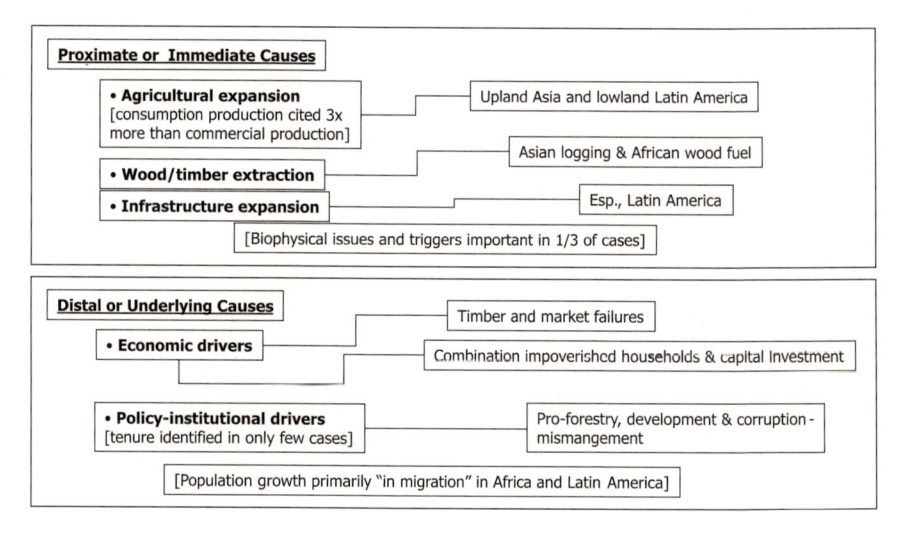

Proximate or Immediate Causes
- **Agricultural expansion** [consumption production cited 3x more than commercial production] — Upland Asia and lowland Latin America
- **Wood/timber extraction** — Asian logging & African wood fuel
- **Infrastructure expansion** — Esp., Latin America

[Biophysical issues and triggers important in 1/3 of cases]

Distal or Underlying Causes
- **Economic drivers** — Timber and market failures; Combination impoverished households & capital investment
- **Policy-institutional drivers** [tenure identified in only few cases] — Pro-forestry, development & corruption - mismanagement

[Population growth primarily "in migration" in Africa and Latin America]

indicate the proximate causes to be some set of increasing agricultural prices, increasing roads, and a shortage of off-farm employment opportunities, while the distal causes involve market and policy failures, and debt and terms of trade. They conclude with the simple lessons that tropical deforestation takes place when and where these forests or the lands on which they grow become profitable to someone or by the ability of some groups to influence policy (e.g., forest access) to enlarge their economic self-interests. Likewise, Geist and Lambin (2001) identify the proximate causes of agricultural, timber, and/or infrastructure expansion into tropical forests, but enlarge the distal factors from economic and policy shifts to population increases driven by in-migration. Significantly, they suggest that the dominant combinations of proximate and distal drivers vary by locations (i.e., Sub-Saharan Africa, Southeast Asia, and Latin America).

Causes
- Spatially explicit understanding of different land-uses and land-covers is in its infancy;
- Tropical deforestation is driven by combination of distal and proximate causes;
- Dominant sets of these causes may be identified by large regions of the world.

3.3 Model-Methods

Surely the most advances to date in land-change and land-cover studies have focused on new or improved methods of detection and observations of land change and models of this change. Worldwide systematic data sets have been developed, reducing the need to rely so heavily on the FAO self-reported country data. These new data sets include 1 km Global Land Cover (Fig. 3.3; DeFries et al. 2000) and croplands (Ramankutty and Foley 1998) as well as longer term land-change data, compiled and evaluated in the HYDE project (Klein Goldewijk 2001). In addition, various compilations of heterogeneous data (historical maps, aerial photography and remote sensing imagery) are being combined to create local and regional land-change assessments. These efforts not only enrich the total data pool, but serve as critical checks on one another.

Various detection and assessment methods have strongly improved our ability to address a variety of land changes underway. For example, detailed studies of grassland burning in portions of the Sahel demonstrate the over- and underestimations of the area annually burned based on the pixel threshold levels used in analysis as well as the seasons in which the imagery is taken (P. Laris, pers. comm.). Satellite data serve more generally as "checks" on ground-based calculations of land change, in some cases seriously challenging the latter, as in the case of cropland coverage in southern China (e.g., Seto et al. 2000). In another example, fine-tuned observations of forest fragmentation and the concomitant enlargement of edge effects increase our ability to address biomass and species losses in this space. Detailed ground studies permit various stages of successional growth to be observed and mapped, providing critical insights for land-change models in which rotational (crop-fallow) cultivation is followed as well as for carbon sequestration calculations (Turner et al. 2001; Watson et al. 2000). And, on the vulnerability front, new techniques and assessment have proven highly robust in projecting the direction of ENSO droughts in southern Africa, providing up to 6–8 weeks lead time in identifying the places to be affected (Eastman and Anyamba 1996).

Finally, major advances are underway in spatially explicit modelling of land changes of various kinds and form various approaches, be they econometric, explanatory, agent based, or scenario-driven (e.g., Angelsen and Kaimowitz 1999; Turner et al. 2001). Increasingly, these models are bringing together previously disparate communities in the human and remote sensing sciences (Veldkamp and Lambin 2001). The modelling community no longer needs to rely on time series generated change rates pro-

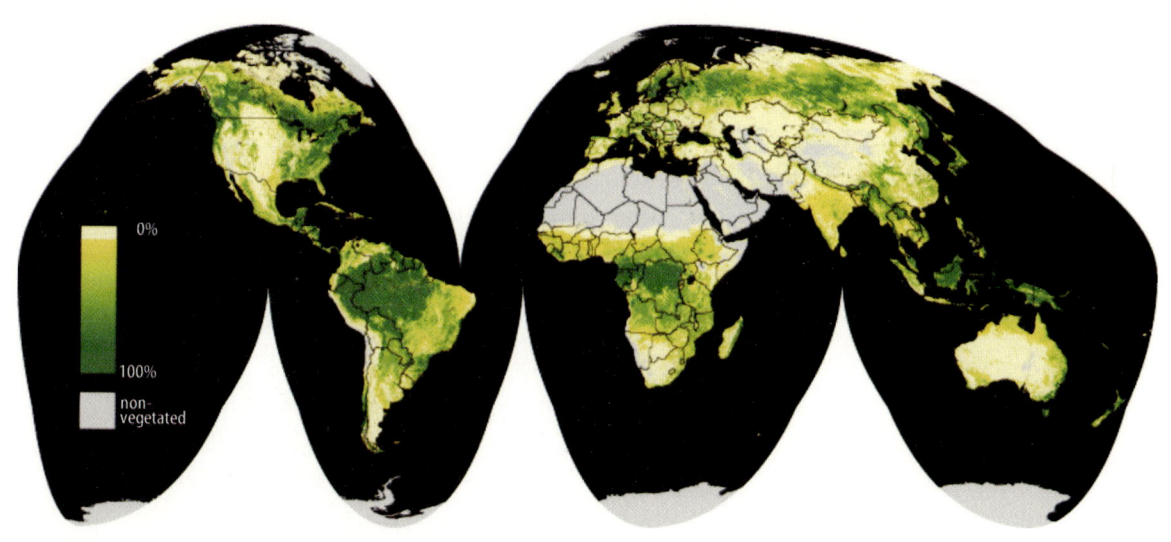

Fig. 3.3. 1 km AVHRR of global tree cover (reproduced from De Fries et al. 2000, with permission from Blackwell Science)

jected into the future or on theoretically based but aspatial models. Rather, the various approaches are joined, linking the robust understanding of behavioural and structural models of decision making with the spatial trends found in time series data, using the two to understand local and regional based land-use/cover changes and to project them into the near future (Fig. 3.4). Such efforts require new measures of calibration and validation suited for spatial modelling (e.g., Walsh et al. 1999), and various developments include the modification of the Kappa statistic and Relative Operating Characteristic (ROC) to apply to land change (Pontius 2000).

Methods and Models

- Systematic global data of land-cover change increase, both current and historical; heterogeneous data sets combined to create local-regional-historical land-change;
- New land-change detection and assessments address grassland burning, forest edges, successional growth, and other such issues;
- New hybrid models of spatially explicit land change explain and project (short term), complete with new measures of calibration and validation.

3.4 Summary and Observations

The advances underway point to the emergence of an "integrated land science" in which the environmental, human, and remote sensing/GIS sciences unite to solve various questions about land-use and land-cover changes and the impacts of these changes on humankind and the environment. This kind of land science is consistent with the integrative restructuring of the IGBP, emerging programs such as the Millennium Ecosystem Assessment, and the move toward sustainability sciences by the IGBP, IHDP, and various national-level global environmental change programmes around the world. The integrated character of this science is such that it invariably requires team-based approaches with high labour and fiscal costs, especially in those cases starting from "ground zero" in terms of teams and data.

The advances outlined in this brief review are substantial and flow from various communities with different ties to global environmental change programmes, illustrating the centrality of land change to the large problems of global change, environment-development, and the transition towards sustainability. It is noteworthy, however, that the land change science lags its counterparts in terms of base detection and firm calculations of use and cover change. The data and estimates underlying the highly studied topic of tropical deforestation, let alone grassland change, are far less rigorous than those employed in studies of the carbon or nitrogen cycles. While perhaps less interesting for much of the land-change community, it is essential that our base understanding of what is changing where and how rapidly improves.

References

Angelsen A, Kaimowitz D (1999) Rethinking the causes of deforestation: Lessons from economic models. The World Bank Research Observer 14:73–98

Behnke R, Scoones I, Kerven C (1993) Range ecology at disequilibrium: New models of natural variability and pastoral adaptation in African savannas. Overseas Development Institute, London

Chameides W, Kasibhatla PS, Yienger J, Levy H II (1994) Growth in contiental-scale metro-agroplexes: Regional ozone pollution and world food production. Science 264:74–77

Daily GC, Söderqvist T, Aniyar S, Arrow K, Dasgupta P, Ehrlich PR, Folke C, Hansson A, Jansson B-O, Kautsky N, Levin S, Lubchenco J, Mäler K-G, Simpson D, Starrett D, Tilman D, Walker B (2000) The value of nature and nature of value. Science 289:395–396

DeFries RS, Hansen MC, Townshend JRG, Janetos AC, Loveland TR (2000) A new global 1 km dataset of percentage tree cover derived from remote sensing. Global Change Biology. 6:247–254

Fig. 3.4.
Cross-tabulation of real and estimated deforestation in southern Campeche and Quintana Roo, Mexico (NASA LCLUC SYPR Project, Clark University). Only legal agricultural lands considered

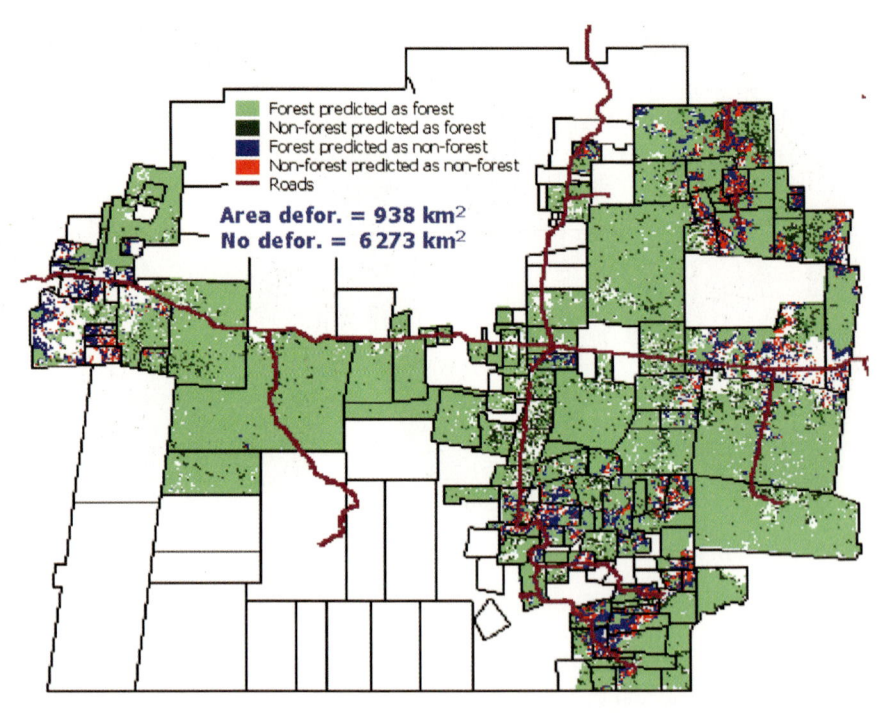

Forest predicted as forest
Non-forest predicted as forest
Forest predicted as non-forest
Non-forest predicted as non-forest
Roads

Area defor. = 938 km^2
No defor. = 6 273 km^2

Eastman JR, Anyamba A (1996) Prototypical patterns of ENSO-drought and drought precursors in Southern Africa. The 13th PECORA Symposium Proccedings, Sioux Falls, South Dakota, 20–22 August 1996, USGS

FAO (1999) State of the world's forests 1999. Rome: Food and Agricultural Organization of the United Nations

Folke C, Jansson A, Larsson J, Costanza R (1997) Ecosystem appropriation by cities. Ambio 26:167–172

Geist H, Lambin E (2001) What drives tropical deforestation? A meta-analysis of proximate and underlying causes of deforestation based on subnational case study evidence. LUCC Report Series No. 4. Louvain-la-Neuve, Be.: LUCC Inernational Project Office

IGBP-IHDP (1995) Land-use and land-cover change science/research plan. IGBP Report No. 35 and HDP Report No. 7, Stockholm: IGBP and Geneva: IHDP

IGBP-IHDP (1999) Land-use and land-cover change (LUCC) implementation strategy. IGBP Report No. 48 and IHDP Report No. 10, Stockholm: IGBP and Bonn: IHDP

Johnson N, Revenga C, Echeverria J (2001) Managing water for people and nature. Science. 292:1071–1072

Kates RW, Clark WC, Corell R, Hall JM, Jaeger CC, Lowe I, McCarthy JJ, Schellenhuber HJ, Bolin B, Dickson NM, Faucheaux S, Gallopin GC, Grübler A, Huntley B, Jäger J, Johda NS, Kasperson RE, Mabogunje A, Matson P, Mooney H, Moore B III, O'Riordan T, Svedin U (2001) Sustainability science. Science 292:641–642

Klein Goldewijk K (2001) Estimating global land-use change over the past 300 years: the HYDE 2.0 database. Global Biogeochem Cy 15: 417–433

Lambin EF, Turner BL II, Geist H, Agbola S, Angelsen A, Bruce JW, Coomes O, Dirzo R, Fischer G, Folke C, George PS, Homewood K, Imbernon J, Lee-mans R, Li X, Moran EF, Mortimore M, Ramakrishnan PS, Richards JF, Skånes H, Steffen W, Stone GD, Svedin U, Veldkamp T, Vogel C, Xu J (2001) Our emerging understanding of the causes of land-use and -cover change. Global Environmental Change 11:261–269

Matson PA, Naylor RL, Ortiz-Monastario I (1998) Integration of environmental, agrinomic, and economc aspects of fertilization management. Science 280:112–115

Mooney HA, Hobbs RJ (eds) (2000) Invasive species in a changing world. Island Press, Washington, DC

Moran EF (n.d.) Progress in the last ten years in the study of land use/cover change and the outlook for the next decade. In: Diekman A, Dietz T, Jaeger CC, Rosa EA (eds) Human dimensions of global change. MIT Press, Cambridge, forthcoming

Nepstad DC, Veríssimo A, Alencar A, Norte C, Lima E, Lefebvre P, Schlesinger P, Potter C, Moutinho P, Mendoza E, Cochrane M, Brooks V (1999) Large-scale impoverishment of Amazonian forests by logging and fire. Nature 398:505–508

Pacala SW, Hurtt GC, Baker D, Peylin P, Houghton RA, Birdsey RA, Heath L, Sundquist ET, Stallard RF, Ciais P, Moorcroft P, Caspersen JP, Shevliakova E, Moore B, Kohlmaier G, Holland E, Gloor M, Harmon ME, Fan S-M, Sarmiento JL, Goodale CL, Schimel D, Field CB (2001) Consistent land and atmosphere-based US carbon sink estimates. Science 292:2316–2320

Pontius R Jr. (2000) Quantification error versus location error in comparison of categorical maps. Photogrammetric Engineering, and Remote Sensing 66:1011–1016

Ramankutty N, Foley JA (1998) Characterizing patterns of global land use: An analysis of global croplands data. Global Biogeochemical Cycles 12:667–685

Rojsraczer S, Sterling SM, Moore NJ (2001) Human appropriation of photosynthesis products. Science 294:2549–2552

Seto KC, Kaufmann RK, Woodcock CE (2000) Landsat reveals China's farmland reserves, but are they vanishing? Nature 406:121

Skole D, Tucker C (1993) Tropical deforestation and habitat fragmentation in the Amazon: Satellite data from 1978 to 1988. Science 44:314–322

Thomas DSG (1997) Science and the desertification debate. Journal of Arid Environments 37:599–608

Turner BL II, Cortina Villar S, Foster D, Geoghegan J, Keys E, Klepeis P, Lawrence D, Macario Mendoza P, Manson S, Ogneva-Himmelberger Y, Plotkin AB, Pérez Salicrup D, Roy Chowdhury R, Savitsky B, Schneider L, Schmook B, Vance C (2001) Deforestation in the southern Yucatán peninsular region: An integrative approach. Forest Ecology and Management 154:353–370

Veldkamp A, Lambin E (eds) (2001) Special issue: Predicting land-use change. Agriculture, Ecosystems and Environment 85

Vitousek PM, Mooney HA, Lubchenco J, Melillo JM (1997) Human domination of Earth's ecosystems. Science 277:494–500

Wackernagel M, Yount JD (2000) Footprints for sustainability: The next steps. Environment, Development and Sustainability 2:21–42

Walsh SJ, Evans TP, Welsh WF, Entwisle B, Rindfuss RR (1999) Scale-dependent relationships between population and environment in northeastern Thailand. Photogramm Eng Rem S 65:95–105

Watson RT, Noble IR, Bolin B, Ravindranath NH, Verardo DJ, Dokken DJ (eds) (2000) Land use, land-use change and forestry. Special Report of the IPCC (Intergovernmental Panel of Climate Change), Cambridge University Press, Cambridge

Chapter 4

Climate Variability and Ocean Ecosystem Dynamics: Implications for Sustainability

Michael J. Fogarty

4.1 Introduction

Climate variability and change will fundamentally alter the physical structure of the oceans with direct implications for marine ecosystems and human societies (IPCC 2001). These changes will be superimposed on other impacts resulting from human activities including fishing, pollution, and habitat loss in coastal areas. Climate change can interact with other human-induced change to alter the fundamental production characteristics of marine systems. Living marine resources have sustained human cultures for millennia as an essential source of protein and as a cornerstone of maritime commerce and trade. However, continuously increasing fishing pressure and demand related to the burgeoning human population have resulted in declines in many previously abundant fish and shellfish populations. It is now clear that humans have the capacity to outstrip the production potential of the world's oceans. Climate change will exacerbate the stress on living marine resources imposed by harvesting and other anthropogenic activities in many regions of the world's oceans.

Total reported yield from marine capture fisheries has levelled off at approximately 85 million t after a period of rapid development over the last half century (FAO 2001). Estimates of the production potential of the seas in coastal and continental shelf systems are on the order of 100 million t. Forty-seven percent of the world's fisheries for which assessments are possible are considered to be fully exploited with no capacity for further development, 18% are considered to be overexploited, 9% are depleted and not currently capable of supporting fisheries, 21% are moderately exploited, and 4% are considered underexploited. It has been estimated that yield could increase by 20 million t with improved management. Consideration of observed yields and estimated production potentials suggests that we are close to the limits of production and yield from capture fisheries in coastal/continental shelf regions of the world ocean and that the potential for changes in production under global change must be carefully considered.

It is currently expected that if primary production does not decrease, the net impact of climate change on yield from global fisheries will be relatively small, with increases in production in mid to higher latitude regions due to a projected increased growing season offsetting losses in other regions (Everett 1995). The important role of climate variability and change on fishery production is clearly evident in the covariation of landings of pink salmon off Alaska and the Pacific Decadal Oscillation Index (Mantua et al. 1997), a measure of temperature conditions in the North Pacific (Fig. 4.1). Changes related to temperature and associated environmental factors (including wind fields and precipitation patterns) have resulted in distinct production regimes in salmon populations.

Although net impacts on fishery production may be neutral, shifts in regional production characteristics are expected to have important local effects on fishery-dependent human communities. Changes in the distribution of key resource species will alter local availability and accessibility, potentially affecting food supply and products for human communities with limited capability to adapt by changing fishing grounds. These impacts are expected to fall disproportionately in the developing world where mobility of fishers is restricted.

Fishery production is a critical issue of global food security. Over 1 billion people are strongly dependent on fish as an integral component of their diet (FAO 2001). Fish contribute a high proportion of the total animal protein intake in Asia (26%) and in Africa (21%) (FAO 2001).

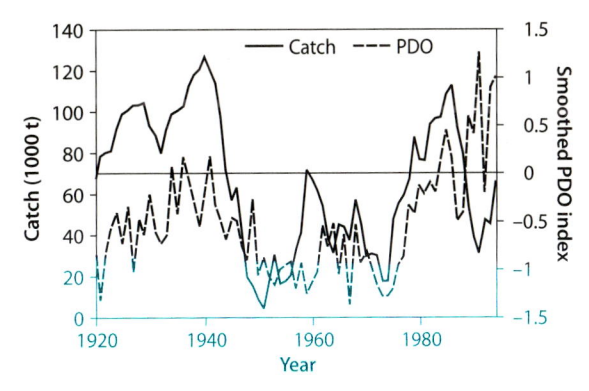

Fig. 4.1. Trends in landings of Alaska pink salmon and the Pacific Decadal Oscillation Index (smoothed using five-year running average)

4.2 Climate Effects on Marine Ecosystems

The atmosphere and oceans are dynamically coupled. Climate change scenarios indicate increases in atmospheric temperatures and regional changes in precipitation patterns. Temperature changes are expected to be greatest at higher latitudes. This differential warming is expected to result in a reduction of latitudinal temperature gradients that may result in a lessening of intensity of important wind fields. Winds drive major ocean currents such as the Gulf Stream that are critical for the redistribution of heat in the oceans and that serve important roles in the dispersal of marine organisms (e.g., Mann and Lauzier 1996).

Changes in water temperature will affect ocean systems in a number of critical ways. Current climate change scenarios indicate an increase in water temperatures within the next century. Increased temperature will have important effects on the distribution of marine species with a general poleward shift in distribution anticipated (Frank et al. 1990; Murawski 1993). Most marine species exhibit distinct thermal preferences with a well-defined optimal temperature. Populations of marine animals at the extremes of their thermal ranges will be adversely affected under current climate change scenarios if redistribution to more favourable conditions is not possible. Populations below their thermal optimum are expected to experience increases in growth rates and production as water temperatures increase. It is expected that increases in production in mid- to high-latitude regions will be a dominant influence in global fishery production due to increases in growth and survival of exploited species (Everett 1995).

Increased temperature will also result in increases in sea level through thermal expansion of sea water and increased melting of ice and snow with resulting runoff into the oceans (Gornitz 1995). Depending on the rate of sea level rise, coastal habitats such as mangroves, sea grass beds, and marshes that serve as important nursery areas for many marine and estuarine species may be adversely affected, resulting in reduced production in coastal areas (Everett 1995).

Increased temperatures are related to coral bleaching events in which the growth and survival of corals are adversely affected (e.g., Wilkinson et al. 1999). Coral reefs are critical centres of biodiversity in marine systems and are highly vulnerable to temperature change.

Increased temperatures will also result in the loss of sea ice. Sea ice is a critical habitat for many species of birds and mammals in polar regions. In the Southern Ocean, sea ice declines have been linked to declines in krill recruitment (Loeb et al. 1997). Krill are a keystone species in the Antarctic food web, and reduction in krill populations can have critical effects on seal and penguin populations. Krill also support an important fishery in the Southern Ocean.

Regional changes in salinity are expected under climate change. Decreased salinity is expected in coastal areas affected by high precipitation and runoff. Increased runoff will intensify buoyancy-driven coastal currents. Conversely, increased salinity is anticipated in offshore areas where higher temperature will lead to higher evaporation rates. Many marine organisms exhibit distinct salinity tolerance levels, and it is anticipated that these changes will contribute to overall changes in distribution patterns of marine species (Everett 1995). Changes in salinity will also affect the density of sea water and hence stratification.

Specific changes can be expected to have different ecological impacts in different regions. For example, stratification of the water column with depth can have positive or negative effects on production. In areas dependent on mixing of nutrient-rich bottom waters with surface waters, stratification can reduce overall productivity. Conversely, in systems with alternative nutrient inputs such as river runoff, stratification can result in increases in productivity if mixing and distribution of phytoplankton cells are constrained to depths where light levels are adequate for photosynthesis.

Upwelling systems are extremely productive regions that support important fisheries for pelagic species such as sardines and anchovies. If the tradewinds and other important wind fields are diminished, reduced upwelling is expected. Decreased upwelling will result in lower levels of system productivity as a result of lowered nutrient regeneration. However, there may be conditions where the temperature differential between land and sea will increase, potentially leading to increased winds and upwelling (Bakun 1990) and higher overall productivity.

During El Niño events, upwelling is diminished with dramatic changes in temperature, productivity and species distribution patterns. An increase in the frequency and intensity of El Niño events has been predicted in some climate models. An interaction between harvesting and El Niño is evident in the records of the Peruvian Anchovetta fishery, once the largest single species fishery in the world. A combination of over-exploitation and the 1973 El Niño led to a sharp decline in the

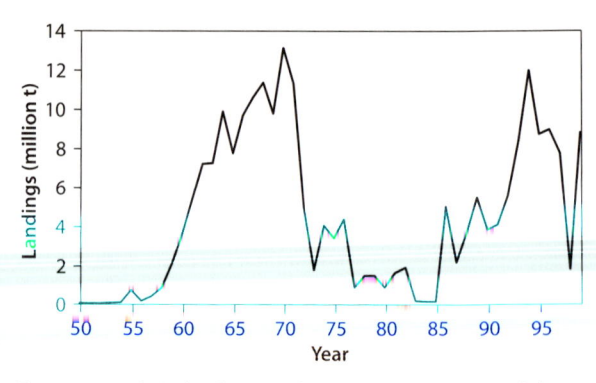

Fig. 4.2. Trends in landings in the Peruvian anchovetta fishery. Strong El Niño events occurred in 1973, 1983, and 1997/98

fishery, with a collapse in the mid-1980s with the strong 1983 El Niño (Fig. 4.2). The population and fishery recovered in the 1990s, but landing dropped precipitously following the recent 1997/1998 El Niño. The fishery rebounded in 1999, suggesting that availability rather than abundance had been affected.

4.3 Implications for Sustainability

Understanding the potential for synergistic interactions between climate and exploitation regimes is critical in devising appropriate management approaches. Populations of exploited marine species are strongly shaped by environmental variability on a broad range of space and time scales. High frequency variation in environmental forcing plays an important role in variability in the growth and survival of young fish and shellfish, while lower frequency forcing on broad spatial scales affects overall levels of productivity on multidecadal time scales. Fishery management strategies must contend with the uncertainties introduced by large-scale variation in the number of young surviving the critical early months of life and must consider the implications of climate-induced changes on broader time horizons.

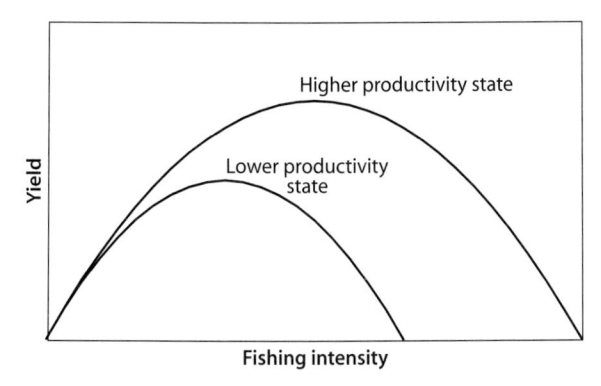

Fig. 4.3. Sustainable yield curves under two environmental regimes. In the high productivity state, yield levels and the resilience to exploitation are higher. Under the lower productivity state, yield and resilience are reduced (reprinted from Fogarty 2001, with permission of Academic Press)

We expect that under a given set of environmental conditions, the yield will be maximised at an intermediate level of fishing intensity (Fig. 4.3). A shift in climatic conditions fundamentally changes the production characteristics of these systems and the vulnerability to exploitation. A shift to less favourable environmental conditions can lead to a synergistic effect where levels of exploitation that were sustainable under more favourable conditions are no longer sustainable, and a population collapse is predicted. Understanding the implications of climate-induced change in production is paramount and will require fundamental adjustments in management strategies in a changing world ocean.

References

Bakun A (1990) Global climate change and intensification of coast ocean upwelling. Science 247:198–201
Everett (1995) Fisheries. In: Watson RT, Zinyowera MC, Moss RH, Dokken DJ (eds) Climate change 1995. Impacts, adaptations and mitigation of climate change: Scientific-technical analyses. Contribution of Working Group II to the Second Assessment Report of the Intergovernmental Panel on Climate Change, pp 511–538
FAO (2001) State of world fisheries and aquaculture
Fogarty MJ (2001) Dynamics of exploited marine fish populations. In: Steele JH, Turekian KK, Thorpe SA (eds) Encyclopedia of ocean sciences. Academic Press, London, pp 774–881
Frank KT, Frank RI, Perry RI, Drinkwater KF (1990) Predicted response of Northwest Atlantic invertebrate and fish stocks to CO_2 induced climate change. Trans Am Fish Soc 119:353–365
Gornitz VM (1995) Sea-level rise: A review of recent past and near future trends. Earth Surf Process Landforms 20:7–20
IPCC (2001) Summary for policy makers. A Report of Working Group I of the Intergovernmental Panel on Climate Change
Loeb V et al. (1997) Effects of sea ice extent and krill or salp dominance on the Antacrctic food web. Nature 387:897–900
Mann KH, Lazier JRN (1996) Dynamics of marine ecosystems: Biological-physical interactions in the oceans. 2nd Edition. Blackwell Science, Cambridge
Mantua NJ, Hare SR, Shang Y, Wallace JM, Francis RC (1997) A Pacific interdecadal climate oscillation with impacts on salmon production. Bull Am Met Soc 78:1069–1079
Murawski SA (1993) Climate change and marine fish distributions: forecasting from historical analogy. Trans Am Fish Soc 122:658–667
Wilkinson C, Linden O, Hodgson G, Cesar H, Rubens J, Strong A (1999) Ecological and socio-economic impacts of 1998 coral mortality in the Indian Ocean: An ENSO impact and a warning of future change? Ambio 28:188–196

Chapter 5

Food in the 21st Century: Global Climate of Disparities

Mahendra Shah

5.1 Food in the 21st Century: Global Climate of Disparities

We are living in a unique and defining moment in history. It is unique with respect to the progress in science and technology that has been achieved in the past half century. Beginning with men on the moon in the 1960s and continuing with the green revolution of the 1970s, the information revolution of the 1980s, and the genetic revolution of the 1990s, the 20th century ended with the mapping of the human genome.

These scientific and technological achievements are formidable accomplishments with significant potential for the future. But today we live in a world of disparities where a fifth of the global population exists in poverty and hunger. The poor are poor because they lack tangible assets, lack formal education and technical skills and have little access to such basic needs as health care and safe shelter. And the poor often face political and social discrimination.

Some 800 million people go hungry every day, and over one billion live on less than a dollar a day. Without social, economic, and scientific progress, a third of the world's expected population of some 9 billion could be living in poverty in the second half of the 21st century. Every minute of every day, 15 children and 15 adults die of hunger in the developing world. During the course of this four-day meeting alone, some 200 000 people will die from a lack of food. This food insecurity affecting over 10% of the world's population is a sad indictment of the world's failure to respond adequately in a time of unprecedented plenty.

The challenge of poverty reduction is not an option but an imperative in a world of interdependence, reciprocity and interpenetration. The progress in science and technology, including the knowledge revolution and environmentally sound management of natural resources, has the potential to reshape and manage the emerging challenges of the 21st century. It is a defining era of knowledge-based decisionmaking that can put the world on a path toward equity and sustainability.

5.2 The Critical Role of Knowledge

It is knowledge that is available, accessible, and affordable that will drive progress in the 21st century, and perhaps become more relevant than capital. Living at a time of rapid changes and challenges, timely action based on credible and comprehensive information is critical.

Information becomes knowledge when it has utility. It needs to be policy relevant, scientifically credible, and available at the right place at the right time. Knowledge empowers and enables. In an open forum, it promotes synergy, transparency, and accountability.

Science and technology are increasingly becoming proprietary, and owning knowledge is becoming the order of the day. Does this mean that those who cannot afford it will be denied the fruits of scientific participation and progress? The number of patent applications has grown from 1 million in 1985 to over 7 million today. Recently a US company even attempted to patent turmeric – an herb valued for centuries in India for its medicinal properties! Fairness, justice, or ethics – which way do we turn?

The internet and other communication developments have ushered in a new era of knowledge sharing. Access to the internet is growing rapidly; in the next 4 years the number of people with access will double to one billion. In the past 10 years, the number of websites increased from 1 000 to 20 million today. But it is a world of disparities: it costs 2 months average wage to access the internet in Bangladesh, in comparison to 1% of the monthly average wage in the United States. Also some 50% of the population in the United States and a third in the European Union are internet users, whereas less than 0.5% of the population in Asia and Africa use the internet. Such a continuing knowledge divide would extenuate the wide inequity gaps and hinder progress towards goals of sustainable development, especially the reduction of hunger and poverty in the world.

5.3 Environment and Sustainable Development

From Stockholm to Rio and now to Johannesburg, in just 30 years we have begun to understand our living environment. Sustaining this environment demands local, national, and international action.

Climate change, air and water pollution, pests and diseases, economic and social turmoil, and even genetically modified crop contamination do not recognise or respect political and geographical boundaries. It is critical that we act based on rational analysis, but at the same time we must be wise in our actions. Without responsible decisionmaking, compromises, and even sacrifice, the children born at the end of the 21st century may face a bleak future indeed. We must begin now to consider our response to global changes and challenges, because the actions taken today will affect the quality of life for us and for generations to come.

The rapid population growth of the past 50 years combined with the explosive increase in consumption is threatening the ability of ecosystems to supply our needs and to absorb the impacts of pollution and waste generated by human activities. The good news is that the rate of population growth is decreasing, and world population is very likely to stabilise towards the close of the 21st century. However the next 50 years are critical, as the number of people continues to increase, especially in Africa and Asia. These two regions will add some 3 billion people to their populations by the year 2050.

Many developed countries and some developing countries such as China are likely to see a significant reduction of their populations. At the same time, new problems of a demographic divide will come to the fore: an aging population in some developed countries and a younger population in developing countries, with an implication of human capital migration; underconsumption vs. overconsumption – synonymous with the poor-rich divide with implications for environmental change and degradation; and a science and technology and communication divide, exacerbating progress towards an equitable and a sustainable world.

On the economic front, the disparities around the world are growing ever wider, while at every meeting and forum the international political and scientific community continues to stress the pressing need to reduce the widening gaps between the haves and have-nots.

5.4 Global Environmental Change

Food production systems interact with land resources, forest ecosystems, and biodiversity, and climate change will affect these systems both positively and negatively. Ensuring soil fertility, genetic diversity, agricultural water resource management, and adapting to the impacts of climate change are critical to enhancing production, while agricultural practices of inefficient fertiliser and pesticide use as well as lack of land and water conservation measures and conversion of forest areas will result in irreversible damage to ecosystems and loss of production potentials.

The rapid land-cover changes, biotic fluxes, and extinction of living species of the past 50 years are worrisome. The disturbing truth is that we do not even know what biodiversity is being lost around the world – in our forests, in the oceans, and on land. China, which once had 10 000 land-race varieties of wheat, now has less than a thousand. No one knows what genetic traits leading to insect and disease resistance, stronger plants, higher yields, or even better tasting crops may have been irrevocably lost.

The need for food for an increasing population is threatening natural resources, as people strive to get the most out of land already in production or push into virgin territory for agricultural land. The damage we are inflicting on the environment is increasingly evident: arable lands lost to erosion, salinity, desertification, and urban sprawl, disappearing forest and threats to biodiversity, and water scarcities.

About 70% of the world's fresh water goes to agriculture, a figure that approaches 90% in countries such as China and India, which rely on extensive irrigation. Though renewable, fresh water is a finite resource, not evenly distributed across countries, regions or even seasons. Two-thirds of the world's population live in areas that receive only one-quarter of the world's annual rainfall, while sparsely populated areas as the Amazon Basin receive a disproportionate share. Because of extensive upstream use, some of the world's major rivers – the Nile and the Ganges, for instance- barely run into the sea any more. The growing water scarcity in the future will pose a serious threat to food security, poverty reduction, human health, and protection of the environment.

Thirty years ago, the world faced a global food shortage that some predicted would lead to catastrophic famines. The danger was averted because an international research effort enabled scientists to develop and farmers to adopt high-yielding varieties of major food crops. The lessons of that green revolution indicate that an integrated biological, environmentally sound, and socially viable strategy has to be at the core of the next precision green revolution.

In the 21st century, we now face another threat – perhaps a more devastating environmental threat of global warming and climate change. There may be uncertainties, but we cannot be complacent, not when the most fundamental of human survival need for food is at risk.

5.5 Global Agro-Ecological Assessment

Given the complex and interlinked components of the food security challenge in the 21st century, it is clear that solutions that deal with one part – for example, crop productivity, land use, water conservation, or forest protection – will not be sufficient. The issues are connected and must be dealt with as an interlinked holistic system to ensure sustainable management of natural resources.

Future land use and agricultural production are not known with certainty:

- What will be the spatial extent and productivity of arable land in each country?
- What will be the availability and adoption of agricultural technology?
- What new improved genetic crop varieties will come available?
- What impact will climate change have on specific countries and regions, on forest ecosystems, on biodiversity and on water resources?

The agro-ecological zone (AEZ) methodology follows an environmental approach: it provides a standardised framework for the characterisation of climate, soil and terrain conditions relevant to agricultural production. Crop modelling and environmental matching procedures are used to identify crop-specific environmental limitations under assumed levels of inputs and management conditions. The main elements in the AEZ framework are shown in Fig. 5.1.

The AEZ approach is a GIS-based modelling framework that combines land evaluation methods with socioeconomic and multiple-criteria analysis to evaluate spatial and dynamic aspects of agriculture. Land and climate resources are assessed to quantify crop production. All important food and fibre crops are considered.

The global AEZ resources database comprises of a digitised overlay of climate (derived from 1901 to 1996 time series climate data set, Climatic Research Unit, University of East Anglia); FAO/UNESCO soil map of the world linking soil associations and attributes, elevation and slope distribution; global land cover data set: crops, forests, woodlands, wetlands; and spatial population distribution by grid cells. A large amount of agronomic farm management data from around the world has also been incorporated.

A total of some 2.2 million-grid cells covering all countries' land resources are delineated. For most countries, a grid cell amounts to a land area of some 5 to 10 thousand ha. For each grid cell, the assessment considers 28 possible crops at three levels of inputs, namely low, intermediate and high. The high level assumes the

Fig. 5.1.
AEZ methodology (reproduced with permission from Land Use Project/IIASA 2001)

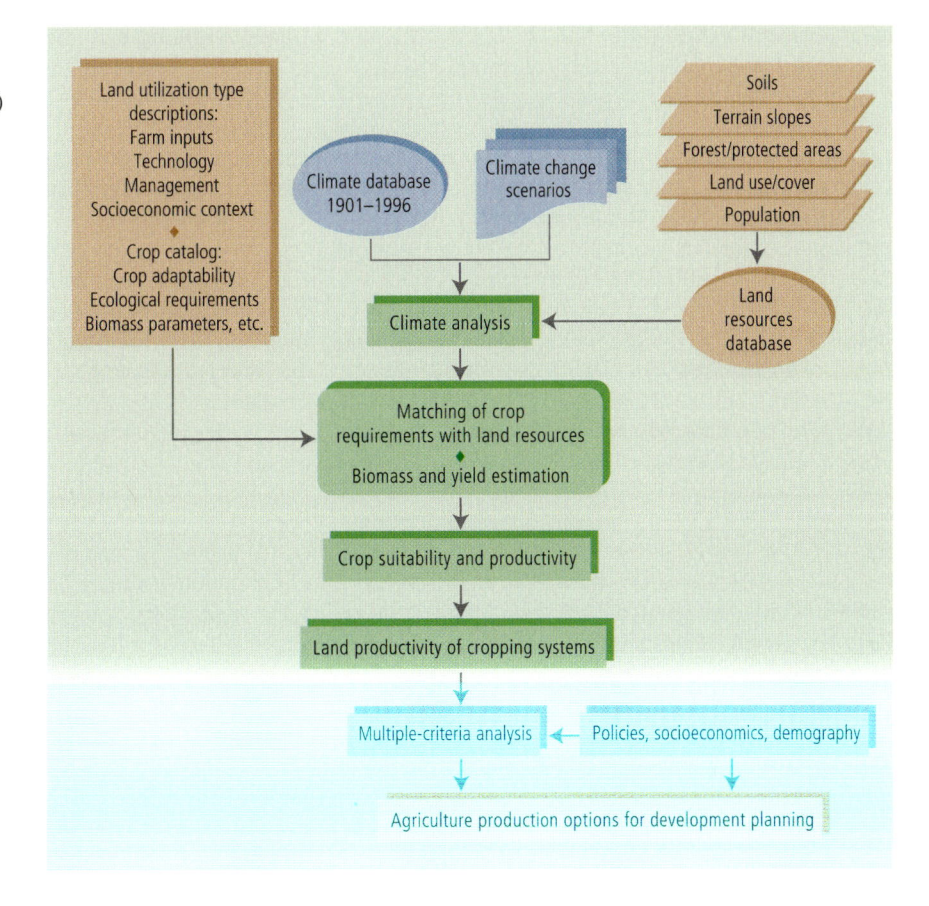

best farming technology, inputs, and management known today. Sustainable natural resources management and precision agriculture are also at the core of this technology. Future developments in new crop varieties and productivity can be incorporated into the scenario approach.

5.6 Global AEZ Findings

Natural resources constraints to crop production. Three-quarters of the global land surface – over 10.5 billion ha – suffer rather severe constraints for rain-fed crop cultivation: 13% is too cold, 27% is too dry, 12% is too steep, and 65% is affected by unfavourable soil conditions, with multiple constraints coinciding in some locations.

Climate change will have positive and negative impacts, as some constraints will be alleviated while others may increase. This detailed information, at the country level, is relevant to agricultural research – including biotechnology – which is focused on the priority and viability of relieving constraints, covering land areas where the returns to research investments could be substantial. For example, agricultural research in Mexico has resulted in the application of biotechnology to increase plant tolerance to aluminium, thus countering soil toxicity problems common in some tropical areas.

Land with cultivation potential. In Asia and Europe, the rain-fed land that is currently cultivated amounts to 90% of the land that is potentially suitable or very suitable for agricultural production. In North America, some 75% of the potentially suitable or very suitable land is currently under cultivation. By contrast, Africa and South and Central America are estimated to have some 1 billion ha of land in excess of the currently cultivated land of some 350 million ha. However, most of this additional cultivable land is concentrated in just seven countries – Angola, Congo, Sudan, Argentina, Bolivia, Brazil, and Colombia.

Cultivation potential in forest ecosystems. About a fifth of the world's land surface – some 3 billion ha – is under forest ecosystems. Eight countries – Russia, Brazil, Canada, the United States, China, Australia, Congo, and Indonesia – account for 60% of the world's forest land. During the past decade, some 127 million ha of forests were cleared, while some 36 million ha were replanted. Africa lost some 53 million ha of forest during this period – primarily from expansion of crop cultivation.

The AEZ results show that some 470 million ha of land in forest ecosystems have crop cultivation potential. However, using this land for agriculture would have serious implications, as forests play a critical role in watershed management and flood control, and serve as carbon sinks and stores of biodiversity.

Irrigation and multiple cropping. Currently about 200 million ha of arable land are irrigated in the developing countries, and some 54% of this is accounted for by China and India. Irrigation contributes some 40% of production on 22% of cultivated land in the developing countries. For cereals, the major component of food consumption, the share of irrigated production in total world production is almost 60%. The AEZ results for cereal production highlight the important role of multiple cropping and irrigation:

- Rain-fed multiple cropping increases production by about 25% in developing countries and some 10% in developed countries, in comparison to a single rain-fed crop per year;
- Rain-fed and irrigated multiple cropping increases production by about 75% in developing countries and some 50% in developed countries in comparison to a single crop per year.

The AEZ methodology and the spatial database provide a comprehensive and detailed basis for ecological assessments and quantification of regional and national crop production impacts of climate change.

5.7 Global Warming and Climate Change

Global warming will affect agro-ecological suitability of specific crops as well as their water requirements. It may also lead to increased pest and disease infestations.

The increasing atmospheric concentration of carbon dioxide will enhance plant photosynthesis and contribute to improved water-use efficiency.

Increased climate variability and extreme events are reported in some countries. In the absence of mitigation and response capacities, losses from damage to the infrastructure and the economy, as well as social turmoil and loss of life could be substantial. And this burden will fall on the poorest people and in the poorest countries.

It is only in poor countries that drought turns to famine and often that results in substantial economic losses, population displacement, suffering and loss of life. The social and economic costs of such occurrences may undo, just in a day or a month, the achievements of years of development efforts. The challenge of integrated mitigation and adaptation to climate change, variability and extreme events will entail incorporating these issues in long-term strategies for development that are equitable and sustainable.

Responses to climate change can be of two broad types. The first type is adaptive measures to reduce the impacts and risks, and maximise the benefits and opportunities of climate change, whatever its cause. The other type of response involves mitigation measures to reduce human contributions to climate change. Both adaptive measures and mitigation measures are necessary elements of a coherent and integrated response to climate change. If future emissions are higher, the impact will be stronger, and vice versa. At the same time, no matter how aggressively emissions are reduced, climate change is a reality for the 21st century, since existing emissions in the atmosphere will remain for decades to come. Thus, adaptation to climate change is inevitable.

If we are to stem global warming, we have no choice but to reduce the rapidly increasing emissions of greenhouse gases such as carbon dioxide. But in doing so, the contribution to and consequences of such emissions as well as the different national development needs and priorities have to be central in reaching economically efficient and environmentally effective agreements.

Here, the good news is that the scientific understanding of global warming is growing by the day, and we are already at the point where uncertainty is no longer an acceptable excuse for inaction. These are the challenges that politicians, policy makers, and scientists must meet, in the interest of everyone everywhere.

The AEZ climate impact assessment is based on a range of projections of the ECHAM4 model of the Max Planck Institute of Meteorology, the HadCM2 model of the Hadley Centre for Climate Prediction and Research, and the CGCM1 model of the Canadian Center for Climate Modeling.

All three climate models predict that global warming will occur and that the heat index will rise. Precipitation is more likely to come in heavy and extreme events. In some cases, the Canadian model predicts drier conditions than do the Max Planck and the Hadley models. There is uncertainty in future climate prediction. Our aim has been to analyse the robust conclusion of various climate model predictions and in this context, an analysis of the AEZ results is also relevant to further improvements and refinements of the various climate models.

The AEZ approach fully accounts for optimal adaptations of crop calendars as well as switching of crop types. Yield increases resulting from higher carbon dioxide concentrations in the atmosphere are also incorporated.

5.8 Impact of Climate Change on Worldwide Cereal Production

The results of changes in cereal production in the 2080s based on the three climate models, for single rain-fed cropping, multiple cropping, and multiple rain-fed and irrigated cropping on currently cultivated land and at a high level of inputs and management, are illustrated in Fig. 5.2. The number of countries affected is shown in the bar charts.

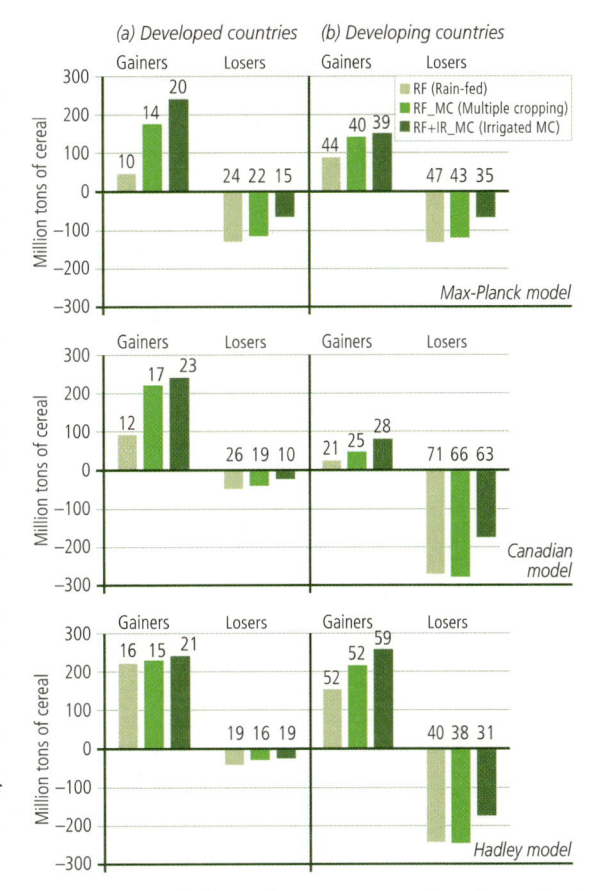

Fig. 5.2. Impact of climate change, 2080s; number of countries is shown above *bar charts* (reproduced with permission from Land Use Project/IIASA 2001)

With climate change, multiple cropping, in comparison to a single crop per year, provides an additional potential production increase of about 120 million t in developed countries for the Max-Planck and Canadian models. In the case of the Hadley model, the increase is much less pronounced at 10 million t. For the developing countries, the Max-Planck and Hadley results show an increase in production of some 55 million t, and the Canadian model shows a production increase of some 15 million t.

When irrigation is considered as well, depending on the climate projection model, further net increases of 10 to 60 million t for the developed countries and some 10 to 40 million t are attained for the developing countries.

For the developed countries, climate change impact for the rain-fed multiple cropping and irrigation results highlight a net gain in cereal production of some 200 million t in the case of all three climate models. Here about half the developed countries gain, and the remainder lose or have little change in cereal production.

For the developing countries, the net gain is some 30 million and 50 million t for the Hadley and the Max Planck models respectively and a net loss of about

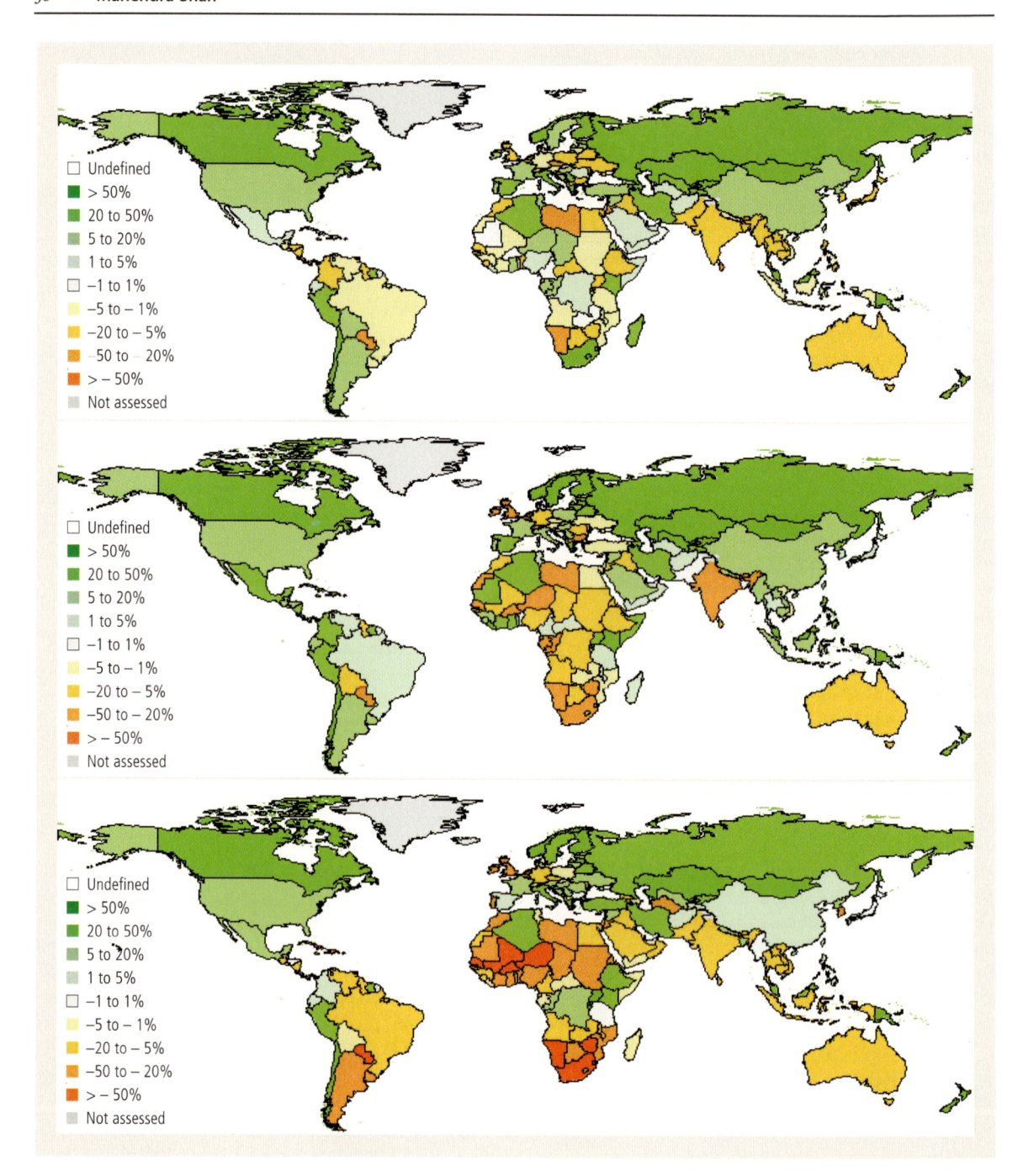

Fig. 5.3. Country-level climate change impacts on cereal production potential on currently cultivated land 2080s (reproduced with permission from Land Use Project/IIASA 2001)

170 millions for the Canadian model. Of the 117 developing countries in the world, 39, 59 and 28 countries gain and 35, 31 and 63 countries lose cereal production due to climate change for the Max Planck, Hadley and the Canadian models respectively.

The results highlight that climate change will benefit the developed countries in terms of net gain in cereal production due to climate change substantially more than the developing countries.

The results in terms of percentage losses and gains for individual countries are shown in map form in Fig. 5.3. Among the developed countries, the winners for all three climate models include Canada, the United States, Spain, France and Italy while losers include the United Kingdom, Germany, Poland and Australia. In the developing regions, India, Thailand, Colombia and many sub Saharan African countries lose production while countries such as China, Mexico, Chile and Kenya gain production.

At the global level, the gain in cereal production due to climate change amounts to some 230 million t for the Max Planck and the Hadley models and only about 20 million t for the Canadian model. However, in spite of this positive global outcome, there is profound concern for many developing countries that lose production due to climate change.

An important feature of the global agro-ecological assessment of climate-change impact is that it is a uniform assessment of all developed and developing countries. This provides for a level comparison of the impacts of climate change. The information is especially relevant to most developing countries and some developed countries that have not yet assessed the impact of climate change on their economies and environment, especially their agriculture sector. Countries such as the United States, the United Kingdom, Canada, France, Germany, and Japan have undertaken scientific assessments of the potential impacts of climate change on their economies and the environment.

Any negotiations, such as the Kyoto Protocol, between the well informed and the less informed will always be constraining. We hope that the results of the AEZ study with a worldwide coverage will contribute to this need for knowledge and information.

5.9 Food Security and Climate Change

Of the 117 developing countries in the world, some 94 countries account for the 792 million undernourished people, as estimated by the Food and Agricultural Organisation of the United Nations. Sixteen of these countries, each with a relatively high per capita GDP of over US$3 000, are not considered here.

Of the remaining 78 countries, 28 countries with a population of 2.2 billion account for 223 million undernourished, and their average daily per capita calorie deficit is about 220 calories. China is among this group of countries. Another 25 countries including India account for 339 million undernourished out of a total population of some 1.5 billion. The average daily per capita deficit of this group of countries is 285 calories. The remaining 23 counties have a total population of some 460 million, of which 220 million are undernourished with a daily per capita deficit of some 360 calories.

At present, the total population of these 78 countries amounts to some 4 billion, and it is projected to increase to over 7 billion by 2050. Currently over half of the populations in most of these countries derive their livelihoods from agriculture. Also, in many of these countries agriculture accounts for 20% to 30% of the total gross domestic product. The current food gap for the undernourished population of these countries is estimated at some 25 million t.

The impact of climate change on potential domestic cereal production in these food insecure countries is shown in Fig. 5.4. Depending on the climate model, some 17 to 37 countries gain cereal production due to climate change. Among these countries, China, with some 140 million undernourished and a corresponding food gap of some 4 million t, gains about 100 million in cereal production due to climate change. In contrast, India, accounting for some 200 million undernourished and an equivalent food gap of about 6 million loses some 30 million t in cereal production due to climate change.

The impact of climate change on cereal production is cause for serious concern in some 25 to 45 "losing" developing countries (Fig. 5.3). These countries have a total combined population of about 1.3 billion to 2.1 billion, of which about a fifth of the population is undernourished. Comparing the decrease of over 60 million t cereal production for the Max Planck model and some 150 million t for the Hadley and the Canadian models, with the current food gap for the undernourished of some

Fig. 5.4.
Food security: Impact of climate change on food production, 2080s (reproduced with permission from Land Use Project/IIASA 2001)

Model	Number of Countries	1995 Population (millions)	Under-nourished (millions)	1995 Cereal Production (million t)	1995 Cereal Gap (million t)	2080s Climate Impact (million t)
Losing						
Max-Planck	27	1661	386	362	−12	−60
Hadley	25	1379	321	277	−10	−156
Canadian	45	2077	396	467	−12	−135
Winning						
Max-Planck	20	1592	210	481	−6	99
Hadley	37	2057	275	598	−8	192
Canadian	17	540	166	100	−6	42

Fig. 5.5.
Climate change impacts and carbon dioxide emissions, 2080s: Fairness and equity (reproduced with permission from Land Use Project/IIASA 2001)

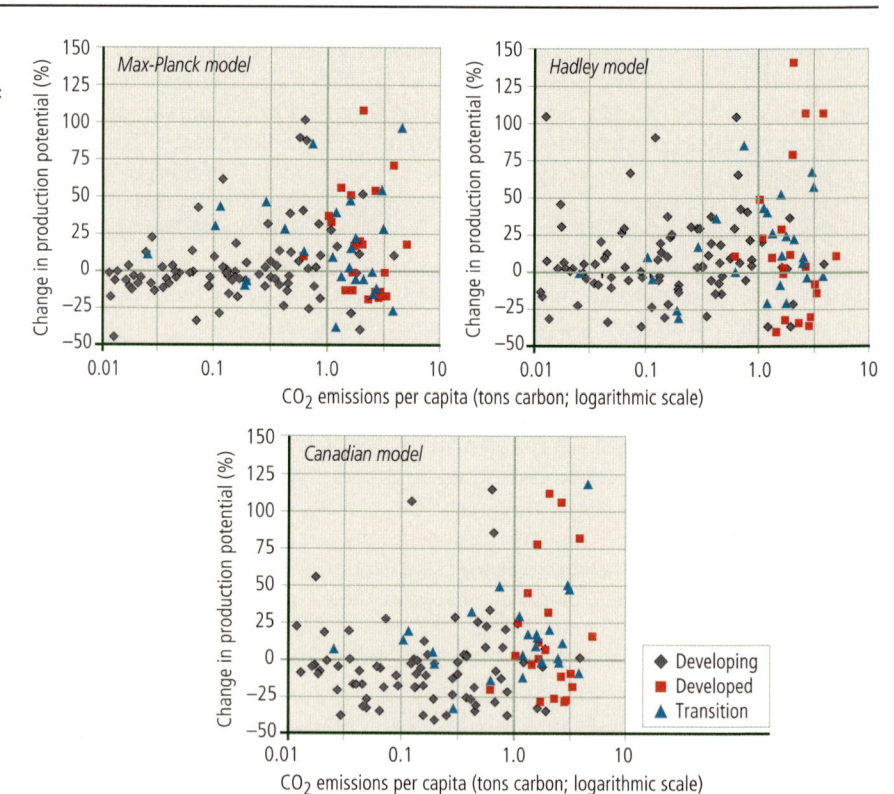

10 to 12 million t, the substantial loss due to climate change in domestic production in the 2080s implies that the number of the undernourished may drastically increase.

Many of these countries are poor, agricultural-based economies. They often lack the foreign exchange to finance food imports. Hence any domestic production losses resulting from climate change will further worsen the prevalence and depth of hunger, and this burden will undoubtedly fall disproportionately on the poorest and the most vulnerable.

5.10 Climate Change Impact: Fairness and Equity?

Global warming raises the issue of fairness, as illustrated in Fig. 5.5. This shows the magnitude of climate change impacts on cereal production potential in relation to average cumulative (1950 to 2000) per capita carbon dioxide emissions. The cumulative emissions over the past 50 years for the developing countries, accounting for more than four-fifths of the world's population, totals less than a quarter of global emissions. Yet, many of these countries will suffer substantially from the impact of climate change on food production.

The world community of nations must fairly and equitably meet the challenge of addressing climate change mitigation policies. This must take into account differences between nations in their past and future emissions, as well as socioeconomic considerations. The timely implementation of economically efficient and environmentally effective international agreements on climate change and national adaptive measures will be critical in the context of achieving worldwide societal goals of equity and sustainable development.

5.11 Concluding Remark

Diverse and disparate as they are, humans are a single species on Earth. They bear responsibility for most of the global changes currently under way. It seems only right that humanity should give a moment of thought for the many other countless living species that also inhabit the Earth and bear the brunt of the impact of human activities. Threatened with extinction, for example, members of the animal world have no voice and they are not often seen. But their plight must not be forgotten as humankind efforts to secure its own future.

General References

Fischer G, Shah M, van Velthuizen H, Nachtergaele F (2001) Global agro-ecological assessment for agriculture in the 21st century. IIASA Summary Report, July 2001

Fischer G, Shah M, van Velthuizen H, Nachtergaele F (2001) Global agro-ecological assessment - methodology and results. IIASA Research Report, forthcoming

Higgins G, Kassam A, Naiken L, Fischer G, Shah M (1983) Potential populations supporting capacities of lands in the developing world. Final report of the Land Resources for Populations of the Future, FAO/IIASA/UNFPA, Rome

Chapter 6
Equity Dimensions of Dam-Based Water Resources Development: Winners and Losers

Madiodio Niasse

6.1 Introduction

This chapter analyses the changes in patterns of access to and control of water resources that result from dam construction. It builds extensively on examples from the Senegal River (the author's personal work) and other watercourses around the world (with an emphasis on some of the river basins studied by the World Commission on Dams[1]). It provides an illustration of the complexity and importance of equity issues in river basin contexts, especially when major changes occur as a result of dam construction. Secondly, the paper elaborates on the role of dams as a means of reallocating water resources. Thirdly, it presents some of the mechanisms that can be considered to improve equity dimensions of dam-based development interventions.

6.2 Why is Equity Relevant in River Basin Development Contexts? An Illustration from the Senegal River

In 1988, I started long-term village-based anthropological research to monitor the socioeconomic aspects of the newly completed Diama and Manantali Dams on the Senegal River. The Senegal River is the second longest in West Africa (1 800 km long). It is formed by the merging of multiple tributaries at Bakel, 800 km from Saint-Louis where the river empties to the Atlantic Ocean. Below Bakel, the river enters a depression, meanders and creates a fertile floodplain, annually inundated when the river spills out of its main channels. Each year, between August and October, the area inundated by the river varies from a few hundred ha in years of severe water deficit to more than 500 000 ha in the wettest periods. This huge floodplain is inhabited by more than half a million people whose production systems are highly dependent on the natural flow regime of the river. The main components of these production systems (agriculture, livestock production and fishing) are practised in such a way that they temporally and spatially rely upon each other. In these systems the river and its floodplain are the backbone of an integrated agro-halio-pastoral resource management. Despite the fact that this system has been seriously disrupted since the late 1980s, its basic principles are still largely adhered to by local communities.

In August/September 1990, OMVS (Organisation pour la Mise en Valeur du Fleuve Sénégal), decided not to generate artificial flood releases that would, combined with the uncontrolled flow of other tributaries, create flood patterns similar to natural pre-dam conditions (Hollis 1996). They stressed the imperative of speeding up the filling of the reservoir of the newly completed upstream Manantali Dam. This decision resulted in a no-flood situation, a man-made drought season. To appease riverine populations' anxieties, OMVS downplayed the impact of the no-flood situation by explaining that the loss of 50 000 ha of recession agriculture normally available in an average year of flooding can be easily compensated by adding 1 000 ha of irrigated land in the off-season. To achieve this, they said the Manantali Dam would be managed in a way that a minimum dry season flow is guaranteed to allow such an increase in the area irrigated. The rationale was that the average yield in recession agriculture (400 kg of sorghum per ha) is 10 times lower than the productivity of irrigated land (more than 4 000 kg of paddy rice per ha) (Salem-Murdock and Niasse 1993).

This reasoning has serious weaknesses that illustrate the need to pay more attention to equity dimensions of dam-based development interventions. First, equating the annual flood to recession agriculture takes a restrictive view of water use, only taking into consideration the most immediate and manifest uses of water resources, and therefore ignores many other social and economic activities more or less dependent on them. In this example, fishing activities, herding, groundwater recharge, and forestry are as dependent on the annual flood as recession agriculture. In addition to the role of the annual floods on forestry and groundwater recharge, it is estimated that the inundated floodplain generates on average 70 kg of fish per ha for an estimated

---

[1] Although I build heavily on the work of the World Commission on Dams (WCD), the ideas expressed here are my own and do not necessary reflect the views of the WCD.

Table 6.1. Household access to recession and irrigation in the middle valley of the Senegal River (SRBMA-IDA unpublished field data)

Reaches of the Middle Valley (upstream to downstream)	Number of households	Percentage having access to recession agriculture	Percentage having access to irrigation
Reach A	45	18	2
Reach B	34	44	21
Reach C	53	45	45
Reach D	38	74	0
Reach E	32	0	59
Reach F	25	8	80
Reach G	37	70	49
Reach H	32	78	81
Reach I	35	6	54
All	331	39	40

400 000 ha of floodplain and 0.35 Tropical Livestock Unit per ha (1 *TLU* = 250 kg) annually. This cannot be compensated for by the proposed increase in the area irrigated.

Secondly, the suggested compensation mechanism (irrigation vs. recession agriculture) seems logical at the aggregated basin-level, but is inoperative at the actual units of production (the households of the middle valley). All households involved in recession agriculture (and which therefore stand to lose from a no-flood situation) do not necessarily have access to irrigation. Table 6.1 (based on a survey conducted in the Senegal River valley in the early 1990s) shows that while the level of access to irrigation (40% of households) and recession agriculture (39%) is very similar at the level of the middle valley, significant differences in access are noted in some other sections of the survey area. For example, none of the households in Reach D has access to irrigation. These households would be net losers from an option that would ensure a year-long regulation of the river flow, and therefore would not allow the release of enough water from the upstream dam to inundate the floodplain. The potential net winners would be households such as the ones in Reach E, as they are not involved in recession agriculture and have a relatively high level of access to irrigation (59% of households).

This example is an illustration of the fact that weighing costs against benefits of dam-based development interventions at a highly aggregated level can become socially destructive, because it fails to address critical equity issues.

6.3 Role of Dams in Water Resources Allocation

By storing fresh water during seasons and years of abundance and making it available when needed, dams are a means to address scarcity and unreliability of water and achieve a dependable water supply for domestic consumption, agricultural and industrial needs, energy supply, navigation and recreational activities. Dams have played these roles so effectively that humanity has heavily invested in them, especially since the beginning of the last century. A few hundred large dams (higher than 15 m) existed in the world by 1900. By the middle of the 20th century there were about 5 000 large dams. The construction rate of large dams peaked between 1970 and 1980. Today, the number of large dams worldwide is estimated at more than 45 000. It is clear that countries which have invested in dams (China, USA, India, Spain, and Japan) have notably improved their resilience to seasonal and interannual variability in rainfall conditions. Equipped with more than 75 000 dams higher than six feet (1.83 m) (including almost 7 000 large dams), the United States can store 60% of its entire river flow in man-made reservoirs.

How do dams affect equity? In the Senegal River valley and in other basins where they are built, dams affect existing resources allocation in many ways. Firstly, transformed or expanded resources may continue to benefit those who previously had access to them. The Tarbela Dam in Pakistan illustrates this: the expansion of the cropped area by 39% and the increase in the cropping intensity (from 105% to 117% in Punjab and 116% to 132% in the Sindh province) have primarily benefited land owners (World Commission on Dams 2000). Secondly, resources generated may be allocated to population groups who previously lacked access: for example when dams expand access to electricity, safe water or irrigated land. In both these scenarios, the allocation of dam-based resources therefore has a positive social impact as it "spreads prosperity"[2].

In many cases however, this allocation implies a resource transfer process. The comparison of pre- and post-dam fish catches in the Tocantins River in Amazonia provides a good illustration (see Table 6.2).

---

[2] Which is what they are supposed to do as rightly pointed out by P.V. Indiresan who mentioned in "Dams and Drinking Water" (The Hindu, Dec. 1, 1999) that "the purpose of building a dam is to spread prosperity".

Table 6.2. Changes in fish catches in various reaches of the Toncantins River affected by the Tucuruí Dam (computed from data in WCD Tucuruí Case Study, p. 81–89)

Reach	Pre-reservoir filling (1981)		Current (1998)		Percentage change
	Fish catches (t yr^{-1})	Percentage	Fish catches (t yr^{-1})	Percentage	
Upstream	315	20.5	959	20.5	+204
Reservoir	319	21.0	3 211	69.0	+906
Downstream	900	58.5	492	10.5	−45
All	1 534	100.0	4 662	100.0	+203

Although the total commercial fish catches rose by up to 200% for the river as a whole, as a result of impoundment by the Tucuruí Dam, people downstream experienced reduced access and a 45% drop in fish catches. In contrast, upstream and reservoir-area residents experienced 200 and 900% increases in fish catches respectively after commissioning of the dam. In terms of allocation of the total fish resources, the share of downstream areas fell from 60% of total catches to only 10%, while the share of the reservoir upstream increased from 20% in the pre-impoundment period to about 70% after filling of the reservoir (Table 6.2). This example shows that even if no physically observable diversion is made, dams can take a resource from some groups and reallocate it to other groups.

Similarly, in many cases, there is a lack of coordination and even total disconnection between groups benefiting from dam projects, such as urban areas receiving electricity, fresh water, etc., and groups adversely affected such as displaced populations and downstream communities. In the case of the Kariba Dam mentioned above, while electricity generated is used by the remote Copperbelt mining companies (700 km from the dam) or urban areas such as Harare (400 km away), most of the displaced peoples' settlements (e.g., Nyaminyami, Siakobvu and Lisutu) still do not have access to electricity, forty years after the completion of the dam (WCD 2000a). In Brazil, even if today about 97% of Pará State and 100% of Maranhão State power demand is supplied by the Tucuruí Dam, rural electrification is still a dream for many farmers in the vicinity of the dam (WCD 2000b). In Indonesia, it was found in 1993 that 72% of the 27 000 people displaced by the Kedung Ombo Dam on the Seran River were worse off than before, while 87 000 families of farmers (440 000 people) in irrigated areas experienced a 35–150% income increase over the pre-project baseline (Gutman 1999). In Guatemala in the mid-1980s, 2 500 Maya Indians displaced (some with the use of force and coercion) to make way for the Chixoy Hydro-electric Dam were poorly resettled and continued to experience tragic hardships, while all the electricity generated went to urban areas (Gutman 1999; Ferradas n.d.).

These examples support the view that dams "take a set of resources – a river and the lands along its banks, generating food and livelihood for local people; and transform them into another set of resources – a reservoir, hydro power and irrigation, providing benefits to people living elsewhere." By doing so, they remove river and land resources from the productive domain of one community in favour of another (Brody 1999).

6.4 Addressing Equity Dimensions of Dam-Based Water Resources Development

6.4.1 The Concept of Equity

By balancing gains and gainers on the one hand and losses and losers on the other, it may often appear that dams have performed well from a social and economic point of view, as the amount of quantifiable benefits and number of beneficiaries may outweigh quantifiable costs and losers. This does not, however, necessarily mean that such dams have performed well from an equity perspective.

Equity refers to the moral acceptability or fairness of the patterns of distribution of positive and negative impacts. It speaks more to the costs of a project than to its benefits, and directs attention to those who are most at risk. It points to the vulnerabilities of isolated, less powerful populations for whom development can all too easily mean loss, which is an unacceptable outcome of dams as long as their purpose remains to spread prosperity. This means that when equity is an objective, the limits and dangers of a "balance sheet" perspective should be recognised, as it hides inequities under aggregated statistics. Where dams are implemented without taking into account their proneness to inequity, powerless and poor communities tend to bear a disproportionately high share of the costs, and a disproportionately low share of the benefits.

6.4.2 Ideas for Improving the Equity Performance of Dam Projects

In a number of cases, initiatives have been taken in the past to alleviate potential inequities resulting from dams or to promote social justice regardless of whether the inequities are associated with the project or preexisted in affected communities.

a Where there is unequal access to resources within communities, the advent of a project such as a dam can offer an opportunity for improving prevailing imbalances. The Tarbela Dam case study notes cases where agriculture and homestead land was provided even to those who were landless as part of resettlement processes. This study also mentions cases where marginal farmers were given land in excess of the original size of their plots. In the Senegal River valley, there are cases where state-led distribution of irrigated land provided women more opportunity to have access to ownership of farmland compared to the traditional system of land allocation in the floodplain. A village monograph in the same region shows that traditional landless caste groups (19% of households) had almost the same level of access to irrigated plots (18% of plots were allocated to them) as families of traditional land owners (who comprised 50% of households and to whom 68% of the irrigation plots were allocated). This has been likened to a "social revolution", given the fact that in a traditional land tenure system all farmlands are under the control of noble caste groups (Horowitz et al. 1994; Niasse 1991).

b The comparison of access to electricity between Zimbabwe and South Africa in the last few decades shows that some aspects of inequities noted earlier are not necessarily inevitable. Almost 40 years after commissioning of the Kariba Dam, only about a fifth of Zimbabwe's households (mainly located in urban areas) have access to electricity. Prohibitive power pricing policies adopted by the government (Bond 2000) continue to exclude poor people from access to electricity. In contrast, as a result of clear government policy, the percentage of households having access to electricity increased from 20 to 50% in South Africa in the 1990s (Bond 2000). Policies therefore make a difference to the way the benefits generated by dams are allocated.

c A number of benefit-sharing mechanisms can be considered to improve equity in the distribution of the wealth created by dams. These include: (*a*) free or preferential electricity rates for jurisdictions affected by dams projects (e.g., Norway), (*b*) property taxes paid to local or regional authorities (e.g., France), (*c*) royalties (e.g., the Urra 1 dam in Colombia, and the Itaipu bi-national project in Brazil and Paraguay), (*d*) equity sharing (e.g., the Minastuck project in Quebec, and of a number of agreements between Hydro-Quebec and indigenous communities) (Milewski et al. 1999; van Wicklin 1999).

d Solidarity mechanisms between project beneficiaries and negatively impacted groups can also be promoted. For example, "Funds for Reservoir Areas Development" are used in Japan to foster solidarity between downstream beneficiaries who finance the fund and displaced people (Inoue 2000).

e In order to address losses facing downstream inhabitants in tropical areas, this solidarity concept can be considered as a national initiative, or be reflected in the benefit-sharing mechanisms mentioned above (preferential electricity rates, royalties and/or trust funds benefiting communities in downstream areas or their local government units). Other forms of support to downstream-affected communities could be priority access to irrigated land or to fishing rights, etc.

f In the case of already existing dams, measures for sharing the wealth or minimising costs can include mechanisms for addressing grievances of the past (e.g., the Gwembe-Tonga Rehabilitation and Development Project in Zambia) or adjusting the operational priorities of the dam to new demands as in the case of artificial flood releases to revive the traditional functions of the floodplain (possibilities exist with Pongolapoort Dam in South Africa and the Manantali Dam on the Senegal River).

6.5 Conclusions

One of the major equity issues arising from the typical impacts of dams is less the general disconnection between those who pay and those who benefit, than the fact that dams often result in impoverishment of those who owned the resources transformed by dams. There is, for example, a sense of injustice when displaced people lose their traditional land and water resources and when downstream people experience reduced access to floodplains and are exposed to waterborne diseases, while benefits generated are directed to other social groups. As values change, this form of injustice is increasingly unaccepted by society, and a number of initiatives have already been taken to improve the equity performance of dam projects.

There are ways of responding to these changing values, because while discrepancies are highly probable in the sharing of risks and in the allocation of benefits of dam projects, they are not inevitable outcomes of any dam. To achieve improved balance and fairness in the distribution of the impacts of new as well as of existing dams, the appropriate legal and institutional framework (land tenure, water law, etc.) needs to be set in place, and a number of approaches to avoid and minimise risk, as well as share benefits, can be considered.

References

Bond P (2000) Paying for Southern African dams. Socio-Economic-Environmental Financing Gaps. Submission to WCD, EMG-GEM-IRN, Cape Town

Brody H (1999) Assessing the project – Social impacts and large dams. Prepared for WCD Thematic Review on Social Impacts of Large Dams: Equity and Distributional Issues

Ferradas C (n.d.) Social impacts of large dams: A Latin America perspective. Contributing paper to WCD Thematic Review on Social Impacts of Large Dams: Equity and Distributional Issues

Gutman P (1999) Some evidences on the overall distributional and equity impacts of large dams. Prepared for WCD Thematic Review on Social Impacts of Large Dams: Equity and Distributional Issues

Hollis GE (1996) Hydrological inputs to management policy for the Senegal River and its floodplain. In: Acreman MC, Hollis GE (eds) Water management and wetlands in sub-Saharan Africa. IUCN. Gland. p 176

Horowitz M, Salem-Murdock M, Niasse M, Magistro J, Nuttal C, Kane O, Grimm C, Sella M (1994) Les Barrages de la Controverse. Le Cas de la Vallee du Fleuve Senegal. Harmattan, Paris, pp 125–210

Inoue T (2000) Institutional frameworks for social acceptability of dam projects in Japan. Japanese National Committee on Large Dams (JANCOLD). Presentation at the WCD Hanoi Consultation (March)

Milewski J, et al. (1999) Dams and benefit sharing. Submission to WCD by Hydro Quebec, Direction Environment. Monreal. Nov.

Niasse M (1991) Les Perimetres Irrigues Villageois vieillissent mal: les paysans se desengagent-ils en même temps que la SAED? In: Crousse, Mathieu, Seck (eds) La Vallee du Fleuve Senegal. Evaluations et perspectives d'une decennie d'amenagements (1980–1990) Karthala. Paris, pp 97–115

Salem-Murdock M, Niasse M (1993) Landuse, labor dynamics, and household production strategies: The Senegal River valley. IDA Working Paper No. 94. Binghamton, N.Y. p 436

van Wicklin W (1999) Sharing benefits for improving resettlers' livelihood. Submission by to WCD. World Bank. Nov.

WCD (2000a) Kariba Case Study

WCD (2000b) Tucurui Case Study, 14

World Commission on Dams (2000) Tarbela case study. WCD. Cape Town

Part IIb

**Out of Breath:
Air Quality in the 21st Century**

Chapter 7

Atmospheric Chemistry in the "Anthropocene"

Paul J. Crutzen

Since the beginning of the agro-industrial period, mankind's use of Earth's resources has grown so much that it seems justified to denominate the past two centuries into the future as a new geological epoch: "The Anthropocene". This transition is also marked by major changes in the chemistry and chemical composition of the atmosphere, such as:

- The atmospheric concentrations of several climatologically important "greenhouse gases" have grown substantially: CO_2 by 30%, CH_4 by more than 100%, and tropospheric ozone regionally in the troposphere by more than 100%. These changes presently exert an additional infrared climate-warming forcing of about 2.5 W m^{-2}, which to a large degree can have contributed to the observed average warming of the planet by 0.6 °C. The increases in the concentrations of these gases were caused by the combustion of fossil fuels as well as deforestation and agricultural activities. The average global surface warming caused by these changes has been estimated with climate models to be in the range 1.4–5.8 °C by the end of this century, the wide span of this "prediction" reflecting both insufficient knowledge about the many positive and negative feedback processes in the complex physical/chemical/biological climate system, as well as insufficient knowledge about future developments in the energy and agricultural sectors and resulting "greenhouse gas" emissions. Thus, the possibility exists that concentrations of atmospheric "greenhouse gases" and resulting climate warming may reach levels that have not existed on Earth since the emergence of *Homo sapiens*;
- The release of SO_2 (about 160 Tg S yr^{-1}; Tg = Teragram = 10^{12} g) by coal and oil burning is at least two times larger than the sum of all natural sulphur emissions. Over industrialised regions, the increase has been more than an order of magnitude, causing acid precipitation, health effects, and regionally poor visibility due to light scattering by sulphate aerosol. Interestingly, this factor has also led to increased reflection of solar radiation to space, causing a cooling effect on climate, and thereby partially counteracting the warming by the "greenhouse gases";
- Release of NO to the atmosphere from fossil fuel and biomass burning is larger than or comparable to natural emissions, over extended regions causing photochemical smog, including high surface ozone concentrations which are harmful to human health and plant productivity. It should be noted here that photochemical smog is not only a phenomenon of the urban/suburban industrial world, but also extends to rural regions in the developing world as a consequence of widespread biomass burning during the dry season in the tropics and subtropics;
- A special class of products from the chemical industry, the chlorofluorocarbons $CFCl_3$ and CF_2Cl_2, which were never produced in nature, have caused the greatest large-scale change in the atmosphere: the "ozone hole", the rapid loss of almost all ozone in the lower stratosphere over Antarctica during spring, due to "cold chemistry" involving surface reactions on particles. The "ozone hole" develops over a time period of a few weeks from 12–22 km, that is in the height region in which maximum ozone concentrations were always found until about 20 years ago. Smaller, but nevertheless, significant ozone depletions are also observed in some years during late winter/spring over the Arctic. It is important to note that these large, chemical ozone losses were not predicted. It was in fact thought that ozone in the polar regions was chemically inert. This experience shows the enormous importance of observations and limitations of model predictions. Especially in these times of increasing pressures of mankind on the environment, this experience raises questions about the stability of the environment and climate system, or parts thereof. Clearly, special human action has led to an instability in stratospheric ozone chemistry, in a region that seemed the least likely. It should be asked whether there might be other breakpoints in the complex Earth System that we are not aware of. How well may models predict these in the extremely complex Earth System with its many positive and negative feedbacks? See Fig. 7.1, which only depicts the complexity of stratospheric chemistry.

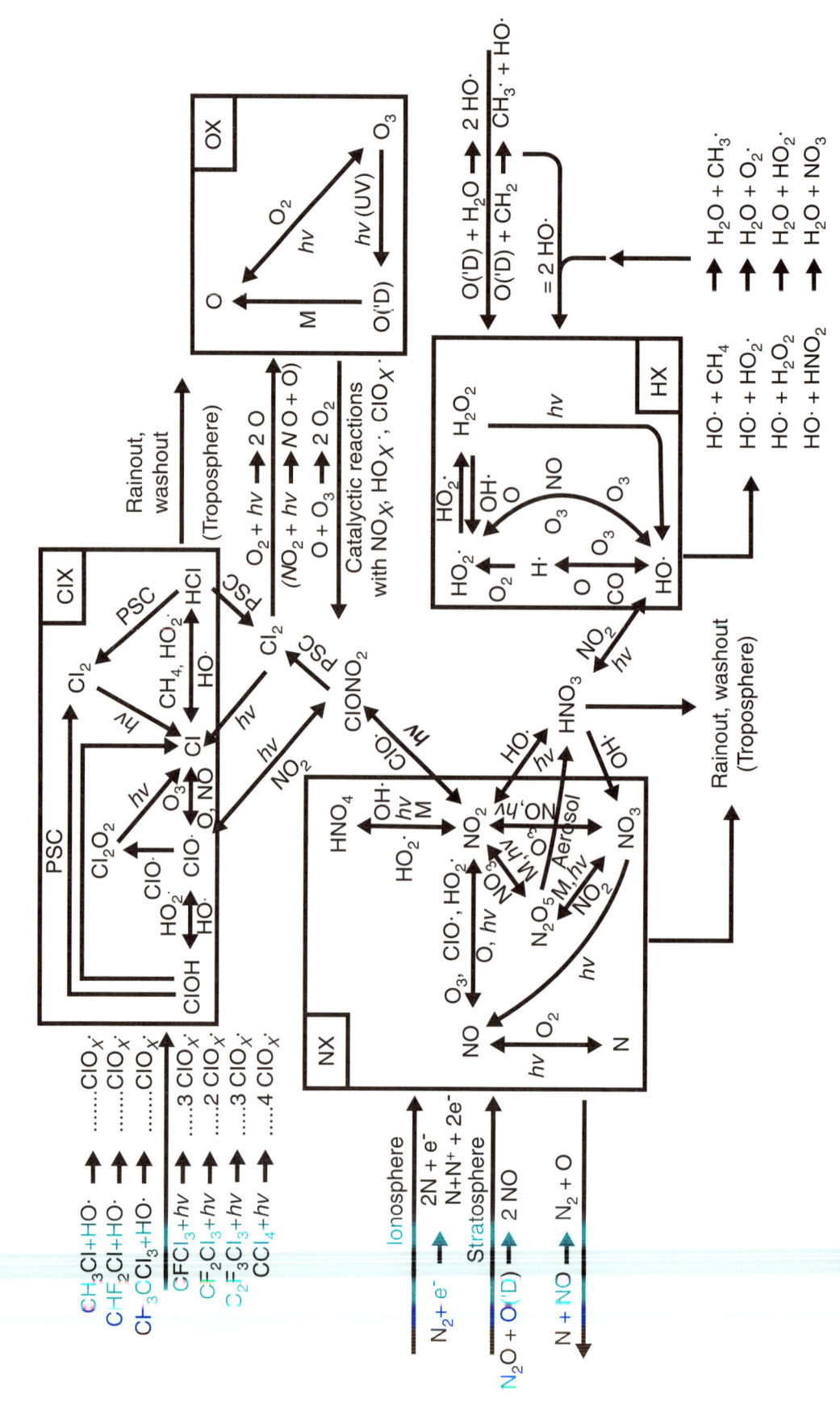

Fig. 7.1. An example of the complexity of Earth System research. The figure shows the reactions that are needed to describe stratospheric ozone chemistry. Not shown are reactions involving Br compounds, which are responsible for additional ozone loss (reproduced with permission from Crutzen (1996) Angew Chem Int Engl 35:1758–1777)

Fortunately, international agreements have been reached to phase out the production of the CFCs and some other halogenated organic compounds since the beginning of 1996. The impact of these restrictions on tropospheric CFC concentrations is already noticeable. Although reductions in stratospheric CFC levels will soon follow, it will take a relatively long time, on the order of some 50 years, before CFC levels will have decreased so much that the "ozone hole" will be filled up. There is, however, the possibility of further delays due to the cooling of the stratosphere by increases in the concentrations of CO_2 or continued increase in water vapour. The international regulations that were reached by the phasing-out of CFC production may be considered an environmental success story. Unfortunately, one can not feel optimistic that international agreements against CO_2 emissions will be reached similarly in a timely fashion.

We will next concentrate our discussions on a few issues involving atmospheric chemistry, which I see as particularly important: the self-cleansing activity of the atmosphere, and how it might be impacted by human activity, and the role of the tropics and subtropics in atmospheric chemistry and how it is disturbed by the large human population in this part of the world.

The self-cleansing propensity (oxidation efficiency) of the atmosphere. About three decades ago it was proposed that hydroxyl (OH) radicals play a key role in removing almost all gases that are emitted to the atmosphere by natural processes and human activities. Hydroxyl radicals are formed by the action of solar ultraviolet radiation on ozone, producing energetic $O(^1D)$ atoms, which possess enough energy to react with water vapour:

R1: $O_3 + h\nu$ (<420 nm) $\longrightarrow O(^1D) + O_2$

R2: $O(^1D) + H_2O \longrightarrow 2\,OH$

The atmospheric concentrations of OH are very low, on average only 4 molecules per 10^{14} molecules of air. Nevertheless, for most gases, their atmospheric average residence time and spatial/temporal variability are determined by how fast they react with OH. For instance, CH_4 reacts only slowly with OH and shows much less variability than isoprene (C_5H_8), which is emitted by vegetation and can only be found in close vicinity to forests. Methane and carbon monoxide are the main gases with which hydroxyl reacts in most of the troposphere. Because both gases have been increasing by 0.5–1% yr^{-1} (in the case of CO only up to the end of the 1980s), a lessening in OH concentrations could be expected. However, other processes work in the opposite way. For instance, due to anthropogenic emissions of NO_x (NO + NO_2) catalysts, hydroxyl concentrations can increase, first because NO reacts with HO_2, reaction R4, (or other peroxy radicals) and because ozone increases, for instance via the reaction chain:

R3: $CO + OH\ (+O_2) \longrightarrow CO_2 + HO_2$

R4: $HO_2 + NO \longrightarrow OH + NO_2$

R5: $NO_2 + h\nu\ (+O_2) \longrightarrow NO + O_3$

Net: $CO + 2\,O_2 \longrightarrow CO_2 + O_3$

From observational studies of CH_3CCl_3, a gas that is (until 1996 has been) produced by the chemical industry and that is largely removed from the atmosphere by reaction with OH, it follows that globally averaged tropospheric OH concentrations have not changed much in time. What will happen in the future depends much on what will happen in the tropical and subtropical regions of the world with the largest and growing fraction of the world population, because it is there that OH concentrations maximise due to high levels of water vapour and solar ultraviolet radiation.

The importance of the tropics and subtropics in atmospheric chemistry. Because of maximum abundance of hydroxyl radicals in the tropics and subtropics, maximum rainfall, and relatively low industrial activity, the tropics should be the cleanest part of our atmosphere. This is, however, not the case. Due to tropical deforestation activities and the burning of large amounts of biomass in various kinds of agricultural activities during the dry season, as well as lack of pollution controls, large amounts of pollutants are emitted into the atmosphere, including parts of the Southern Hemisphere. It is thus estimated that each year between some 2 000 to 5 000 Tg of biomass carbon is burned, releasing large amounts of photochemically active gases like CO, hydrocarbons and NO_x into the atmosphere, a mixture very similar to that which produces photochemical smog in the city plumes of the industrialised world. Thus high concentrations of ozone are observed in the dry season in many parts of the developed world, in Africa, South America and Asia. This affects the oxidation efficiency of the atmosphere; however, by how much, and in what direction, are not known. Besides the effects, just mentioned, there are other factors that influence the oxidising efficiency of the atmosphere. For instance, because of tropical deforestation, which is as high as 1% yr^{-1} in some regions, reactive hydrocarbon emissions (such as isoprene) are declining. This will probably lead to an increase in OH concentrations both in the forested regions, where reactions with reactive hydrocarbons act as strong sinks for OH, and at larger scales where CO, an oxidation product of the hydrocarbons, will tend to decrease, and thus causing OH to increase.

A recently recognised, and maybe the greatest, impact on atmospheric chemistry and climate may well result from large anthropogenic emissions of particulate matter in both industrial and especially developing countries. Already now these are exerting a global average radiative (cooling) forcing of about 1.5 W m^{-2} on climate, mostly due to back-scattering of solar radiation to space, close to the current infrared warming forcing of about 2.5 W m^{-2} by the "greenhouse gases". Past studies were especially concerned with the effects of sulphate aerosol resulting from coal and oil burning in the developed world. Most of these studies have neglected the potentially very large contributions that can be caused by fossil fuel and biomass burning in the developing world. Already now, the emission of SO_2 to the atmosphere in the Asian countries is about equal to those from Europe and North America combined. In about 20 years, it will be 3 times larger. Because of lack of appropriate pollution controls, this is causing heavy air pollution in these countries. To the "classic" air pollution emissions must be added those resulting from extensive biomass burning in primitive agriculture, and for cooking and heating purposes. As a consequence, especially during the dry season, high concentrations of smoke cover extensive regions of the developing world. During a field experiment (INDOEX), which was conducted February–April, 1999 in India, the Maldives and the Indian Ocean, the observed particulate loading of the atmosphere was so high, it caused a reduction in solar radiation fluxes reaching the Indian Ocean surface north of the Intertropical Convergence Zone (the meteorological equator) by up to about 20 W m^{-2} (or about 10% of the solar irradiation), almost ten times larger than the global "greenhouse top of the atmosphere forcing." This may have a very strong impact on regional climate and the water cycle, an impact which will still strongly grow in coming decades. Contrary to the cooling forcing on climate resulting from fossil fuel derived sulphate, current emissions from fossil fuel and biomass burning in the developing world are characterised by large emissions of sunlight-absorbing black carbon (soot). Its impact on regional or global climate, thus has a quality which is quite distinct from that by the emissions in the developed world.

Future changes in atmospheric chemistry and climate will to a large extent depend on what is going to happen in the developing world. Especially, quantitative information about the production of the various kinds of particulate matter, their distribution and optical and microphysical properties is very uncertain, calling for strong enhancements in future research efforts.

Chapter 8

Fires, Haze and Acid Rain:
The Social and Political Framework of Air Pollution in ASEAN and Asia

Simon S. C. Tay

8.0 Introduction: Air Pollution and Asia in Context

Air pollution has many different sources, dimensions and driving factors. In Europe, North America and North East Asia, industrial pollution and the phenomenon of acid rain are common concerns. In the ASEAN region, the discussion of transboundary air pollution necessarily comes against the background of the "haze" pollution experienced in the region in 1994, 1997 and 1998 that resulted from fires in Indonesia.

While each of these examples of air pollution is unique and different in some respects, has common points of reference in its social and political framework. These issues also share common points in their interconnection between national, regional and global environmental challenges. It is on these common points of reference that this paper hopes to elaborate by focusing on the particular examples of (1) Indonesian fires and the haze; and (2) acid rain in Northeast Asia. They will illustrate the need to strengthen the social and political frameworks, as well as to increase non-governmental participation and recognition of cross-border and global concerns.

8.1 Part I: The Problem in Perspective

8.1.1 The Haze: Summary of a Recurring Disaster

8.1.1.1 Local and Global Effects

The 1997 and 1998 fires ravaged the Indonesian islands of Sumatra and Kalimantan, their epicentre. Conservative estimates are that between 800 000 and 1.7 million ha of forests and bush were burnt. A study by the World Wide Fund for Nature (WWF) puts the figure at two 2 million ha. The haze smoke and particulate pollution from the fires affected Indonesia, Malaysia and Singapore, and to a lesser extent, parts of Brunei, Thailand and the Philippines.

The haze pollution exacted a high economic cost on the region. A preliminary study estimates that regional economies in total suffered over US$ 4.5 billion losses in health care, tourism and airlines because of the 1997 fires and haze. These estimates are preliminary estimates, as they do not include global losses, such as the release of climate-change gases or losses to biodiversity, nor do they include possible long-term health effects. The WWF estimates that losses from the 1997 fires and haze for Indonesia and the region exceed US$ 20 billion after lost productivity, timber, health effects, and other costs are tallied.

During the 1997 fires, the Pollution Standard Index (PSI) in the worst-hit areas often exceeded the 300 PSI mark (graded as hazardous for health) and reached a high of 839 PSI in Kuching, in the East Malaysian state of Sarawak (Crowell and Morgan 1997). The estimated number of people in the region exposed to the harms of the haze ranges between 20 million (Dudley 1997) and 70 million (Crowell and Morgan 1997). Due to unusually dry weather associated with the 1997/1998 El Niño, the wet monsoon season ended early in January 1998, allowing the fires to spread soon after, thus indirectly also causing food shortages, famine and a loss of human life. The long-term exposure to the haze and smoke may possibly have serious impacts on human life and health, especially if the fires recur annually.

The fires and haze also have global impacts on world biodiversity and climate change. As one of the twelve "mega-biodiversity" countries in the world, Indonesia holds the second largest tract of tropical forests in the world, with some 92–109 million ha. The effects of the fire have yet to be fully measured, but it is apparent that flora, fauna and ecosystems have suffered, with fires razing protected areas and reserves in East Kalimantan. The fires will also potentially add to the problem of global warming, given the large amounts of carbon dioxide released.

8.1.1.2 Causes

Widespread forest fires on the islands of South East Asia have been an on-going problem for many years (Wirawan 1993). Some Indonesian officials have been quick to blame the El Niño weather phenomenon (Nicholls 1993) as the main cause of the 1997 and 1998 fires. El Niño brings longer and drier dry seasons, which leave the forests vulnerable to fire. But while drought may create the conditions, the Southeast Asian fires – like all fires – need ignition.

Clearly, traditional practices and climate have contributed to the Southeast Asian fires, but to single them out suggests that the fires are almost inevitable and excusable natural disasters. This does not provide a complete picture. A closer look at the causes of the fires of 1997 and 1998 would reveal causes rooted in human agency.

In 1997 and 1998, Indonesian officials reported that eighty percent of the fires could be attributed to land clearing by large plantation owners, primarily in the timber (especially plywood) and oil palm production businesses, after the valuable timber has been extracted. Additionally, the large plywood industry in Indonesia is facing a rapid decline in their source of virgin logs and is exploiting previously logged lands for raw material. Fire was used not because of tradition, but because it is a cheap alternative to other, less pollutive methods.

A full discussion of logging and deforestation is beyond the scope of this paper (but see Dudley et al. 1995 and Abramovitz 1998 for overviews). It is clear, however, that the region's logging and developmental policies might contribute to the outbreak of fires. Criticisms of government policies that have favoured logging and development projects often take on political dimensions. They arise from the blindness of top-down development planning, but also from cronyism and corrupt patron-client relations in the allocation of natural resources. International trade and investment also have a role to play in the fires as production for export and industry is often held in partnership with foreign investors, such as Japan and the US as well as ASEAN neighbours.

The fires in Southeast Asia, therefore, are not so much a part of natural history and tradition as a new phenomenon. Although fires have always been part of the natural equation of forest ecology, human intervention has tipped the balance dangerously in favour of fires (e.g., Dudley 1997; Ambramovitz 1998). The fires in Southeast Asia in 1997 and 1998 were a disaster resulting from deliberate policy, man-made activities, large corporations in pursuit of larger profits, and the failure of authorities to regulate sufficiently and effectively. For this reason, the SEC International Policy Dialogue concluded that the fires "related to the use and clearance of forest and other land, mainly by big business, engaged in logging and palm oil, and could have been foreseen, given climatic forecasts and patterns of forest use".

8.1.2 Northeast Asian Acid Rain: The Smog of Growth

8.1.2.1 *Effects*

Northeast Asia's rapid industrialisation and urbanisation have led to growing air pollution problems. There is now a growing awareness of the impacts of long-range transport and deposition of pollutants across national boundaries, as the various energy use and development scenarios imply much higher levels of acid deposition over a much wider area. Economic scenarios indicate increases in deposition rates that would overwhelm the soil's capability to cope by 2020 in large parts of eastern China, the Korean Peninsula and Japan. If true, the region would face ecological disaster from acid rain, unless major changes in emission control policies are effected (Carmichael and Ardnt 1997).

8.1.2.2 *Causes*

Modelling studies demonstrate that a significant proportion of the acid deposition in southern Japan and Korea arises from sulphur dioxide and nitrogen dioxide emissions in China. An important source of emissions is from the burning of fossil fuels by the energy, industry and transportation sectors. In 1990, coal supplied 48%, and oil fuels 35% of the region's primary energy needs (Streets 1997). Power plants are the largest contributors to long-range transport, while residential and industrial emissions primarily affect local urban acid precipitation (Sinton 1997).

Inefficient energy production and use are another important source of high emissions. For example, in 1990, it took China 20 times as much primary energy to produce a dollar of economic product as compared to Japan.

The distribution of impacts over Northeast Asia is uneven. Some areas in southeast China regularly experience average precipitation with pH below 3.5 and many areas below 4.0. Long-term measurements in Ryouri, Japan show a decrease in pH levels from 5.2 in the late 1970s to below 4.7 at the present, a five-fold increase in acidity. Prevailing northwesterly winds from high-pressure systems over mainland Asia carry dry air-masses over the sea, which as they absorb moisture, become unstable and precipitate as they reach Japan or Taiwan. Sulphur compounds emitted on the continent are incorporated in this process. Source-receptor relationships vary by season. A country (or region) can change from being downwind of another country's sulphur emissions, to being the upwind source of acid deposition. Even the small fraction of acid rain that does cross China's borders and reaches the land of other countries in Northeast Asia makes a large contribution to total deposition of acid rain in these countries. Modelling studies are beginning to clarify how the various processes during transport affect deposition and untangle source – receptor relationships (Ardnt et al. 1995; Foell et al. 1995).

8.2 Part 2: International Principles and Practice

8.2.1 International Law and Practice in Other Regions

The international laws on transboundary pollution are clear in principle: a state is legally responsible for all transboundary pollution emanating from areas under its jurisdiction or control, where significant harm results to the other state.

It is said to be part of customary international law, binding on all states. The consensus arises from two sources. First, the venerable Trail Smelter Arbitration between Canada and the USA and second, the soft law instruments of the Stockholm and Rio Declaration, which embody similar wordings as in the Trail Smelter: namely that "no state has the right to use or permit the use of its territory in such a manner as to cause injury by fumes in or to the territory of another or the properties or persons therein, when the case is of serious consequence".

There is reason to doubt that the international norm on transboundary pollution, often referred to as Principle 21, is fully and faithfully applied in practice, despite being embodied in some national laws (e.g., the US Clean Air Act) and often cited as a cornerstone of international environmental law.

Other than the Trail Smelter case, the absence of suits, despite notable incidents of transboundary pollution such as the Russian Chernobyl nuclear reactor incident in 1985 or the Sandoz Spill in the Rhine in 1986 (McClatchey 1996), indicate a dearth of adjudication.

With the exception of treaties that deal with "ultra hazardous" activities like oil spills and nuclear accidents, few regional and bilateral treaties seem to set out strict liability regimes in similar terms to the international law principles or provide for institutional means to enforce those legal principles of responsibility (Merrill 1997; Scnieder 1997).

Regional and bilateral treaties prefer to focus on efforts to establish specific pollution limitations or to foster cooperative action between different states instead of liability. A vast majority of the treaties promise cooperative action between states in areas such as: advance notification and consultation, exchanges of information and research, or a general undertaking to take "appropriate measures".

It has been suggested that these treaties, which emphasise cooperative action, without liability or setting specific limits, have been ineffective (Merrill 1997). However, the Convention on Long-Range Transboundary Air Pollution (LRTAP) and its protocols have been effective.

8.2.2 The ASEAN Way and Environmental Cooperation

8.2.2.1 *Why ASEAN?*

At the national level, Indonesia's continued political and economic crisis coupled with widespread corruption and ineffective legal and administrative frameworks have rendered it unable to resolve the fires that occur almost wholly within its borders. The global community's response has been mainly piecemeal and made uncoordinated efforts that aimed at fire fighting, rather than dealing with the systemic problems.

In the gap of both national and international inaction, hopes have come to be pinned on the regional institution, ASEAN, in expectation that it will serve as a useful middle ground between the national and international levels. ASEAN has indeed responded to the Indonesian fires and haze by undertaking a number of plans for cooperation and action. These plans have yet to prove effective, as they have depended heavily on the usual methods used in different areas of ASEAN cooperation, known as the ASEAN way.

8.2.2.2 *The ASEAN Way*

Before its present problems, ASEAN enjoyed a reputation as one of the most credible regional organisations, outside of the European Community. While ASEAN enjoyed political and economic success, its poor environmental record has come under criticism from environmentalists and NGOs. This is especially with respect to tropical deforestation and the lack of conservation efforts.

At the 1992 Earth Summit, ASEAN was better known for taking a pro-developing country stand. However since 1978, ASEAN has embarked on environmental programmes. These have evolved into the ASEAN Strategic Plan of Action (1994–1998) (Koh 1996). The strategies cover a broad range of environmental concerns and adopt many of the approaches recommended in Agenda 21 and seek to integrate environmental and developmental concerns in the decision-making process of governments through such mechanisms.

The effectiveness of such measures, however, suffers from ASEAN's preference for noninterference in the domestic affairs of member states, and for nonbinding plans, instead of treaties. There is also a preference for relying on national institutions and actions rather than a stronger central bureaucracy with powers of initiative and more resources.

8.2.2.3 ASEAN Plans on Atmospheric Transboundary Pollution

Following the haze in 1994, ASEAN environment ministers agreed on a Cooperation Plan on Transboundary Pollution in June 1995. At the national level, each country undertook to establish focal points and enhance national capabilities to deal with forest fires. They also agreed to share knowledge and technology on the prevention and mitigation of forest fires, and establish a mechanism for cooperation in combating forest fires.

A common air quality index and a regional fire danger rating system were to be developed for the region. ASEAN institutions like the Specialized Meteorological Centre were tasked to develop ways of predicting the tracts and spread of smoke haze. The Cooperation Plan also envisaged seeking support from countries outside the region with knowledge in fire management systems, like New Zealand and the USA, as well as institutions such as the International Tropical Timber Organization.

The occurrence of the 1997 fires and haze indicated the failure of the Cooperation Plan. This was due to a lack of implementation. The ASEAN environment ministers then agreed in December 1997 (ASEAN Ministerial Meeting on Haze, Singapore) to a further Regional Haze Action Plan.

In comparison to other treaties discussed earlier, the ASEAN Cooperation and Action Plans do not create liability regimes to hold one state responsible to another. Nor do the plans set out specific pollution limits. Rather, they must be characterised as efforts to foster cooperative action. This may be expected, given the norms of the ASEAN way and the reality of relations of other members with Indonesia. Given the inaction and ineffectiveness seen in the years after the 1994 Cooperation Plan and the recurrence of fires and haze in 1997 and 1998, there are, however, many doubts.

Upon review in April 1998, the Action Plan agreed to establish two sub-regional firefighting arrangements for Kalimantan and for the Sumatra/Riau provinces in Indonesia. Indonesia also announced that it would lay down the framework for an ASEAN research and training centre for land and forest fire management in Central Kalimantan. Notably, the ASEAN Environment Ministers met monthly to review progress on the Action Plan during the flare-up of fires in early 1998.

In subsequent years, without a prolonged recurrence of the haze, the intensity of attention given to the issue has again waned. Also, efforts by think tanks and NGOs in the region to bring attention to the issue of fires have not been well received by the political leadership.

In 2001, experts' suggestions that an early return of the El Niño phenomenon may lead to the fires and haze in 2001–2002 have placed the issue back on ASEAN's environmental agenda.

8.2.3 Cooperation in Northeast Asia

There is no equivalent to ASEAN in Northeast Asia. The methods of cooperation in Northeast Asia have taken on various forms and involved a wide range of state and non-state actors, including experts, international financial agencies, businesses and governments. For example, some expert or epistemic networks have been concerned with the standardising monitoring techniques, cooperating on modelling atmospheric transport and improving overall understanding of acid deposition and its impacts. International banks and development agencies have begun exploring ways to finance or otherwise facilitate the transfer of cleaner technologies. Businesses with environmental technologies are seeking partnerships and markets for their products in other countries. Governments have begun considering how to incorporate regional environmental concerns in their domestic and international policies.

There are a number of ongoing cooperative programmes on transboundary air pollution. Prominent among these are the Northeast Asian Conference on Environmental Cooperation (NEAC), the Meeting of Senior Officials on Environmental Cooperation in Northeast Asia (SOM), the Acid Deposition Network in East Asia (EANET), the China-Japan-Korea Tripartite Environment Ministers Meeting, and the Expert Meeting for Long-Range Transboundary Air Pollutants in Northeast Asia.

Most of these efforts are in their early stages and focus on the exchange of information, the promotion of research and raising prospects of cooperation. In some cases, intergovernmental institutions like the Asian Development Bank are brought in. In others, a non-regional country like the USA or Europe is also involved. Almost all of these are moving slowly, if at all. Questions of liability are not raised between the countries.

8.3 Part 3: Assessing and Improving Regional Cooperation on the Environment

8.3.1 Assessment and Prospects for Improvement

Efforts at regional cooperation in Asia on these environmental issues are quite limited in their reach and implementation. Institutions remain weak, especially in NE Asia, and modes of discussion and behaviour between governments remain more formal than effective.

There are a number of different general approaches to improving cooperation in both regions. What follows is, however, more a "toolbox" of possible options than a single solution and contains alternative suggestions, which might, to a degree, be mutually exclusive or inconsistent. Additionally, there are possible solutions that lie beyond the purview of ASEAN. Alternatives may arise from international institutions, individual companies, or consumers. These, however, lie beyond the scope of the present paper.

8.3.2 How to Improve ASEAN?

8.3.2.1 Treaty Making

Given the varied environments that ASEAN countries function in, harmonising of limits (Gerardin 1996) may be a difficult task. Instead, states can set their own specific limits, which will be enforced by an intergovernmental panel. This approach is used by a number of agreements on transboundary pollution, for example The North American Free Trade Agreement (NAFTA) between Canada, Mexico and the United States (Esty 1995).

ASEAN Plans could still be strengthened by widening the review process and holding them on a regular basis. Such reviews are the most common mechanism for encouraging state compliance in environmental and other international treaties.

8.3.2.2 Reconciling Norms

Some of the problems facing ASEAN's efforts to take effective action arise from the ASEAN norm of nonintervention and the "ASEAN way". While this norm of nonintervention might seem to be a primary reason for ASEAN inaction and neighbourly silence, there is an urgent need to recognise the situation as one that falls outside the norm. ASEAN's efforts in recognising the issue's importance and promising cooperative actions are an indication of this. More can be done, however, as ASEAN members should consciously and publicly recognise that the ASEAN norm against noninterference does not preclude lawful actions regarding transboundary pollution, which by its very nature is not a purely domestic affair.

ASEAN can also develop practical cooperation measures such as the joint mobilisation of fire-fighting resources. While political sensitivities make such cooperation difficult, they also serve as confidence-building measures.

8.3.2.3 Strengthening Institutions

ASEAN members need to recognise and account for the human, economic and other costs of the fires and haze. This would help the Indonesian political system to recognise the real costs of the fires and provide a basis for affected states to make the economic commitment and provide justification for the expenditures to prevent or mitigate fires. It is common sense that cooperation will only succeed if, for all parties, the benefits exceed the costs (Merrill 1997).

ASEAN's general capacity for environmental policy-making and assisting compliance needs to be strengthened. ASEAN should have sufficient resources and knowledgeable personnel to add value to national efforts to deal with the transboundary pollution. These assets might assist and help smooth matters if and when bilateral discussions between involved states become strained.

8.3.2.4 Linking Environment and Economic Policy

ASEAN countries have emphasised trade and other economic activity without sufficient recognition of its full environmental costs. Linking the environment to economic development would not impede economic activity but direct ASEAN towards sustainable development practices.

Links between environment and economics for ASEAN countries can be positive and cooperative, using "carrots" rather than "sticks". Assistance and incentives for Indonesia from fellow ASEAN members or from outside the region would help in this regard. Such assistance would not be charity but rather due recognition of the economic costs and ill effects caused by fires and haze.

8.3.2.5 Involving People and the Private Sector

NGOs have played a considerable role in protecting the environment (Wapner 1996). The pressure of ASEAN's inability has increased their acceptance of NGOs and NGO involvement.

The privatisation of environmental conflicts devolves them from the interstate level to legal proceedings at the level of municipal laws. This approach might lessen the possibility of political embarrassment and intergovernmental confrontation, risks which are antithetical to the ASEAN way.

Private individuals and organisations can also help foster and enforce compliance within the relevant industries in Indonesia through internal economic measures and through externally created incentives, such as voluntary eco-labels (Staffin 1996). The other financial methods for increasing compliance amongst industries that can be used are debt-for-nature swaps, private lending, insurance coverage and the use of ethical funds.

8.3.3 Northeast Asian Cooperation

Some experts suggest that for the immediate and medium term, the prospects of cooperation on acid rain are best where they coincide with the national interests of China. There is also need to develop stronger national environmental regulatory agencies, with the authority to set environmental goals and enforce laws, together with supporting policies and institutions that emphasise pollution prevention and continuous improvement, increased transparency and improved cooperation.

8.3.3.1 Transfer of Efficient and Clean Technologies

Despite having different capacities and capabilities for developing cleaner production technologies, the combination of using efficient technologies and cleaner technologies such as natural gas or renewables, for example the sun or wind, can have a great reduction in pollution levels. Multilateral financial organisations such as the Asian Development Bank, World Bank and the Global Environmental Facility have for several years provided financial assistance for such technology transfers. For example, one of the first bilateral agreements in Northeast Asia to deal explicitly with the acid rain problem, signed in March 1994, called for increased Japanese environmental technology transfers to Beijing. Japan, through its Green Aid Plan to China, has also provided substantial amounts of foreign aid, some of it aimed directly at reducing SO_2 emissions. Government cooperation can facilitate these technological transfers by providing protection for intellectual property rights and reducing trade barriers.

Despite the availability of cleaner technologies, it appears that it will take more time and financial assistance before they can be adopted on a large scale in China or North Korea. Improving energy efficiency is, therefore, a key strategy, and one that has and should continue to be supported through international cooperation.

8.3.3.2 Laws and Harmonisation

China took an important step domestically towards solving the acid rain problem in 1995, when China's National People's Congress amended the Atmospheric Pollution Control Law of 1987.

Similarly, the harmonisation of standards and goals of regulations is an important step to be taken, as regional cooperation can reduce the costs of capacity building for each nation. Environmental management is extremely information and knowledge intensive. Although the most appropriate policy instruments must vary among countries to fit the different socioeconomic and ecological contexts, there is also scope for improved exchange of experiences with different policies.

8.3.3.3 People and the Private Sector

Participation of a non-state actors has been and must continue to be a critical part of the process. For example, civil society groups remain important in creating internal pressures for change that place greater value on environmental protection. The private sector will also have a large role in implementing and reaping the benefits of efficiency programmes and the transfer of cleaner technologies. Consumer behaviour changes will make green marketing more and more important to business.

8.4 Conclusion: Strengthening the Social and Political Frameworks

The suggestions offered by this paper are more of a toolbox or menu of possibilities, and not all are necessary to improve our response to the issue.

One common factor in all of these possibilities is that what some see as a problem of air pollution is in fact intertwined with much deeper social, political and economic factors. The connections to industrial, agricultural and development policy are indeed complex.

The approach to problems of transboundary pollution can range from the litigious – seeking to hold countries strictly liable – to the technical – setting and enforcing pollution limits – to the cooperative. Government decisionmakers are the key variable to success of various plans undertaken. The important proviso is that their answer should result in an approach that is not only acceptable, but also one that helps solve the problems at hand.

But the tools of persuasion in the current situation may extend beyond cooperative stances or agreements or other technical subjects. To get attention and priority to the issue, the issue of liability may need to be raised. In this way, legal norms will help support the correctness of countries suffering from the transboundary pollution.

In this, I would join others in arguing that the role of civil society and NGOs is key. The non-state sector must be recognised as a strong factor that might be able to make the plans more effective. This would include the positive role that the private sector can play. For in looking at how institutions and regional cooperation works, the driving factors are not simply legal, institutional or technical.

They relate to social and political frameworks by which people understand and express their concerns over the impacts they experience as human beings. It is this common point of reference that the ASEAN plans share with similar plans from other regions, and it is on this point that the ASEAN way shows perhaps the most considerable difference. The non-state sector of people and private sector companies can also make a significant difference in Northeast Asia.

To include civil society, NGOs and the private sector in a meaningful way will potentially change the traditional ASEAN or Asian ways of handling issues.

Issues of a transboundary nature like the Indonesian fires and haze and the Northeast Asian acid rain that underscore the difficulties in addressing this environmental challenge are, especially, interstate cooperation, legal recourse for transboundary harms and the participation of civil society groups and non-governmental organisations.

Social and political frameworks in Asia must be strengthened and greater room allowed for non-governmental participation and attention to cross-border and global concerns.

General References

IGES (2001) Regional/sub regional environmental cooperation in Asia, Feb 2001, (Kazuo Kato and Wakana Takahasi)
Noda PJ (ed) (2002) Cross-sectoral partnerships in enhancing human security. Institute of Southeast Asian Studies and Japan centre for International Exchange, especially Louis Lebel's Acid Rain in Northeast Asia (Chap 3)
Tay SSC (1999) South East Asian fires: The challenge to international environmental law and sustainable development. Georgetown International Environmental Law Review, Winter 1999

Specific References

Abramovitz JN (1998) Taking a stand: Cultivating a new relationship to the World's forests
Arndt R et al. (1995) Long-range transport and deposition of sulfur in Asia. Water, Air and Soil Pollution 85:2283–2288
Carmichael GR, Arndt R (1997) Baseline assessment of acid deposition in Northeast Asia. Energy, Security and Environment in Northeast Asia Project, The Nautilus Institute
Crowell T, Morgan P (1997) Twelve months of turning points. AsiaWeek, Dec. 26, 1997
Dudley N (1997) The year the World caught fire. WWF International Discussion Paper, Gland, Switzerland, WWF, Dec 1997
Dudley N, et al. (1995) Bad harvest?: The timber trade and the degradation of the World's forests
Esty DC (1995) Making trade and environmental policies work together: Lessons from NAFTA. In: Cameron J, Demaret P, Gerardin D (eds) Trade and the environment: The search for balance, vol I. London
Foell W, Green C, Amann M, Bhattacharya S, Carmichael G, Chadwick M, Hettelingh J-P, Hordijk L, Shah J, Shresta R, Streets D, Zhao D (1995) Energy use, emissions, and air pollution reduction strategies in Asia. Water, Air and Soil Pollution 85:2277–2282
Gerardin D (1996) Lessons from the European Community. In: Tay SSC, Esty D (eds) Asian dragons and green trade
Koh KL (ed) (1996) Selected ASEAN documents on the APCEL document series
McClatchey DF (1996) Chernobyl and Sandoz one decade later 1986–1996. GA. J. Int'l & Comp. L, vol. 25:659
Merrill T (1997) Golden rules for transboundary pollution. Duke Law Journal 46:967
Nicholls N (1993) ENSO, drought and flooding rain. In: Brookfield H, Byron Y (eds) South East Asia's environmental future. p 154
Scnieider J (1997) World public order of the environment: Towards an international ecological law and organization. pp. 168–171
Sinton JE (1997) China's view of acid rain in Northeast Asia and regional cooperation strategies for mitigation. ESENA
Staffin EB (1996) Trade barrier or trade boon? A critical evaluation of environmental labeling and its role in the "greening" of world trade, 21 Colum. J Envtl L 205
Streets DG (1997) Energy and acid rain projections for Northeast Asia. Energy, Security and Environment in Northeast Asia Project, The Nautilus Institute
Tay SSC (1998) South East Asian fires: A haze over ASEAN and international environmental law. Review of European Community and International Law, UK
Wapner P (1996) Environmental activism and World civic politics. State University of New York
Wirawan N (1993) The hazard of Fire. In: Brookfield H, Byron Y (eds) South East Asia's environmental future. p 242

Part IIc

**Managing Planetary Metabolism?
The Global Carbon Cycle**

Chapter 9

Carbon and the Science-Policy Nexus: The Kyoto Challenge

Robert T. Watson · Ian R. Noble

The overwhelming majority of experts in both developed and developing countries recognise that while scientific uncertainties exist, there is little doubt that the Earth's climate has warmed over the past 100 years in response to human activities (combustion of fossil fuels and land-use changes) and that further human-induced changes in climate are inevitable (Fig. 9.1). The question is not whether climate will change further in the future in response to human activities, but rather by how much, where and when.

It is also clear that climate change will, in many parts of the world, adversely effect socioeconomic sectors, including water resources, agriculture, forestry, fisheries and human settlements, ecological systems (particularly coral reefs), and human health (particularly through an increased exposure to vector-borne diseases). Indeed, more people will be adversely affected by climate change than be benefited, and the greater the rate and magnitude of change the more adverse the consequences.

The good news is, however, that significant reductions in net greenhouse gas emissions are technically feasible due to an extensive array of technologies in the energy supply, energy demand and agricultural and forestry sectors, many at little or no cost to society. However, realising these emissions reductions involves the development and implementation of supporting policies to overcome barriers to the diffusion of these technologies into the marketplace, increased funding for research and development, and effective technology transfer.

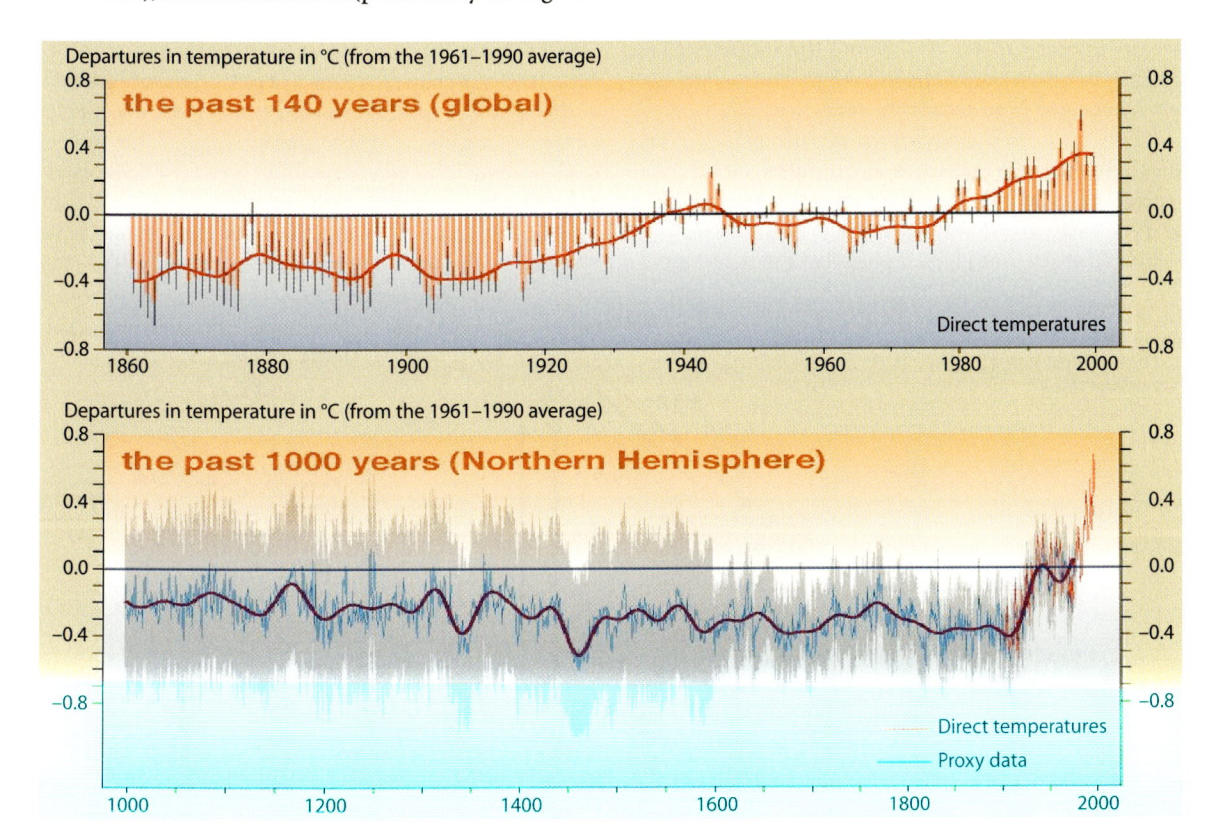

Fig. 9.1. Variations of the Earth's surface temperature over the last 140 years and the last millennium (reproduced with permission from Albritton et al. 2001, © IPCC 2001)

The short-term challenge for industrialised countries is to achieve the Kyoto targets, i.e., a reduction in overall emissions of six greenhouse gases (or families of gases) by an average of 5.2% below 1990 levels in 2008–2012. The longer-term challenge is to meet the objectives of Article 2 of the UN Framework Convention on Climate Change, i.e., stabilisation of greenhouse concentrations in the atmosphere at a level that would prevent dangerous anthropogenic interference with the climate system, with specific attention being paid to food security, ecological systems and sustainable economic development.

This paper briefly discusses the current understanding of the global carbon cycle and then the carbon cycle in the context of the Kyoto Protocol.

9.1 The Global Carbon Cycle

Figure 9.2 summarises the main pools and fluxes in the global carbon cycle in the 1990s. Each year approximately 60 Gt C are exchanged, in each direction, between the atmosphere and terrestrial ecosystems, and another 90 Gt C, in each direction, between the atmosphere and the oceans. Currently, both the oceans and terrestrial systems show net uptakes of carbon. Terrestrial systems have a net uptake of about 1.4 Gt C yr^{-1} (uptake of 3.0 Gt C yr^{-1} offset by about 1.6 Gt C yr^{-1} in emissions from land-use changes, primarily in the tropics) and oceans uptake 1.7 Gt C yr^{-1}. Before the industrial era, uptakes balanced emissions in both of these exchanges when averaged over decades to centuries.

There have been longer-term variations in the balance between these exchanges, and the atmospheric concentration of CO_2 has varied between 180 and 280 ppm over the past 420 000 years in a regular pattern (Fig. 9.3). Falkowski et al. (2000) have argued that the tight bounding of atmospheric concentrations of CO_2 results from the interplay between ocean and terrestrial sources and sinks. Oceans dominate the global carbon cycle over time scales of decades to millennia. The most important processes are the solubility pump, whereby CO_2 is taken up in the cold waters of high latitudes and transported towards the equator, and the biological pump, whereby phytoplankton remove carbon from the upper ocean and transport a significant fraction of this to the deeper ocean. Higher concentrations of CO_2 enhance the solubility pump, although higher temperatures may partially offset this and may also increase ocean stratification which, in turn, slows the transfer of carbon to the deep ocean. The biological pump is also sensitive to nutrient transport from terrestrial systems; thus the ocean and terrestrial carbon cycles are intimately linked. The overall controlling mechanisms probably involve physical and chemical reorganisation of the ocean and changes in nutrient inventories.

9.2 Human Perturbation of the Carbon Cycle

Human activities, primarily through the combustion of fossil fuels and land-use changes, have and are continuing to perturb the carbon cycle, increasing the atmospheric concentration of carbon dioxide and methane.

From about 1850 to 1990, human activities are estimated to have added about 336 Gt C to the atmosphere, 212 Gt C from the burning of fossil fuels and 124 Gt C from land-use change. A total of 144 Gt C have remained in the atmosphere, and models estimate that the oceans have taken up about 107 Gt C, which implies that terrestrial systems have taken up about 85 Gt C (Noble et al. 2000 and Prentice et al. 2001).

Since the industrial era, these perturbations have resulted in the atmospheric concentration of CO_2 in-

Fig. 9.2.
Summary of the main pools and fluxes in the global carbon cycle in the 1990s

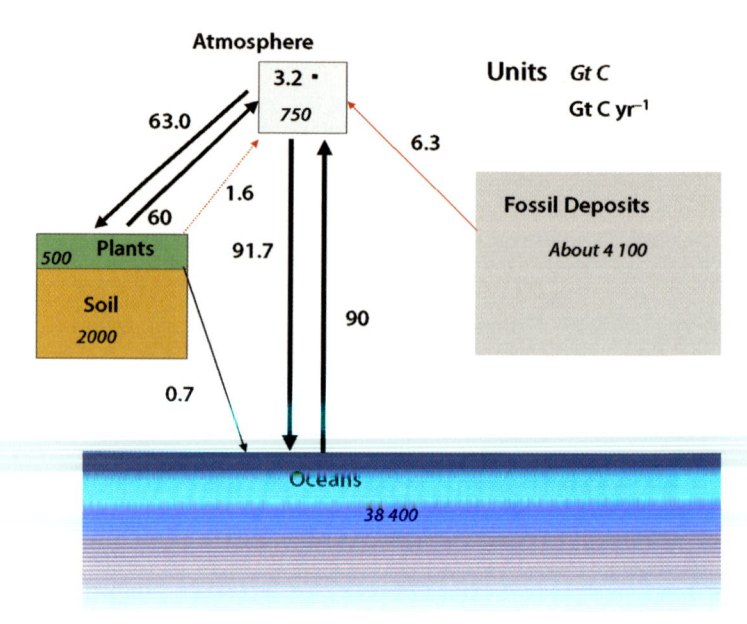

Fig. 9.3.
Over the last 420 000 years, CO_2 concentrations have fluctuated in a regular way between upper and lower limits of 280 and 180 ppm respectively. It is only recently that an increase has exceeded the upper limit, and this is predicted to increase still further

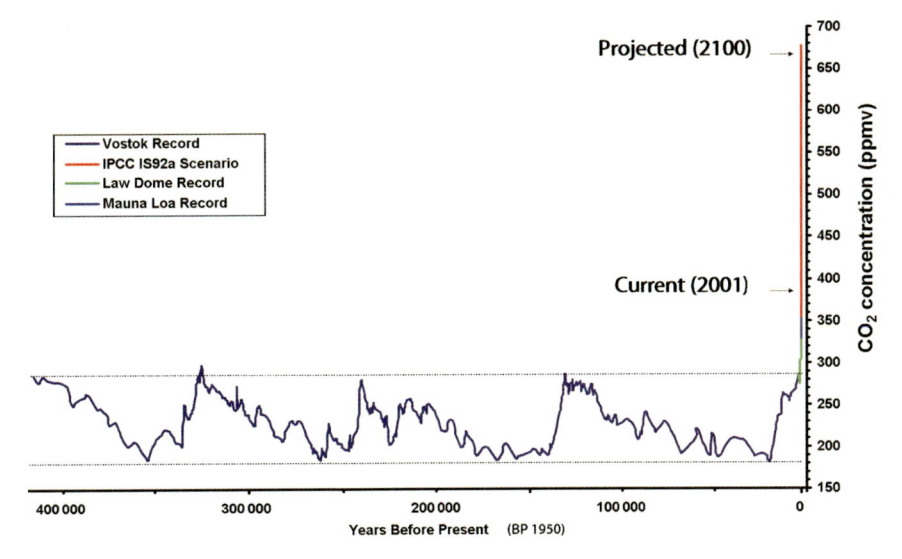

creasing to 369 ppm, greatly exceeding the 280 ppm upper bound that has existed for at least the past 420 000 years.

9.3 The Terrestrial Sink

Since 1850, the terrestrial biosphere has been a net source of 39 Gt C, made up of an uptake of 85 Gt C and emissions of 124 Gt C from land cover change. However, over the past several decades, the terrestrial biosphere appears to have become a net sink (Table 9.1). In the 1980s, the net terrestrial uptake (uptake less emissions from land-use change), as estimated by established modelling and isotope measurements, was 0.2 ±0.7 Gt C yr^{-1}, and this increased to about 1.4 Gt C yr^{-1} in the 1990s. In the 1980s, emissions from land-use change were approximately 1.7 Gt C yr^{-1}, which implies that the uptake by terrestrial ecosystems was about 1.9 Gt C yr^{-1}. Emissions from land-use change during the 1990s are not yet well established, but if they were similar to those in the 1980s,

uptake by terrestrial ecosystems appears to have increased to about 3 Gt C yr^{-1} (Prentice et al. 2001).

These estimates have major implications for the goals of the UNFCCC and the Kyoto Protocol. They lead to questions of what is causing this increase in uptake by terrestrial ecosystems, where is it occurring, by how much does it vary from year to year, and for how long will it continue? The answers to each significantly affect our best approach to achieving the objective of the UNFCCC of preventing dangerous anthropogenic change to the climate system.

9.4 Location of the Terrestrial Sink

There are two approaches to determining the location of the terrestrial sink. The first is based on forest and other inventories, which are of variable quality and coverage across the globe. The second is by atmospheric transport inversion analysis (inverse models), in which circulation and biological models are used to work backwards from measured concentrations of CO_2 (or other trace gases) and measured emissions data to estimate where carbon is being sequestered or released by terrestrial and ocean systems. The results from inverse models are still unreliable, depending very much on the geographic resolution selected, the particular models used and the mathematical techniques applied.

An original estimate by Fan et al. (1998) that the coterminous USA was a terrestrial net sink of 1.4 Gt C yr^{-1} from 1988–1992 has led to a series of alternate analyses. The overall results suggest that the USA is a sink but possibly of lesser magnitude (averaging about 0.6 Gt C yr^{-1} according to a series of inverse analyses by Pacala et al. 2001). However, analyses by Bousquet et al. (2000) show that the exchange of carbon between the atmosphere and North America as a whole is very variable, ranging between a sink of 1 Gt C yr^{-1} to a source of a similar magnitude. The long-term average for North America appears to

Table 9.1. Sinks for carbon in the coterminous USA 1980–1990 (Gt C yr^{-1}) (from Pacala et al. 2001)

Category	Low estimate	High estimate
Forest trees	0.11	0.15
Other forest processes	0.03	0.15
Wood products	0.03	0.07
Woody encroachment	0.12	0.13
Cropland soils	0.00	0.04
Reservoirs (alluvium, colluvium)	0.01	0.04
Net exports of food, wood etc.[a]	0.04	0.09
Exported by rivers[a]	0.03	0.04
Total	0.37	0.71

[a] These both create a corresponding source elsewhere.

be close to zero (Bousquet et al. 2000). Most models point to a significant sink in Eurasia probably of more than 1 Gt C yr^{-1} on average but again with significant variation. Net carbon emissions from the tropics are estimated to fluctuate by 0.3 Gt C yr^{-1} with an average of about zero, implying that the emissions of about 1.6 Gt C yr^{-1} resulting from land-use changes are balanced by an uptake of a comparable amount. In summary, of the 3 Gt C yr^{-1} uptake by terrestrial ecosystems, about half is taken up by Northern Hemisphere ecosystems with the major contribution fluctuating between Eurasia and North America, and the other half in tropical ecosystems.

9.5 Variability of the Terrestrial Sink

The variation in uptake by the terrestrial sink from year to year is significant, especially when compared with the targets associated with the Kyoto Protocol (0.2 Gt C yr^{-1} below 1990 levels and about 0.7 Gt C yr^{-1} below the IPCC IS 92 a or SRES A2 emissions scenarios). Annual net uptake by terrestrial ecosystems is estimated (via inverse analyses) to vary by as much as 3 Gt C yr^{-1} about the mean value (Bousquet et al. 2000), and over a five-year commitment period the cumulative variation could be 5 Gt C or more. Most of the variation derives from variations in net emissions from terrestrial ecosystems and is strongly linked with the El Niño phenomena. In strong El Niño years, global terrestrial uptake falls and the rate of increase in atmospheric CO_2 can be double that of the long-term average. These variations are not uniformly distributed across the globe with the net uptake across the USA, for example, varying from about +1 to –1 Gt C yr^{-1} from 1989–1990 to 1994–1995. If further studies confirm these conclusions, they imply significant challenges in designing a measuring, verification and compliance system for a ratified Kyoto Protocol.

9.6 Origin of the Terrestrial Sink

Recent studies have attempted to identify the processes contributing most to the terrestrial uptake. The best information is from the USA where Pacala et al. (2001) reviewed a number of studies to develop Table 9.1. Forest processes and wood products account for over half of the terrestrial sink with another major contribution from woody encroachment (also called woody weeds, or vegetation thickening), which is the invasion of grasslands or open woodlands by woody tree and/or shrub species after changes in grazing and fire management practices. The total uptake is higher than previous inventory based estimates, (Birdsey and Heath 1995; Houghton et al. 1999; Brown and Schroeder 1999), although less complete, but is compatible with the inversion analyses described above.

The question remains as to what the causes of the net terrestrial sink are. Schimel et al. (2001) used three models to estimate the uptake due to CO_2 fertilisation in the USA and concluded that only about 0.08 Gt C yr^{-1} of net uptake was due to this process. They also concluded that net uptakes due to other processes such as nitrogen fertilisation and climate warming were also of a similar magnitude or smaller. If these estimates are correct, they imply that a significant fraction of the forest uptake is due to changes in forest age structure. Significant areas of farmland were abandoned or converted to forest during the 20th century, and these forests, along with woody encroachment, make a significant contribution to the net sink (Schimel et al. 2001; estimate 0.2 Gt C yr^{-1}). The USA in its submission to the UNFCCC before COP6 used an estimate that forest management in the USA contributed a net sink of 0.288 Gt C yr^{-1}.

The balance between CO_2 fertilisation, nitrogen deposition, climate warming, and age structure change due to forest regrowth will vary from region to region across the globe. The outcome is that on average about 3 Gt C yr^{-1} are taken up by terrestrial ecosystems, but this can vary by ±(2–3) Gt C yr^{-1}. This uptake plays a significant role in reducing the rate of increase in CO_2 in the atmosphere and is matched by a similar (slightly smaller) uptake into ocean systems. Identification of the relative contributions of the different processes, including the geographic distribution, to the terrestrial sink remains important as the UNFCCC and the Kyoto Protocol seek that nations take actions to reduce the human induced emissions and increase the human induced uptakes of greenhouse gases. The definition of human induced remains open to debate, but at the COP7 meeting in Marrakech the parties agreed to exclude "(*i*) elevated carbon dioxide concentrations above their preindustrial level; (*ii*) indirect nitrogen deposition; and (*iii*) the dynamic effects of age structure resulting from activities and practices before the reference year." This will prove a significant scientific and accounting challenge.

9.7 The Future of the Terrestrial Carbon Cycle

Observational evidence suggests that terrestrial carbon fluxes may be showing a trend of increasing uptake. Several models of the terrestrial carbon fluxes have been developed independently and when compared produce a similar result (Fig. 9.4). They show that the net terrestrial sink will increase until about the middle of the 21st century, peaking at 3–7 Gt C yr^{-1} uptake, after which it will either level off or decline. Some models predict that the uptake could decline to zero or even become a net source again by the end of this century (IPCC Synthesis Report, Fig. 5.5).

Two of the important elements of the Kyoto Protocol that affect the carbon cycle will be briefly discussed, i.e., (*i*) carbon trading among industrialised countries (Arti-

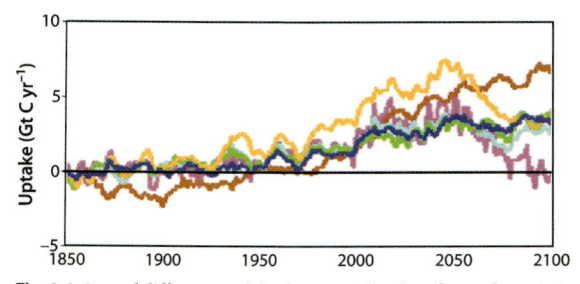

Fig. 9.4. Several different models of terrestrial carbon fluxes show similar results – that the net terrestrial sink will increase until about the middle of the 21st century peaking at 3 to 7 GT C yr^{-1} uptake, after which it will either level off or decline (adapted from Cramer et al. 2001)

cles 6 and 17), and between industrialised countries and developing countries (Article 12), and (*ii*) the use of land-use, land-use change and forestry (LULUCF) activities to mitigate greenhouse gas emissions (Articles 3.3, 3.4 and 12).

International carbon trading through Article 17 (emission rights trading) can significantly reduce the cost of compliance with the Kyoto Protocol targets for industrialised countries. This occurs because of the presence of "hot air" in the Russian Federation, i.e., the Kyoto targets for the Russian Federation are significantly greater than the projected emissions in 2010 due to the economic downturn in the Russian Federation. Thus the Russian Federation can sell its surplus emissions to other industrialised countries. In the absence of international trading, the costs of complying with the Kyoto Protocol for industrialised countries are projected to range from about $150–600 per t C with GDP losses ranging from 0.2–2%, whereas with full trading among industrialised countries, the costs of compliance are reduced to $15–150 per t C, with GDP losses ranging from 0.1–1% (Fig. 9.5). These estimated costs could be further reduced with use of the Clean Development Mechanism (Article 12 allows project-based trading between industrialised countries and developing countries), carbon sequestration through LULUCF activities, non-carbon dioxide greenhouse gases, inclusion of ancillary benefits, and efficient tax recycling. If all cost reduction activities could be realised, then GDP growth rates are projected to slow only by a few hundreds of a percent per year.

Biological mitigation can occur by three strategies: (*a*) conservation of existing carbon pools, (*b*) sequestration by increasing the size of carbon pools, and (*c*) substitution of sustainably produced biological products.

To operationalise key provisions associated with LULUCF activities within the Kyoto Protocol, a number of issues had to be addressed:

- *Definitions of a forest, afforestation, reforestation and deforestation:*

  The current agreement (COP7) allows each Party to define a forest by choosing a canopy cover between 10 and 30%, a minimum tree height between 2–5 m, and a spatial extent between 0.05–1 ha. However, Parties have yet to decide whether biome specific definitions may be used after the first commitment period.

  The definitions of afforestation, reforestation and deforestation all use land-use change as a criteria, with afforestation being on land that has not been forested for at least 50 years, and reforestation on land that has not been forested since 1990. These definitions have the effect of precluding credit for activities that involve the clearing of forests (after 1989) to subsequently claim a credit for reforestation.

- *How does one address issues such as the harvesting-regeneration cycle, and aggradation and degradation in forested systems?*

  The harvesting-regeneration cycle and the issue of aggradation and degradation of forests will be dealt with through forest management under Article 3.4.

- *Can the indirect effects of human activities, e.g., carbon dioxide and nitrogen fertilisation be credited or can only the direct effects of human activities, e.g., planting trees be credited?*

  Article 3.3 states that net changes in greenhouse gas emissions from sources and removals by sinks resulting from "*direct human-induced*" land-use change and forestry activities, limited to afforestation, reforestation, and deforestation since 1990, measured as verifiable changes in stocks in each commitment period shall be used to meet the commitments in this Article of each Party included in Annex I.

  Therefore, a key question for the scientific community is whether the "*direct*" effects of human-induced activities, i.e., the growth increment due to "normal" forest growth following afforestation or reforestation can be separated out from the growth increment due to "*indirect*" human-induced activities, e.g., that due to carbon dioxide and nitrogen fertilisation, changes in age structure and climate change. The IPCC Special Report on LULUCF concluded that for activities that involve land-use changes (e.g., conversion of grassland/pasture to forest) it may be very difficult, if not impossible, to distinguish with present scientific tools that portion of the observed stock change that is directly human-induced from that portion that is caused by indirect or natural factors, such as that due to climatic variability. Emissions and removals from natural causes such as El Niño may be large during any one commitment period compared with an industrialised country's commitment.

- *Which activities are eligible under Article 3.4: whether to limit Article 3.4 credits during the first commitment period; whether "business-as-usual" carbon uptake can be credited, and whether the "direct" effects of human activities can be separated from the "indirect" effects?*

  Article 3.4 states that Parties will "… decide upon modalities, rules and guidelines as to how and which additional human-induced activities related to changes in greenhouse gas emissions by sources and removals by sinks in the agricultural soils and the land-use change

Fig. 9.5.
Projections of GDP losses and marginal cost in Annex II countries in 2010 from global models, with and without carbon emissions trading (data from Banuri et al. 2001, IPCC)

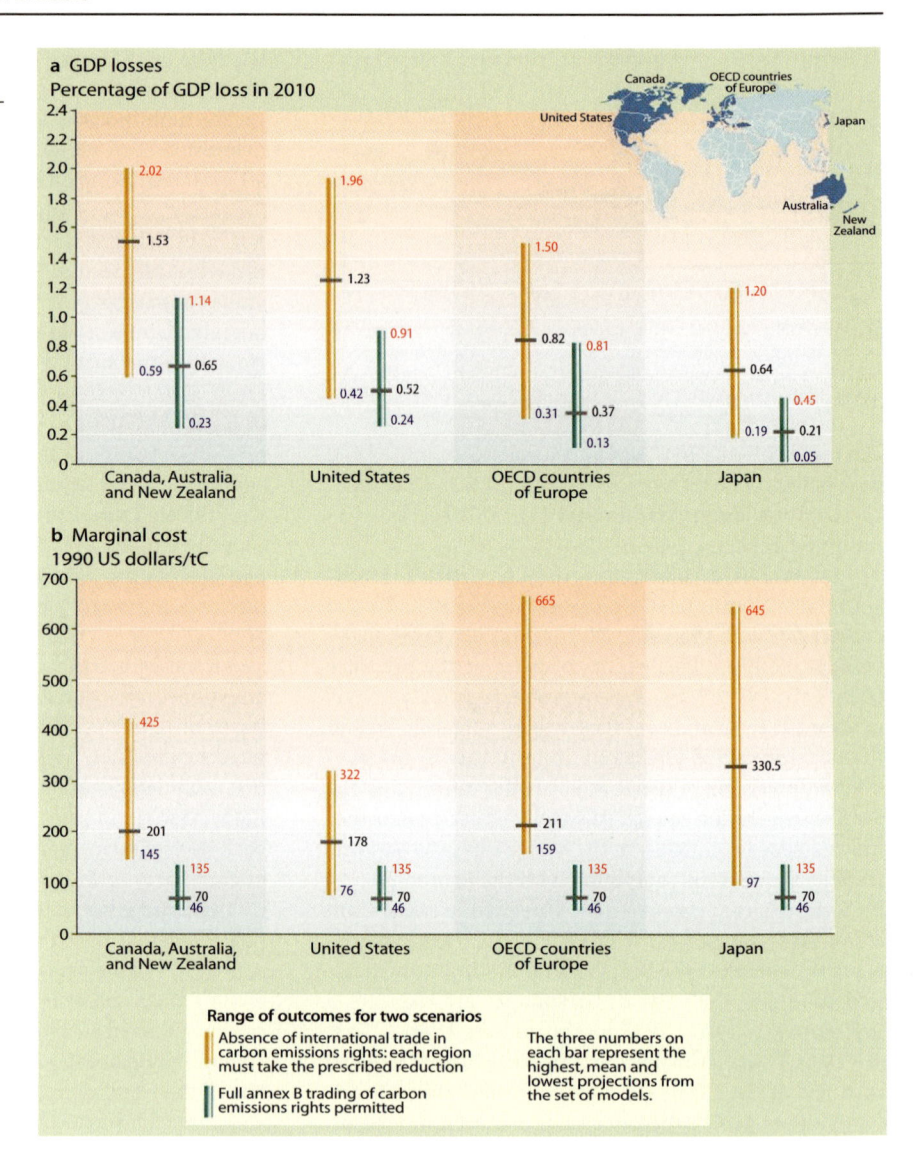

a GDP losses
Percentage of GDP loss in 2010

Canada, Australia, and New Zealand: 2.02 / 1.53 / 0.59; 1.14 / 0.65 / 0.23
United States: 1.96 / 1.23 / 0.42; 0.91 / 0.52 / 0.24
OECD countries of Europe: 1.50 / 0.82 / 0.31; 0.81 / 0.37 / 0.13
Japan: 1.20 / 0.64 / 0.19; 0.45 / 0.21 / 0.05

b Marginal cost
1990 US dollars/tC

Canada, Australia, and New Zealand: 425 / 201 / 145; 135 / 70 / 46
United States: 322 / 178 / 76; 135 / 70 / 46
OECD countries of Europe: 665 / 211 / 159; 135 / 70 / 46
Japan: 645 / 330.5 / 97; 135 / 70 / 46

Range of outcomes for two scenarios
- Absence of international trade in carbon emissions rights: each region must take the prescribed reduction
- Full annex B trading of carbon emissions rights permitted

The three numbers on each bar represent the highest, mean and lowest projections from the set of models.

and forestry categories shall be added to, or subtracted from, the assigned amounts for Parties included in Annex I, taking into account uncertainties, transparency in reporting, verifiability, and the methodological work of the IPCC …" Such a decision shall apply in the second and subsequent commitment periods, but a Party may choose to apply such a decision on these additional human-induced activities for its first commitment period, provided that these activities have taken place since 1990.

The key question is whether these activities must commence after 1990 or whether activities initiated before 1990, but that are continued after 1990, are eligible. If the latter interpretation were to be used, then industrialised countries would automatically have an annual credit of about 1.5 Gt C without enacting any additional activities, i.e., the current net uptake of the terrestrial biosphere in industrialised countries. This credit would far exceed their obligations under the Kyoto Protocol, which amount to a decrease of about 200 Mt C yr^{-1} in 2010 relative to 1990. The issues of which activities will be eligible, whether credits could be obtained from pre-1990 Article 3.4 activities and whether the "indirect" effects had to be subtracted from the "direct" effects, were all finessed by the Parties for the first commitment period by simply allocating each country a maximum allowable credit for forest management activities under Article 3.4 and by measuring credits for other Article 3.4 activities against a baseline year (1990). It should be noted that for activities that involve land-management changes (e.g., tillage to no-till agriculture), it should be feasible to distinguish between the direct and indirect human-induced components, but not to separate out natural factors through the use of control plots and modelling.

- *How does one address the issues of permanence?*

The issue of the permanence of sinks, which are reversible through human activities, disturbances, or environmental change, including climate change, is a more critical issue than for activities in other sectors, e.g., the energy sector. A pragmatic solution is to ensure that any credit for enhanced carbon stocks on a particular plot of land due to Article 3.3 or 3.4 activities is balanced by accounting for any subsequent reductions in those carbon stocks on the same plot of land, regardless of the cause.

- *Which carbon stocks need to be, and can be, measured?*

  The Parties agreed that all carbon stocks that are decreasing and all carbon stocks for which credit is being claimed must be measured. However, if it can be shown that a specific stock is increasing but is difficult or costly to measure, it does not need to be measured.

  Technical methods sufficient to serve the requirements of the protocol exist for aboveground stocks (aboveground biomass, including litter) and most likely for belowground stocks (below-ground biomass and soil carbon).

  Industrialised countries generally have the technologies available, but few currently apply them routinely, for monitoring, whereas developing countries may require assistance to develop the necessary capacities and cover costs.

  Given that methods and research results are highly transferable between countries, rapid improvement in monitoring capabilities should be expected.

- *Which, if any, LULUCF activities are eligible under Article 12 the Clean Development Mechanism (CDM)?*

  Parties agreed that afforestation and reforestation activities are eligible under Article 12, the Clean Development Mechanism, but credits for avoided deforestation or Article 3.4 type activities will not be eligible for the first commitment period. In addition, credit to Annex I countries for LULUCF activities under the CDM are limited to 1% of their 1990 emissions.

  Given the significant potential of avoiding carbon emissions by slowing deforestation, and the concurrent benefits to biodiversity and water resources, the development of accounting rules that address issues such as permanence, baselines, leakage, sustainability criteria and national sovereignty rights under the CDM are urgently needed.

A key question is what the potential is for LULUCF activities to sequester carbon under Articles 3.3, 3.4 and 12 during the first commitment period. The following estimates do not include any contribution from the current terrestrial uptake. There are significant uncertainties in the following "best" estimates, primarily because of assumptions made regarding areas affected, carbon sequestration rates and the rate of adoption of improved management techniques by foresters and farmers.

The potential carbon credit for industrialised countries through afforestation and reforestation activities during the first commitment period is relatively small, i.e., 20–30 Mt C yr^{-1}, assuming current rates of afforestation and reforestation continue. This compares to a potential debit of about 90 Mt C yr^{-1} if current rates of deforestation continue.

The potential for carbon credits under the CDM from afforestation and reforestation projects during the first commitment period is between 350 and 400 Mt C yr^{-1}, however, the total credits allowable to industrialised countries is limited to 1% of their 1990 emissions, i.e., about 40 Mt C. The potential carbon credits for avoided deforestation (not allowed during the first commitment period) equal the rate of deforestation, i.e., about 1.6 Gt C yr^{-1}.

The potential carbon credits for industrialised countries through Article 3.4 activities during the first commitment period is about 260 Mt C yr^{-1} from improved management (forest, cropland, rangeland and agroforestry) and about 25 Mt C yr^{-1} from land-use change (cropland to grasslands) (Fig. 9.6a). The potential carbon credits for Article 3.4 activities in developing countries during the first commitment period is about 300 Mt C yr^{-1} from improved management (forest, cropland, rangeland and agroforestry), and about 400 Mt C yr^{-1} from land-use change (primarily the conversion of degraded lands to agroforestry) (Fig. 9.6b).

There is also significant potential for using modern biomass to displace fossil fuels as a source of energy, but these activities are not accounted for under the LULUCF Articles, except for the standing biomass in the plantations.

The IPCC also estimated the potential global carbon uptake using LULUCF activities over the next 50 years for activities started after 1990, not including any contribution from the current terrestrial uptake. The estimated global potential is on the order of 100 Gt C (cumulative) by 2050, equivalent to about 10 to 20% of projected fossil-fuel emissions during that period, although there are substantial uncertainties associated with this estimate. Realisation

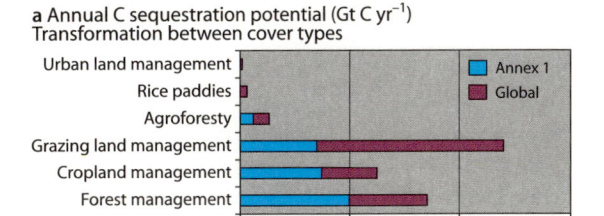

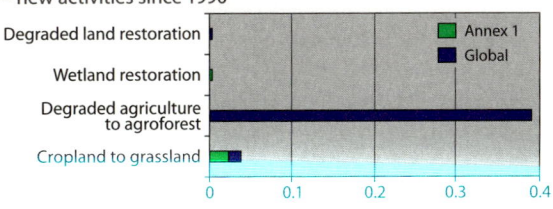

Fig. 9.6. Potential carbon credits through transformation between land-cover type and improved land management for industrialised countries and developing countries (contains a best estimate of the rate of uptake of these activities by 2010 (vary between 3–80%) – current text would inhibit investment under Article 3.4 because forest management because is discounted 85%)

of this potential depends upon land and water availability as well as the rates of adoption of land management practices. The largest biological potential for atmospheric carbon mitigation is in subtropical and tropical regions.

Two questions are often asked: (*i*) Is it possible that sequestered carbon can be released back into the atmosphere due to changes in climate; and (*ii*) is it possible that LULUCF activities can have negative climatic or socioeconomic effects? With respect to the first question, some of the LULUCF sequestered carbon could be released back to the atmosphere due to changes in climate, but this is unlikely for many decades. Even then there would still be more carbon across all carbon pools than without the LULUCF activities, and any carbon released can be debited through an appropriate accounting system. With respect to the second question, there may be a few instances, e.g., in boreal forests at high latitudes, where the benefits of carbon sequestration may be partially or fully offset by changes in surface albedo. However, there is little incentive to plant forests for LULUCF credits in this region, as growth rates are very low. In addition, there is no doubt that LULUCF activities and projects can have a broad range of positive, and potentially negative, environmental, social and economic impacts on biodiversity, forests, soils, water resources, food, fibre, fuel, employment, health, poverty, and equity. Hence, a system of criteria and indicators could be valuable to compare sustainable development impacts across LULUCF alternatives. In particular, slowing deforestation has multiple environmental and social benefits in most regions.

In summary:

- Biological mitigation can occur through three types of activities: (*a*) conservation of existing carbon pools, (*b*) sequestration by increasing the size of carbon pools, and (*c*) substitution of sustainably produced biological products;
- LULUCF activities will result in the sequestration of carbon in three main pools, above- and belowground biomass and soils – for decades to centuries, for which monitoring systems can be put in place to monitor all three pools of carbon with the precision required to implement the Kyoto Protocol;
- The magnitude of trading under the CDM for LULUCF activities will be very limited given the absence of the US from the Kyoto Protocol, and the significant allowances made for sinks under Articles 3.3 and 3.4;
- LULUCF activities can play a critical role in limiting the build-up of carbon dioxide in the atmosphere, especially in the short-term, but stabilising the atmospheric concentration of carbon dioxide (Article 2 of the Convention) will require significant emissions reductions globally, which can only be achieved by either reducing energy related emissions or by capturing and storing of energy related emissions;
- LULUCF activities buy time to transform energy systems to lower greenhouse gas emitting systems but will allow more fossil carbon to be transferred to the more labile biological pools, hence avoiding a tonne of carbon emissions is better than creating a tonne of sinks.

References

Albritton DL et al. (2001) Summary for policymakers. In: Houghton JT, Ding Yihui, Griggs DJ, Noguer M, van der Linden PJ, Dai X, Maskell K, Johnson CA (eds) Climate change 2001: The scientific basis. Contribution of Working Group I to the Third Intergovernmental Panel on Climate Change. Cambridge University Press, Cambridge New York

Banuri T et al. (2001) Technical summary. In: Metz B, Davidson O, Swart R, Pan J (eds) Climate change 2001: Mitigation. Contribution of Working Group III to the Third Assessment Report of the Intergovernmental Panel on Climate Change. Cambridge University Press, Cambridge New York

Birdsey RA, Heath LS (1995) In: Productivity of America's forest ecosystems. Joyce LA (ed). Forest Service General Technical Report RM-GTR-271, US Forest Service, Fort Collins, CO, 1995, pp 56–70

Bousquet P, Peylin P, Ciais P, Le Quere C, Friedlingstein P, Tans PP (2000) Regional changes in carbon dioxide fluxes of land and oceans since 1980. Science 290:1342–1346

Brown SL, Schroeder PE (1999) Spatial patterns of aboveground production and mortality of woody biomass for eastern US forests. Ecol Appl 9:968–980

Cramer W, Bondeau A, Woodward FI, Prentice IC, Betts RA, Brovkin V, Cox PM, Fisher V, Foley JA, Friend AD, Kucharik C, Lomas MR, Ramankutty N, Sitch S, Smith B, White A, Young-Molling C (2001) Global response of terrestrial ecosystem structure and function to CO_2 and climate change: results from six dynamic global vegetation models. Global Change Biology 7:357–374

Falkowski P, Scholes RJ, Boyle E, Canadell J, Canfield D, Elser J, Gruber N, Hibbard K, Högberg P, Linder S, Mackenzie FT, Moore III B, Pedersen T, Rosenthal Y, Seitzinger S, Smetacek V, Steffen W (2000) The global carbon cycle: A test of our knowledge of Earth as a system. Science 290:291–296

Fan S, Gloor M, Mahlman J, Pacala S, Sarmiento J, Takahashi T, Tans P (1998) A large terrestrial carbon sink in North America implied by atmospheric and oceanic carbon dioxide data and models. Science 282 (5388):442–446

Houghton RA, Hackler JL, Lawrence KT (1999) The US carbon budget: contributions from land-use change. Science 285:574–578

Noble I, Apps M, Houghton R, Lashof D, Makundi W, Murdiyarso D, Murray B, Sombroek W, Valentini R (2000) Implications of different definitions and generic issues. In: Watson RT, Noble IR, Bolin B, Ravindranath NH, Verardo DJ, Dokken DJ (eds) Land use, land-use change and forestry. A special report of the IPCC. Cambridge University Press, Cambridge New York, 377 pp

Pacala SW, Hurtt GC, Baker D, Peylin P, Houghton RA, Birdsey RA, Heath L, Sundquist ET, Stallard RF, Ciais P, Moorcroft P, Caspersen JP, Shevliakova E, Moore B, Kohlmaier G, Holland E, Gloor M, Harmon ME, Fan SM, Sarmiento JL, Goodale CL, Schimel D, Field CB (2001) Consistent land- and atmosphere-based US carbon sink estimates. Science 292:2316–2320

Prentice IC, Farquhar GD, Fasham MJR, Goulden ML, Heimann M, Jaramillo VJ, Kheshgi HS, Le Quéré C, Scholes RJ, Wallace DWR (2001) The carbon cycle and atmospheric carbon dioxide. In: Houghton JT, Ding Y, Griggs DJ, Noguer M, van der Linden PJ, Dai X, Maskell K, Johnson CA (eds) Climate change 2001: The scientific basis. Contribution of Working Group I to the Third Assessment Report of the Intergovernmental Panel on Climate Change. Cambridge University Press, Cambridge New York, 881 pp

Schimel DS, House JI, Hibbard KA, Bousquet P, Ciais P, Peylin P, Braswell BH, Apps MJ, Baker D, Bondeau A, Canadell J, Churkina G, Cramer W, Denning AS, Field CB, Friedlingstein P, Goodale C, Heimann M, Houghton RA, Melillo JM, Moore B, Murdiyarso D, Noble I, Pacala SW, Prentice IC, Raupach MR, Rayner PJ, Scholes RJ, Steffen WL, Wirth C (2001) Recent patterns and mechanisms of carbon exchange by terrestrial ecosystems. Nature 414:169–172

Chapter 10

Industry Response to the CO_2 Challenge

Charles C. Nicholson

10.1 Introduction

This paper presents a view on how industry is meeting the issue of climate change.

It is interesting to discuss the topic of climate change at a conference in Holland, a country where nowhere is more than 150 kilometres from the sea, and where the highest point is 321 metres, in South West Limburg. Indeed, the struggle with the sea and taking advantage of its opportunities, has done much to shape the history of this land.

It is also a land of wind power; there are eleven types of windmills, and at one point there were some ten thousand gracing the landscape. As a nineteenth century British travel guide commented: "The Dutchman may be said to have made the wind his slave." No doubt many wish that today's wind turbines had similar elegance.

It is also fitting to recognise the strenuous efforts that Dutch Minister Pronk and his colleagues have made to try to find a way through one of the most difficult challenges the world has set itself. We must hope he will be successful.

Most global change scientists probably live with the issue of climate change on a regular basis and have done so for some time. But for many in business, it is a new issue; it competes with others for management attention, and it is not always easy to see what to do. This paper comes from the perspective of those who have decided they must move forward, and in particular from the direct experience of my own company, BP, which I think connects well with the wider public debate.

We have not concluded that all scientific uncertainty has been removed, but we are sufficiently persuaded by the thoughtful work of scientists that the risk of climate change is real. The latest IPCC reports published earlier this year provide important added confirmation, a point that is addressed later.

The challenge for business, if you judge the prospect of global warming to be real, and you recognise the concerns of governments and society, is what then to do. As a business, and particularly an energy business, we do not believe we have the choice of inaction.

10.2 The Energy Context

Today some 65% of the world's energy demand is met by oil and gas (Fig. 10.1), and as we look ahead that proportion is likely to increase to satisfy the growing requirement not least in the developing countries (Fig. 10.2). It is likely that the demand for oil will rise by some 20% by 2010, and gas by 25% or more. The supply is there, since the industry has become ever more skilful at replacing reserves more economically, and today there are at least 40 years of oil supply available at the current rate of consumption, and 60 years worth of gas.

So demand for energy will continue to grow, driven largely by population growth and economic activity. The growth will be most evident in the developing world. But policy makers can influence the choice of fuel, coal and nuclear for example, and accelerate or retard the rate at which new technology is introduced. Finally, if a bias develops to low carbon, clean fuels, this will favour natural gas and renewables.

However, we must recognise that the increasing demand for hydrocarbon fuels carries with it the risk of increasing carbon emissions, and as this shows by the year 2020, the majority of these will be coming from developing countries (Fig. 10.3). So, the challenge is not lack of resource. It is how resources are used, and the environmental challenge which comes with growing consumption.

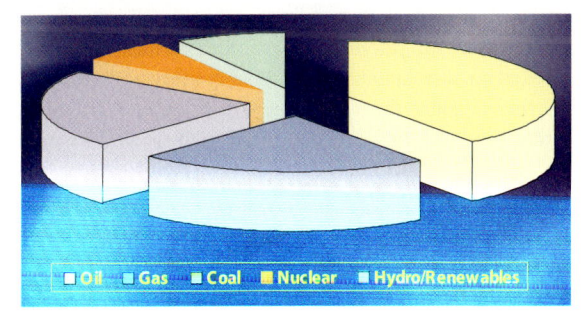

Fig. 10.1. Long run energy demand world consumption 1999 (source: Energy Information Administration (EIA) Energy Outlook 2001)

Fig. 10.2.
Long run energy demand world consumption by region (1970–2020; source: Energy Information Administration (EIA) Energy Outlook 2001)

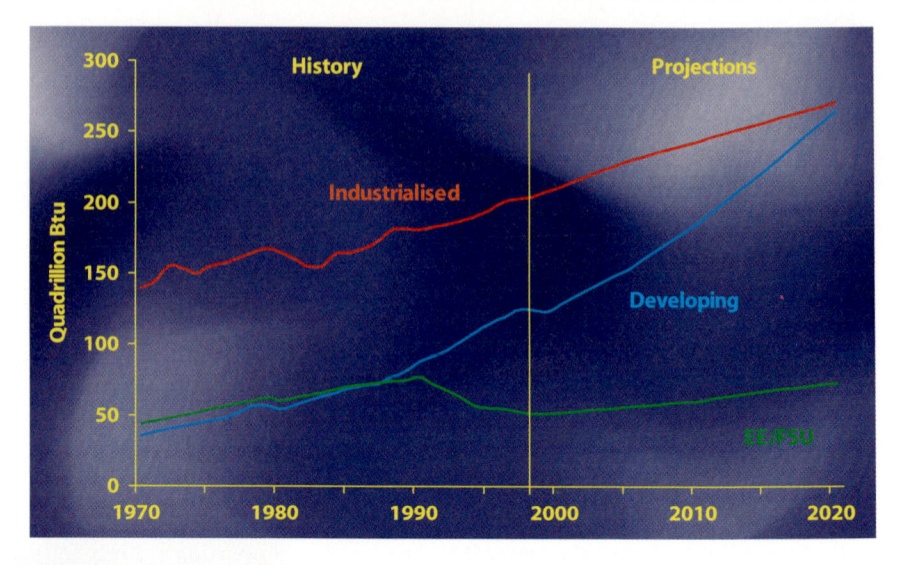

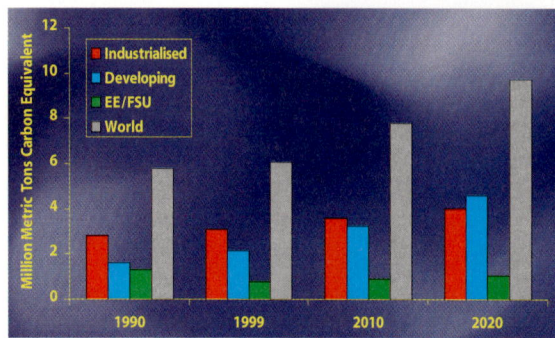

Fig. 10.3. Long run energy demand world carbon emissions by region – by 2020 most of the world's emissions will come from developing countries (source: Energy Information Administration (EIA) Energy Outlook 2001)

10.3 The Industry Response

10.3.1 Reducing GHG Emissions

In the context of global change, I will focus particularly on our company's response to the CO_2 challenge. Three years ago, we set a target of reducing our emissions of greenhouse gases by 10%, and we said very clearly that this would rank with our financial and other business objectives. Our judgement was, and remains, that we needed to start down the road of finding the answers, putting in place the responses which we had available, while developing the longer-term initiatives which would be needed. We did not believe it was right to delay, and we were very clear that if we focused the creative talents of our people on the challenge, we would deliver a set of responses which both addressed the requirement to reduce emissions and improved our business performance.

We have succeeded; by the end of 2000 we had achieved a reduction of 5%, or about 4 million t of CO_2 equivalent, and we should be able to achieve the other 5% in about three years time. In doing this, the choices are really quite simple. You can choose the fuel you use, you can use it more efficiently or not use it, or you can capture and sequester the carbon.

Our focus to date has been on our own operations, our refineries, production platforms, chemical plants and so on. And we have delivered some significant early wins. By adapting the process of producing gas in the western USA, we have saved over 430 000 t of gas. At our Texas City plant, attention to some of the linkages between different systems has produced gains in energy efficiency equivalent to a saving of 150 000 t a year of CO_2. Similar results are evident at our Grangemouth refinery and chemical complex in Scotland. And over the last four years, we have virtually stopped flaring and concentrated on capturing and reusing the gas instead.

In a short time, this is an impressive list, not constrained by national boundaries or political inhibition, and in addition there is a clear business benefit. We have added some $600 million in value to the company for only a modest outlay. The trick has been to gain the attention and support of our staff. Many of them have their own convictions that we must tackle this issue now, but even some sceptics can see the sense of the sort of things we have done.

There was a choice. We could hand out reduction targets to each of BP's 150 business units worldwide and require that they take whatever steps they needed to deliver. But that would have led to some very uneven costs and considerable inefficiency. And we believed we must demonstrate that we could deliver our goal at a reasonable cost. So we decided to use the mechanism of the market to find out where the reduction in emissions could be delivered at the least cost to the Group.

In January 2000, we introduced a pilot emissions trading system to a number of our larger units; 2.7 Mt were traded in 2000, at an average price of $7.60 per t. With the success of that, a year later we extended it across the company. It is now well established, and to date we have traded some 4 million t. Without a doubt, it has helped us to generate a much better awareness and engagement across the company. It has led to some very interesting innovations. Critically it has given us a clearer view of the value of carbon investment, and in addition we are showing a positive return.

10.3.2 Future Innovations

It is not all plain sailing. Our success to date has been substantially achieved by a focus on reducing flaring and energy efficiency. But as we look ahead, we need to expand our focus, to look at alternative ways of doing things, to develop innovations in technology and the way it is used. We are confident this can be done, but to deliver success will need co-operation with others: with scientists, with policy makers and those who work with them, with civil society groups; with others in business, and above all with our customers and the consumer. The good news is that there is a natural drive to a lower carbon economy through the push for greater efficiency, through improved products, and through advances in technology.

As mentioned earlier, there is likely to be a growth in energy demand. But we should remember that over the last decade there has been a significant improvement in the efficiency of energy use. For every 1% growth in GDP total energy consumption has risen by only 0.4%. That is a worldwide figure – an aggregate of different trends in different parts of the world. In the US the ratio is 0.5, in Europe 0.3 and in China a remarkable 0.12. In total China has halved its energy use per unit of GDP over the last decade – and I suggest that is a real measure of the value of technology in delivering both economic growth and environmental performance. It also seems clear from several studies in both the US and Europe that the scope for greater efficiency of energy use is considerable at virtually no cost. The economic and environmental advantages would be considerable.

There has also been a reduction in the amount of carbon produced from every unit of energy consumed, down worldwide by 5%, from a ratio of 0.8 in 1990 to 0.75 in the year 2000, with a particular improvement in China in the last two years as a result of the shift in the energy mix away from coal and in favour of natural gas, and improvements in energy efficiency.

But we need more; we need better options to address the growth in energy demand, never forgetting that energy underpins our well-being. Without energy, people cannot aspire to improve their education, their jobs, their health, the prospects for their children, indeed their custody of the environment. With the likely growth in the world's population and the spread of prosperity and purchasing power, the demand for energy is clearly there. We have to find ways to supply that while addressing the environmental challenge at the same time.

Many look on the issue as a trade-off between economics and the environment. Either you must satisfy the economic goal to achieve the environmental one, or the other way round. I do not think we can accept that, nor do I believe it is a constructive way forward if we face people with stark and unpalatable choices. And it tends to polarise the issue when what we all need is co-operation. We have to find a way of providing our customers with a product or service they ask for, but in a way which reflects the values which society places on the environment. This may not be easy, but we have made a lot of progress, and as we look ahead I believe we can be optimistic about our ability to manage the transition. I base this on our own experience and that of other companies with whom we work.

There is clearly a time scale here (Fig. 10.4) and while as I said earlier, there is a drive towards a lower carbon economy, the pace of this will depend on the level of

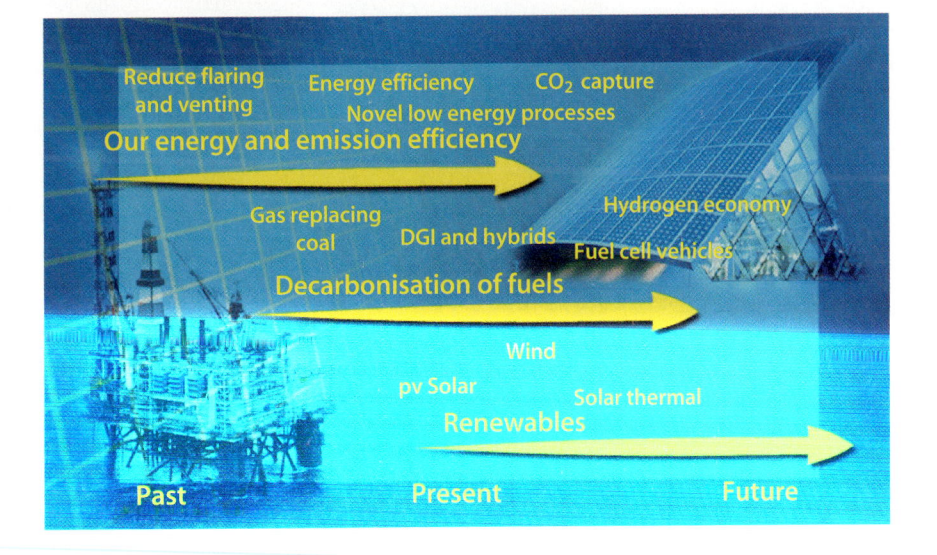

Fig. 10.4.
Options and timescale for achieving lower carbon energy

engagement and the policies that are adopted. Figure 10.4 shows both the way we are considering today (as discussed above), and the way ahead. The three options remain, the choice of fuel, how well it is used, and capturing CO_2 to prevent it from entering the atmosphere.

In BP we now have a database of some 600 emission reduction projects that we are regularly refining in terms of outlay, timeframe, and feasibility. Our 600 or so gas turbines deployed throughout the business probably offer the largest potential for incremental improvement. However, the search for emission efficiency and better knowledge of the value of carbon is leading us to step back and look more fundamentally at some of the key dynamics of our activities.

There are two good examples from a number we are examining, as we look at the changing shape of fuel demand and how it is used. Firstly, it takes a lot of energy to operate our offshore production platforms. The motive power has traditionally been provided by turbines driven by fuel from the field below. Now we are looking at the option of saving that fuel and importing electricity from the shore. It is early and it may not work out, at any rate in the shorter term, but it is an example of the innovative way we need to think. Secondly, with the growth in demand for natural gas, and with it the need for transported gas, we need to set a new standard of efficiency for compression, which is a very energy intensive process.

Carbon capture and sequestration have been mentioned a couple of times. It does and must offer great promise. There are several technical options, but cost is still not aligned. We are thus involved in a number of projects to deliver the prospect. With ten other companies we have a joint research project based in Alaska looking at pre and post combustion decarbonisation for either underground storage or to assist enhanced recovery from the producing field. Interestingly, this project has funding from both the US Department of Energy and the EU. I hope the future will bring further examples of cooperation on the climate issue. In Colorado, we are investigating the use of CO_2 to mix with nitrogen and recover methane from a non-mineable coal formation. In Scotland, we have a project looking at capturing the annual emissions of CO_2 from our Grangemouth complex for possible use in delivering added production from a North Sea field. Thus the carbon would not only be barred from the atmosphere, it would have its own economic value.

I believe this is a promising approach, but we and others, have more work to do to improve the process, reduce the cost and gain acceptance. There should be little quarrel with the notion of storing carbon in secure underground formations, but we recognise that there are concerns that must be addressed before this becomes widely acknowledged as the best option.

10.3.3 Renewable Energy

The growth of renewable energy is still dependent on government subsidies and targets. In the developing world, there are clearly large opportunities where the supply of electricity is not available. But the funds and the know-how are often lacking. In the developed world, the costs are uncompetitive, although wind power is perhaps the nearest to being so. A number of governments have set targets to achieve a proportion of electricity from renewables within a decade, but incentives will be needed to overcome the economic and structural barriers. At the moment the contribution from renewables, excluding nuclear and hydropower, is less than 1%. It will therefore take considerable effort to make significant inroads in the next ten to twenty years.

In BP we have invested around $200 million over the last five years in photovoltaics and the development of

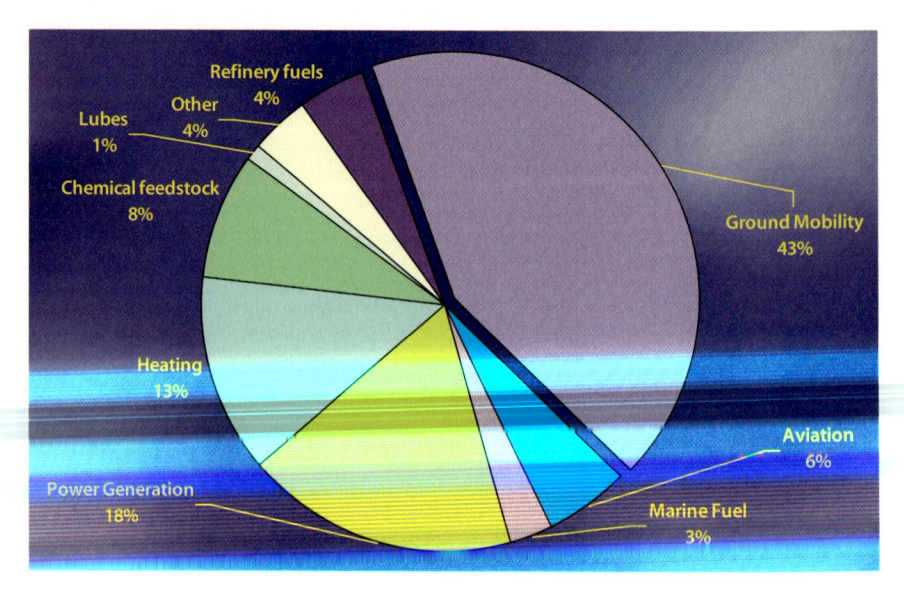

Fig. 10.5.
Global oil consumption 2000: the transport sector was the largest consumer

- Refinery fuels 4%
- Other 4%
- Lubes 1%
- Chemical feedstock 8%
- Heating 13%
- Power Generation 18%
- Ground Mobility 43%
- Aviation 6%
- Marine Fuel 3%

solar power. And we now have one of the largest solar power businesses in the world, growing at about 30% a year. We will be making significant additional investments in renewable energies over the next 5 years. Although the prospect is there, we must recognise the time scale and the support needed to make real progress.

And we are adding to that a programme of research into the potential of hydrogen, which could be the ultimate clean fuel. Again we have the challenge of moving from test bed success to delivering a product that will satisfy the customers' needs for safety, convenience and a competitive price. We believe this will happen, so we are already active in a range of development projects, especially in the area of transport.

10.3.4 Transport Fuels

Over half the world's consumption of oil goes to keep our cars, trucks, aeroplanes and ships on the move, with road transport claiming the lion's share (Fig. 10.5). Most of the growth that lies ahead will come from the developing economies of Asia (Fig. 10.6). The mobility that transport provides is much prized. People need to travel and wish to travel, and aspire to where they are currently constrained from doing so. This is a fairly intractable problem, but we are working with others on this.

In the meantime, we have launched a programme to offer customers a new choice of cleaner fuels, gasoline and diesel without lead, sulphur or benzene. That offer is now available in more than 60 cities worldwide, and should reach 90 by the end of 2001. We aim to remove lead from all BP gasoline by January 2002, and to make 40% of all BP products cleaner fuels by 2005.

If we look at the alternative forms of propulsion (Fig. 10.7), it is clear that pronouncements on the early demise of the internal combustion engine are premature. There is now significant collaboration between those who make road vehicles and those like us who supply the fuel. The result is that cleaner, more efficient vehicles, including hybrid cars and bus fleets running on gas, will become increasingly common.

10.4 Relations between Business and the Scientific Community

Although BP has a very strong focus on climate change, it is not exclusive. The issues of biodiversity and water are also preoccupations, and we see clearly how they link together (Fig. 10.8). Our approach to biodiversity has both a focus on our own operations and involves work with others such as the IUCN to improve both knowledge and practice in the world. On water, which we recognise as a precious commodity with key links to climate and which we both produce and use, we have taken considerable care to economise.

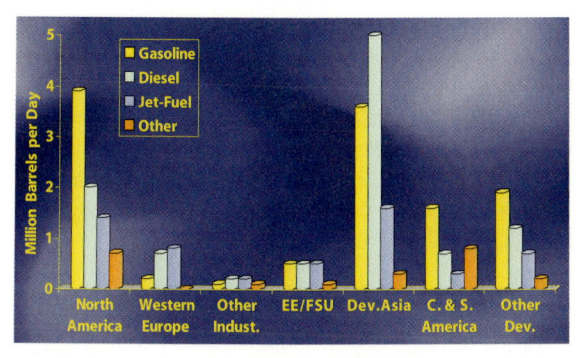

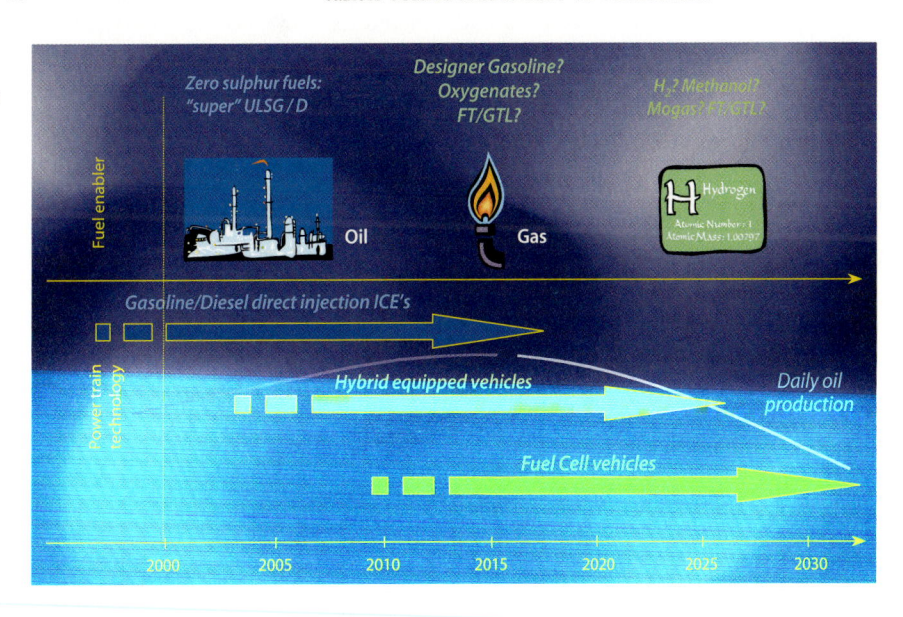

Fig. 10.6. Projected transport fuel consumption by region 1999–2020. Most of the growth in the next few decades will come from the developing countries of Asia (source: Energy Information Administration (EIA) International Energy Outlook 2001)

Fig. 10.7. Anticipated technology developments between now and 2030

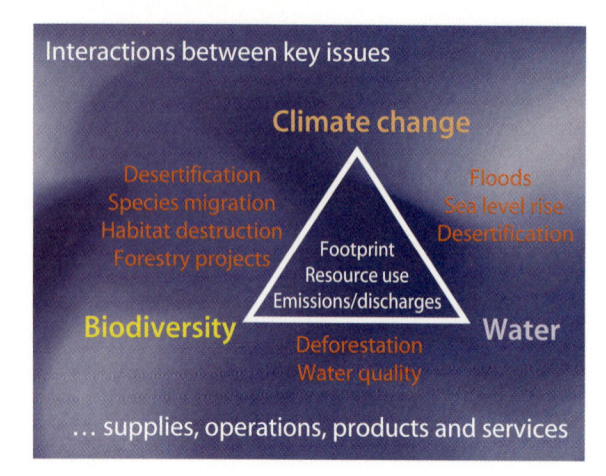

Fig. 10.8. Interactions between key global scientific and environmental issues

This paper is mostly from the vantage point of BP, but much that is said here could have come from colleagues in other companies. You would not expect that all in business would have reached the same conclusion on the significance of the issue and how they should address it, and this anyway varies from company to company or by sector. And in this, industry is no different from other groups in society.

However, we are active in a number of industry organisations – the World Business Council for Sustainable Development (WBCSD), the European Round Table, the International Climate Change Partnership and the Pew Centre, for example. Many of those companies which are taking the lead on carbon management work together in these groups. In the WBCSD, we have done a lot of work on developing a protocol for measuring and reporting on greenhouse gases. This important work has gained wide support from others outside business.

A number of us in the auto and oil industries are also working to understand how the need and demand for mobility could be delivered in ways that are more sustainable and involve reduced, or preferably zero, greenhouse gas emissions. We have also put a lot of time into working on the Clean Development Mechanism, which we see as an important way of engaging the developing countries.

In the European Round Table we produced a report setting out the achievements of a number of its members and outlining how business and governments could work together to put in place better answers. As mentioned above, there are also several collaborative efforts, between the auto and oil industries for next generation vehicles and fuels, and on carbon capture and sequestration.

In the process of developing an emissions trading system in the UK, we now have over 100 companies involved, compared with 25 two years ago. Increasingly, businesses are recognising that this is a time to take precautionary action, to start down the road of 'learning by doing', and that applies not only to those who are visibly committed. Some have chosen to 'do', but not to talk, or at least to ascribe their actions to climate change.

There are some things that would help. Signals are important – constructive signals which recognise that this is not an easy issue; an approach which recognises the time scale needed to engage, to find the good answers that I have covered earlier; an approach which recognises that there is no single answer, no silver bullet, but many, and they all need to be encouraged. We need to think more in terms of incentives not penalties, of policies and approaches that work with the grain of the market, not against it.

We have found that markets operate well within BP, and we and others are working to create national and larger trading schemes, because we believe the same logic applies. We will find the best answers through market incentive and exchange. This is not simply presumption. Work done in this field shows that trading could reduce the sort of targets agreed at Kyoto by $20 bn a year to less than $40 bn, or to $9 bn if it was extended to non-Annex 1 countries.

We also need help from the scientific community. They first brought us the challenge, and I hope they will help us to resolve it. Certainly, if I use the term in its widest application, we are very dependent on their collaboration and findings to help us forward.

We are not just sitting back and waiting for results. In BP, we have a number of important activities together of which I will mention four:

1. With the Carbon Management Institute at Princeton, we are working on the issues surrounding sequestration, both terrestrial and in reservoirs, and on hydrogen;
2. At Cambridge University, we have commissioned work to better understand the flow of fluids, including CO_2, in reservoirs. This work has very wide applications, part of which has extended to the flow of air in buildings;
3. With CalTech and Berkeley work is in hand on methane chemistry, hydrogen and other product related issues;
4. Imperial College in London is looking closely at a range of environmental technologies and their applications.

This is all critical work as we look to develop future answers. As we look to the challenge ahead, we can see where the solutions ought to lie, and also see that the costs are steep. We must endeavour to provide the answers in a way that supports the other objectives that people and societies have, for instance growth and quality of life. I do not believe we will get agreement or the best results for climate change unless we do this. And scientists can help, particularly those who carry the burden of understanding and interpreting the science of climate change. For those of us who have to deliver the answers there would be enormous value in a better understanding of what could happen and where; of greater predictability and evidence for the likely speed of change. At the same time, we fully appreciate the difficulties scientists face and the importance of assembling evidence that is as unassailable as possible.

However, in our business we have to make long-term investments in significant projects, and need the tools to help us weigh the various risks, be they political, economic or climate related. The changes discussed in this volume will have great significance for us and for the ability of business to respond successfully. For example, higher ambient temperatures would affect our calculations on compression of natural gas. Changing temperatures could affect the demand for heating or cooling, depending on where the impact occurred.

Perhaps most difficult of all is where we should be aiming. Quite understandably, at this point no one has been able to define an acceptable level of carbon concentration in the atmosphere. The excellent work of the IPCC has provided a set of pathways, and these led to the discussions and agreement at COP 4 in Kyoto. Taking this forward has since proved difficult, and while I cannot predict how the forward political process will develop, we are clear that there is a need for some kind of multilateral framework to take forward the process of an inclusive, cost effective response to a global issue. An agreed global target would provide a focus, which we currently do not have. It would allow us to move to targets for all those who are participating, with these targets expressed in ways that address the efficient use of energy and ensure that the cost of meeting the target is minimised. We should recognise the time scale, and the need to experiment, learn and adapt as necessary.

The best way forward is to foster and encourage a constructive response by all concerned. We need a judicious mix of government policies and measures; we need instruments designed to encourage action, and to offer business and industry the flexibility they need to develop the right approaches to suit specific circumstances.

10.5 Summary of the Key Points

- The global demand for energy will continue to grow for the foreseeable future;
- There is no absolute shortage of hydrocarbons;
- Evolving attitudes to environment and climate change will alter the primary energy mix with an increasing demand for gas;
- Most incremental demand will be in the developing world, especially in Asia;
- The highest growth rates will be for transport fuels;
- The introduction of clean fuels is ahead of legislation, particularly in northwest Europe;
- Industry is increasingly recognising that we are likely to work in a carbon constrained future and is facing up to this challenge;
- Many of us are building good experience of actions around the conservation and efficient use of energy;
- There is good activity around future options for reducing emissions, but more time and effort is needed to deliver these technically and economically;
- Proper motivation is a key to business success. It is important to work with the grain of the market, mobilise support and find the best answers on a global basis;
- There is no single answer; we need them all and the willingness to experiment and adapt;
- We need the skills of the scientific community to help us develop the answers, and its understanding of climate science to help us plan;
- We need an international policy framework, and a global target for atmospheric concentration would provide a focus which is currently lacking.

Part IId

Summary: Global Change and the Challenge for the Future

Chapter 11

Global Change and the Challenge for the Future

Peter Tyson

In attempting to summarise the papers presented on the first day of the conference, only big-picture findings concerning a decade of global change research activities will be considered. Over the last decade, significant advances have been made in the understanding of the natural temporal variability of the Earth System, its spatial variability and its predictability. It is against this knowledge base that the effects and consequences of human-driven change must be assessed.

Most people are familiar with the outstanding work done by the IPCC in the last few years. The importance of the IPCC cannot be overestimated. However, its very success is helping to perpetuate the notion among nonexperts that global change is largely a question of climate change. It is not. It is much more than this. Many Earth System changes can have significant consequences without involving any changes in climate. It is important that people do not fall into the trap of confusing global change with climate change. Global change is significantly more than a component of climate change in the Earth System.

Human activities are significantly influencing the functioning of the Earth System in many ways. Anthropogenic changes are clearly identifiable beyond natural variability and are equal to some of the great forces of nature in their extent and impact.

- Nearly 50% of the land surface has been transformed by direct human action, with significant consequences for biodiversity, nutrient cycling, soil structure and biology, and climate;
- More nitrogen is now fixed synthetically and applied as fertilisers in agriculture than is fixed naturally in all terrestrial ecosystems;
- More than half of all accessible fresh water is used directly or indirectly by humankind, and ancient and often nonrenewable underground water resources are being depleted rapidly in many areas.

All these changes and many others are altering, sometimes subtly, sometimes overtly, the relationship between humans and their environment. Major anthropogenic changes are recent; they are profound and many are accelerating. They operate in conjunction with constantly changing natural variability. They are cascading through the Earth's environment in ways that are difficult to understand and often impossible to predict. Global change is real, it is happening now and it is accelerating.

Some of the important messages arising from the deliberations of today are that:

- The Earth is a system that life itself helps to control. Biological processes interact strongly with physical and chemical processes to create the planetary environment, but biology plays a much stronger role than previously thought in keeping Earth's environment within habitable limits.
- The human enterprise drives multiple, interacting processes that cascade through the Earth System in complex ways. Global change cannot be understood in terms of a simple cause-effect paradigm. Cascading effects of human activities interact with each other and with local- and regional-scale changes in multidimensional ways. One of the most ubiquitous features of the planetary machinery is the suite of linkages that bind processes in one region to consequences in others thousands of kilometres away.
- The Earth's dynamics are characterised by critical thresholds and abrupt changes. Human activities could inadvertently trigger changes with catastrophic consequences for the Earth System. The Earth System has operated in different quasi-stable states, with abrupt changes occurring between them over the last half million years. Human activities clearly have the potential to switch the Earth System to alternative modes of operation that may prove irreversible.
- In terms of some key environmental parameters, the Earth System has moved well outside the range of the natural variability exhibited over the last half million years at least. The *nature* of changes now occurring *simultaneously* in the Earth System, their *magnitudes* and *rates of change* are unprecedented. Many accept that the Earth is currently operating in a no-analogue state.

In presentations to follow, these ideas will be given further attention, as will consideration of the risks of living with global change, questions of vulnerability and the need to provide the scientific basis for the sustainable development of the Earth and its peoples.

Part III
Advances in Understanding

Part IIIa

Global Biogeochemistry: Understanding the Metabolic System of the Planet

Chapter 12

Ocean Biogeochemistry: A Sea of Change

David M. Karl

12.1 Introduction

A comprehensive understanding of the global carbon cycle is required to address contemporary scientific issues related to the atmospheric accumulation of greenhouse gases and their cumulative effects on global environmental change. Consequently, detailed *in situ* investigations of terrestrial and marine ecosystems are necessary prerequisites for developing a predictive capability for environmental variability and the effects of human-induced perturbations. These investigations need to address the interdisciplinary connections between physics, chemistry and biology, and in each ecosystem, address broad questions regarding the distribution, abundance, diversity, and control of key plant, animal and microbe populations and their interactions with their habitats. Ideally, these field studies should be conducted at strategic sites that are representative of large biomes or in regions that are likely to exhibit substantial interannual variability over large areas. Furthermore, these field investigations should be conducted for at least several decades, in order to distinguish natural variability from that induced by human activities.

In spite of their recognised importance, systematic, long-term biogeochemical observations of oceanic habitats are rare. In response to a growing awareness of the ocean's role in climate and global change research, and the need for comprehensive oceanic time-series measurements, the International Geosphere-Biosphere Programme (IGBP) was established in 1986. One of the essential core components of IGBP, the Joint Global Ocean Flux Study (JGOFS) project, was established in 1987 to improve our understanding of the oceanic carbon cycle and to quantify the exchange of carbon with the atmosphere, the seafloor and the continental boundaries.

One of the enduring legacies of JGOFS will be the long-term time-series stations that were established during the programme and that will continue into this millennium. These stations fulfil a crucial goal within JGOFS. They study the changes in the ocean carbon and nutrient cycles on interannual and, soon, decadal time scales. Each year, nature presents a unique set of physical forcings and biological initial conditions. Each year, the oceans respond in a unique way. By studying the variations between years, we gain a greater understanding of how these ecosystems function. By using natural variability as our guide, we can gain insight into the relationship between biogeochemistry and climate, an insight that will prove invaluable as we try to predict the future fate of the carbon cycle and the climate.

12.2 The Oceanic Carbon Cycle

The large and dynamic oceanic reservoir of carbon, approximately 4×10^{19} g distributed unequally among dissolved and particulate constituents with various redox states (reduction-oxidation), plays an important role in global biogeochemical cycles. The two largest pools are dissolved inorganic carbon and the less oxidised pool of mostly uncharacterised dissolved organic carbon. A chemical disequilibrium between DIC and organic matter is produced and maintained by numerous biological processes. The reversible, usually biologically-mediated interconversions between dissolved and particulate carbon pools in the sea collectively define the oceanic carbon cycle.

Primary conversion of inorganic carbon to organic matter is generally restricted to the euphotic zone of the world's ocean through the process of photosynthesis. The supply of reduced carbon and energy required to support deep water metabolic processes is ultimately derived from the upper ocean and is transported down by advection and diffusion of dissolved organic matter, gravitational settling of particulate matter and by the vertical migrations of pelagic animals and phytoplankton. Each of these individual processes, collectively termed the "biological pump", is controlled by a distinct set of environmental factors, and therefore, the relative contribution of each process may be expected to vary with changes in habitat or with water depth for a given habitat.

Each year, the biological pump removes an estimated 10 Gt C (1 Gt = 10^{15} g) from the surface waters of the world ocean, a value that is equivalent to ~15% of the annual global ocean primary production. This export of organic matter is sustained by a continuous reflux of major nutrients; over broad time and space scales the supply of nutrients from below is balanced by the downward flux of particulate matter. These coupled delivery and removal processes control the ability of the ocean to sequester atmospheric carbon.

12.3 Ocean Time-Series Programmes

When designing a time-series field programme, care must be given to the uncompromised collection of complementary data, including dissolved substrates, dissolved gases, particulate matter, and other parameters. In addition to the availability of numerous sampling devices, there is also a variety of potential sampling platforms, including research vessels, towed and towed-undulating vehicles, submersibles, remotely operated vehicles, autonomous underwater vehicles, moorings, drifters, and Earth-orbiting satellites. Each platform has its own unique capabilities and limitations. Integrated measurement systems including multiple sampling platforms and fast response sensors are likely to emerge as the method of choice in future investigations of biogeochemical processes in the sea (Fig. 12.1).

In 1988, we began a systematic examination of microbial and biogeochemical processes in what was, at that time, considered to be a temporally stable habitat – the North Pacific subtropical gyre (Fig. 12.2). After the first decade of approximately monthly research cruises, it was concluded that this sampling frequency was too

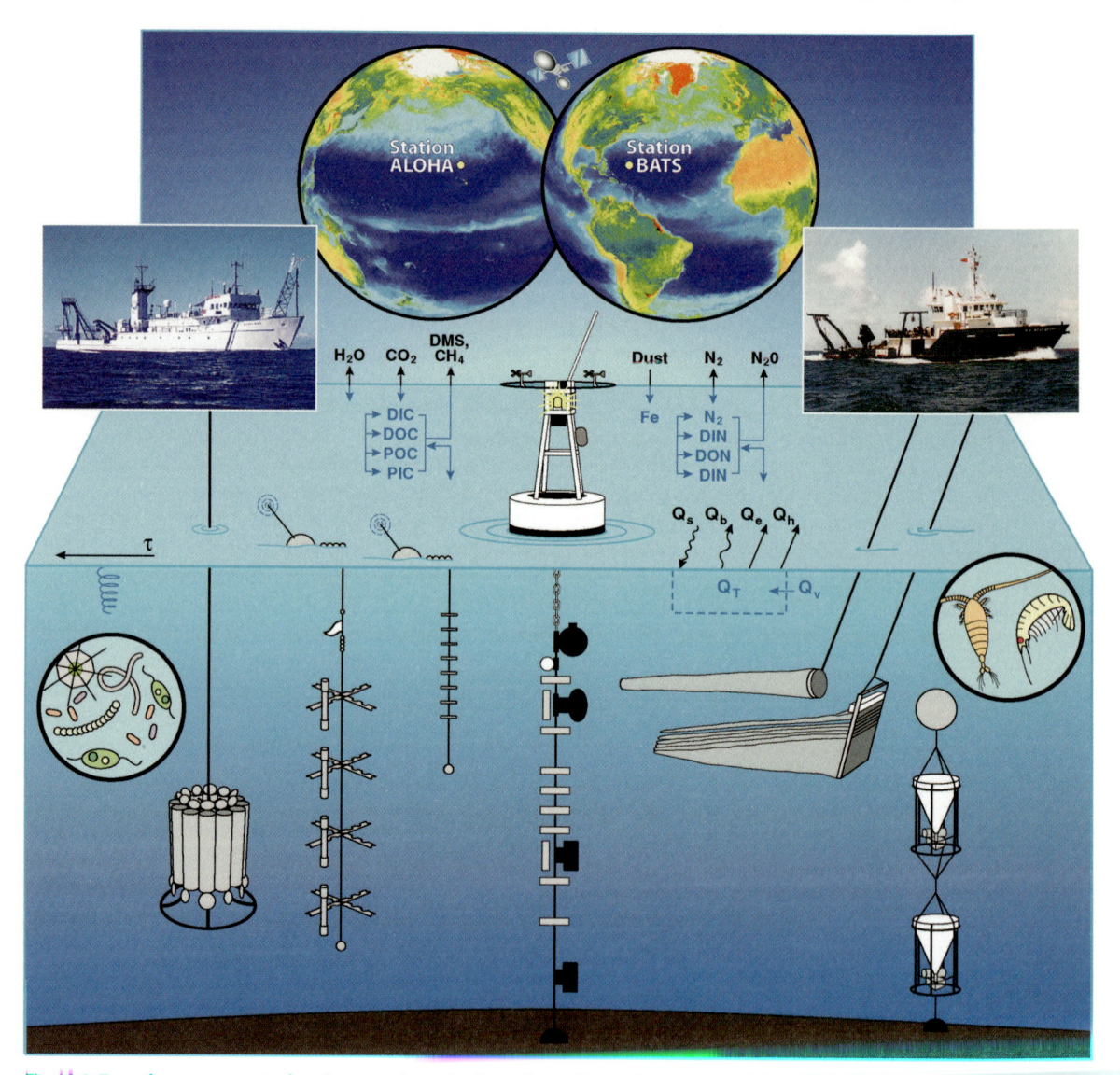

Fig. 12.1. From the macroscope to the microscope, investigations of ocean biogeochemistry at the U.S. JGOFS time-series programs. *Top:* Satellite view of the global biosphere based on observations from the SeaWiFS mission (September 1997–August 1998); imagery courtesy of NASA-Goddard Space Flight Center and the Orbital Sciences corporation (Gene Feldman; see http://seawifs.gsfc.nasa.gov/SEAWIFS.html) *Left-centre* and *right-centre:* The research vessels Moana Wave and Weatherbird II at work in the North Pacific and North Atlantic Oceans. *Bottom:* Examples of the variety of ship-based and remote sampling techniques used in the two time-series programs to obtain relevant information on the cycles of carbon and associated bioelement cycles in the sea. Several important components of these cycles, including dissolved and particulate inorganic and organic pools are shown at centre along with selected air-sea exchange processes. Shown in the *enclosed circles* are microscopic images of nanoplankton (*left*) and mesozooplankton (*right*) (reproduced with permission from the Oceanography Society 2001)

Chapter 12 · Ocean Biogeochemistry: A Sea of Change

Fig. 12.2.
Vital HOT statistics including sea map, key contacts and photo of the now retired R/V Moana Wave returning to port after a successful HOT cruise

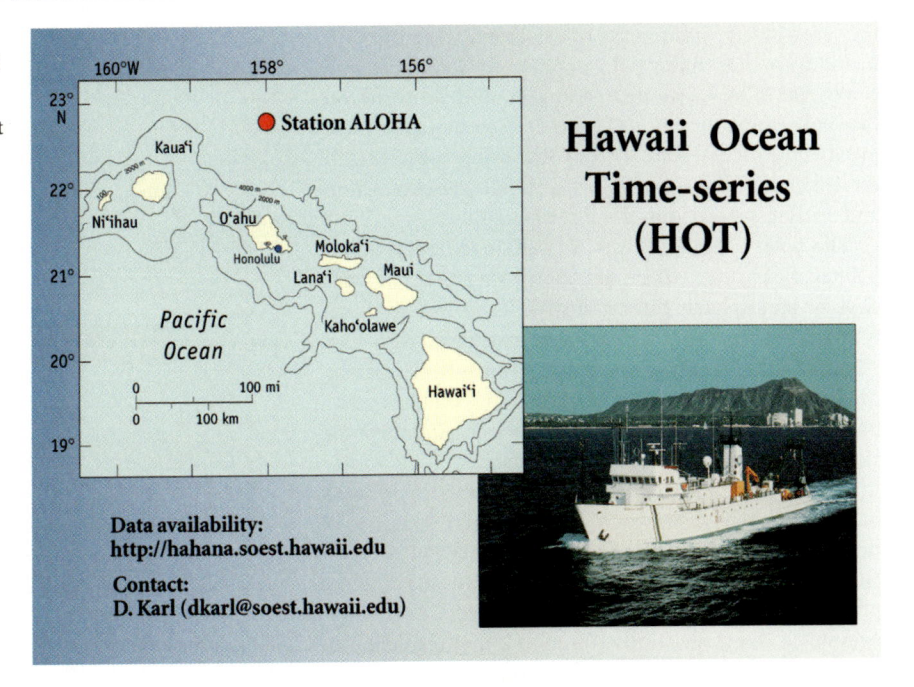

Hawaii Ocean Time-series (HOT)

Data availability:
http://hahana.soest.hawaii.edu

Contact:
D. Karl (dkarl@soest.hawaii.edu)

coarse in time to fully resolve even the most important physical-biological interactions (Karl 1999). The greater the number of time periods and space scales that are involved, the greater the measurement intensity to achieve even a basic understanding. As Stommel (1963) cautioned, "Where so much is known, we dare not proceed blindly – the risk of obtaining insignificant results is too great." Undersampling is, unfortunately, a sobering fact of life in oceanography.

The Hawaii Ocean Time-series (HOT) programme has provided unique insight into the functioning of Earth's largest biome (Karl et al. 2001). The knowledge gained challenges some of the most closely held assumptions in ocean biogeochemistry and provides clues to the mechanisms through which ocean biology may influence global climate. The physical and biological processes at HOT show strong interannual and decadal variability. The El Niño Southern Oscillation (ENSO) has a strong influence on mixing and stratification, and both sites show signs of decadal variations in mixing and biogeochemistry. Nitrogen fixation is an unexpectedly strong component of the major nutrient cycles (Karl et al. 1977). The rate of nitrogen fixation is probably strongly linked to the supply of iron, a critical micronutrient for the enzyme nitrogenase. In the Pacific, the reorganisation of the ecosystem that accompanies the rise of nitrogen fixation appears to occur on decadal time scales as a shift between nitrogen and phosphate limitation.

Finally, ecosystem structure matters. The simple characterisation of biogeochemistry as the flow of nutrients and energy through simplified foodwebs hides a massive complexity of organisms and processes. Changes in the composition of the biological community have a strong effect on how it functions. The "bacteria" are no longer a homogeneous mix of heterotrophic consumers;

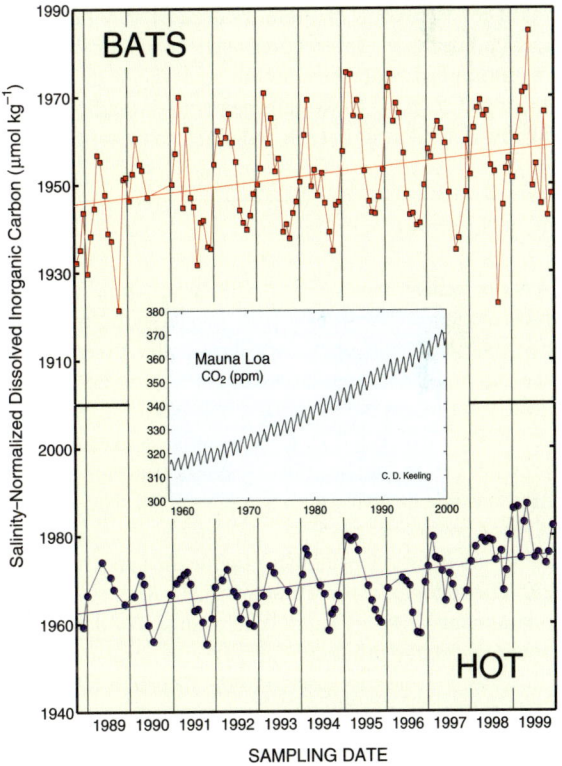

Fig. 12.3. The initial 11-year record of mixed-layer dissolved inorganic carbon (DIC), normalised to a salinity of 35.0 (N-DIC), for the BATS (*upper curve*) and HOT (*lower curve*) open ocean time-series sites. The larger annual amplitude at *BATS* is a manifestation of the greater seasonal temperature range and seasonal variation in the magnitude of greater organic matter production, both of which contribute to N-DIC dynamics. The significant increasing N-DIC trend with time over the 11-yr period (BATS = +1.23 µmol kg^{-1} yr^{-1}, HOT = +1.18 µmol kg^{-1} yr^{-1}) is the ocean's response to the increasing burden of atmospheric carbon dioxide as shown by the centre inset using the C. D. Keeling atmospheric data set from Mauna Loa, Hawaii

they are a diverse community of autotrophs, heterotrophs, including archea, many with unique biochemistries and ecological roles. Large phytoplankton and zooplankton may only be a tiny fraction of the overall biomass, but these parts of the foodweb are disproportionately important in the creation of the sinking particles that transfer carbon and nutrients away from the surface ocean.

The open ocean HOT site, as well as the open ocean habitat at our sister station near Bermuda (BATS), is a net sink for atmospheric carbon dioxide. The concentrations of dissolved inorganic carbon are increasing steadily as a result of the increasing burden of atmospheric carbon dioxide (Fig. 12.3). The uptake of carbon is due to a combination of biological and solubility effects, and these all vary through time. The uptake of carbon is also intimately linked to the mix of unique processes at this and other open ocean stations. This linkage provides a number of clues to the feedback between ocean biogeochemistry and global climate. The understanding of these processes and their inclusion in global carbon models presents a challenge for the future. As that challenge is met, the time-series stations will provide a crucial benchmark for validation of these models, as they become a focal point for the interaction between physical-chemical-biological processes, including large-scale climate forcing (Fig. 12.4). As we gain new understanding, existing biogeochemical models will need to be refined.

12.4 Summary

A major achievement of the JGOFS programme is an improved understanding of the time-varying fluxes of carbon and associated biogenic elements, both within the ocean and the exchanges between the ocean and atmosphere. These accomplishments derive from a network of ocean time-series stations located in representative biogeochemical provinces from low-latitude, subtropical ocean gyres to high latitude coastal and oceanic regions. The variable physical forcing mechanisms and concomitant ecological responses in these key biomes provide the opportunity for a cross-site comparison of fundamental patterns, seasonal/interannual variability and secular change. The decoupling of organic matter production, export and remineralisation processes in time and space, and the detection of decade-scale, climate-driven ecosystem perturbations and feedbacks combine to reveal a time-varying, biogeochemi-cal complexity that is just now becoming evident in our independent and collective ocean time-series data sets.

Ultimately, oceanic productivity, fishery yields and the net marine sequestration of atmospheric greenhouse gases are all controlled by the structure and function of planktonic communities. Detailed palaeoceanographic studies have documented abrupt changes in these processes over time scales ranging from centuries to millennia. Most of these major shifts in oceanic productiv-

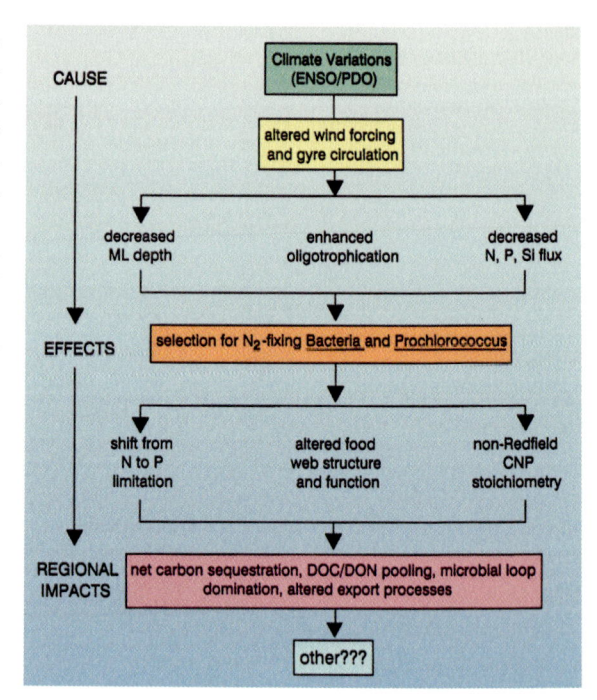

Fig. 12.4. Hypothetical view of the effects of climate variability on ecosystem structure and function in the North Pacific Subtropical Gyre based, in part, on results obtained during the decade-long HOT research programme. Changes in the stratification of the surface ocean has affected nutrient and trace element budgets and has selected for N_2-fixing bacteria and *Prochlorococcus*, resulting in a domain shift from predominantly Eucarya to predominantly Bacteria. This has resulted in numerous observed and potential biogeochemical consequences, including non-Redfield CNP stoichiometry and, therefore, a net sequestration of atmospheric carbon dioxide (redrawn from Karl 1999)

ity and biodiversity are attributable to changes in Earth's climate, manifested through large-scale ocean-atmosphere interactions. By comparison, contemporary biodiversity and plankton community dynamics are generally considered to be "static", in part due to the lack of a suitable time frame of reference, and the absence of oceanic data to document ecosystem change over relatively short time scales (decades to centuries). The establishment of a global network of ocean time-series stations for repeated physical and biogeochemical measurements represents one mechanism for the ecosystem surveillance that is necessary to record and eventually understand these complex oceanic processes.

References

Karl DM (1999) A sea of change: Biogeochemical variability in the North Pacific subtropical gyre. Ecosystems 2:181–214

Karl DM, Letelier R, Tupas L, Dore J, Christian J, Hebel D (1977) The role of nitrogen fixation in biogeochemical cycling in the subtropical North Pacific Ocean. Nature 388:533–538

Karl DM, Dore JE, Lukas R, Michaels AF, Bates NR, Knap A (2001) Building the long-term picture: The U.S. JGOFS Time-series programs. Oceanography 14:6–17

Stommel H (1963) Varieties of oceanographic experience. Science 139:572–576

Chapter 13

The Past, Present and Future of Carbon on Land

Robert (Bob) J. Scholes

In the Earth System, where ultimately everything is connected to nearly everything else, the establishment of cause and effect is often difficult. This does not mean that all processes are equally important or equally sensitive to human alteration. In the context of the global carbon cycle, the pools of carbon on land in the present era are much bigger than those in the atmosphere, but much smaller than those in the ocean, and the annual exchanges between land and atmosphere are slightly larger than those between atmosphere and ocean. The large, periodic variations in the size of the atmospheric pool during the past million years have been accounted as a net transfer of carbon from land to ocean, via the atmosphere, with the implication that because it is the final sink, the ocean and its dynamics control the process. Understanding the medium- to long-term processes in the carbon cycle has relevance to predicting the outcome of the current large excursion of atmospheric carbon dioxide beyond its historical bounds during the Pleistocene, and in particular for the time course of the present large net terrestrial carbon sink.

An understanding of the dynamics of the carbon cycle is impossible without considering its linkages to other cycles: notably those of water, nitrogen and phosphorus. These cycles intersect at discrete pools or processes, which then become critical control points for the whole system. One such point is the control by phosphorus and other trace elements on biological nitrogen fixation on both the land and ocean, which in turn limits the fixation of carbon. A second is the balance of silica and calcium in the ocean, which is influenced by the land, and controls the sinking of carbon out of the oceanic mixed layer.

Two hypotheses are discussed: firstly, that the lower bound of the Pleistocene atmospheric carbon range is determined by the terrestrial ecosystem carbon dioxide compensation point, and secondly, that the transfer of key nutrients from land to sea drives and controls the rate of the decline in atmospheric carbon dioxide from the upper bound to the lower bound.

13.1 What Controls the Behaviour of Biospheric Carbon?

The terrestrial biosphere (i.e., soils and plants) absorbed, on average, approximately 3 Pg C yr^{-1} more carbon dioxide per year during the decade of the 1990s than it emitted to the atmosphere (Prentice et al. 2001). This is a very substantial amount in relation to the total net emissions of carbon dioxide by human activities, including the burning of fossil fuels (6.3 ±0.4 Pg C yr^{-1}), the clearing of tropical forests (1.7 Pg C yr^{-1}; uncertainty range 0.4 to 2.5), and the absorption of carbon dioxide by the oceans (1.7 ±0.5 Pg C yr^{-1}). If the land were not acting as a 'sink' for atmospheric carbon dioxide, the carbon dioxide concentration of the atmosphere would be rising at about twice the current rate, and global climate change would be substantially accelerated.

The land carbon sink is not constant between years (it is profoundly influenced by the distribution of rainfall and temperature around the world in a given year, which is in turn affected by the El-Niño Southern Oscillation phenomenon, large volcanic eruptions and other global-scale variations). Several known processes contribute to the land sink, but the relative proportions of their contributions are very uncertain. The main ones are:

- Regrowth of forests following deforestation in the nineteenth and early twentieth centuries, mainly in Northern Hemisphere temperate regions;
- Fertilisation of plant growth by human-caused increases in carbon dioxide and nitrogen;
- The climate change which has already occurred.

The land sink is believed to have increased in magnitude since about 1950, and is projected to continue to do so – but for how long is uncertain (Schimel et al. 2001). The future of the land and ocean carbon sinks is of great importance in determining 'safe' trajectories for future human-caused carbon emissions.

This paper addresses the influence of other elemental cycles on the carbon cycle, and how the interactions are affected by human action and climate change. It further explores how elements that originate on land may alter the carbon sink strength of the ocean, and presents some ideas regarding the linked carbon dioxide and climate record of the past 400 000 years.

13.2 The Carbon Cycle is Constrained by Other Elemental Cycles

The carbon cycle can be visualised as a cog in gearbox, meshed with other cogs at several points. The rate of rotation of the 'carbon cog' is governed by, among other things, the rate at which the other cogs can turn. In this analogy, the other cogs are the cycles of the approximately twenty other elements necessary for various life forms on Earth. Some of these elements are sufficiently abundant in the biosphere that the cogs are essentially 'free-wheeling'. Others are in restricted supply (perhaps only in certain locations, or at some times, or for particular critical organisms or processes) and may limit the rate of turnover of carbon. Thus, for instance, plants with unlimited water and nutrition may increase their photosynthetic rate by up to 40%, if the CO_2 concentration is raised from 350 to 700 ppm; but under more natural conditions, a 10–15% response is typical (Mooney et al. 1999). Figure 13.1 illustrates some of the key points in the carbon cycle where other cycles may interact.

It has long been known that the ratios of elements within given biological tissues remains within a more or less narrow range. This has been referred to as 'ecosystem stoichiometry', and it implies that for the carbon stock in biomass to rise, the biological stock of other elements must rise proportionately. If the supply of one of these elements limits carbon uptake, then an increase in its supply would increase carbon storage. This oversimplification ('Liebig's Law') needs some *caveats*. In terrestrial ecosystems, the stoichiometric ratios exhibit a much wider range than in the ocean plankton ecosystems for which the concept of fixed elemental ratios was first developed. This is because land plants have more scope to alter the proportions of their constituent tissues than do unicellular plankton, and the plant community itself can change greatly in response to nutrient limitation, disturbance or climate change – for instance, from a grassland to a forest, or vice versa. Therefore, there is considerable 'slop' in the meshing of the metaphorical elemental cycle cogs, particularly on land.

Secondly, ecologists accept that single-factor limitation is the exception rather than the rule in natural ecosystems (Rastetter and Shaver 1992). Adaptation and natural selection lead most organisms to be at or close to limitation by several factors at once. Furthermore, the availability of one element, such as phosphorus, can alter the availability of another, such as nitrogen. Limitation may be alternated rapidly between elements, and different ecosystem components may be limited at the same time by different elements.

13.3 The Nitrogen-Carbon Link

There are three main points of linkage between the carbon and nitrogen cycles in terrestrial ecosystems (Scholes et al. 1999). Similar points apply in aquatic systems as well. Firstly, the availability of nitrogen to plants

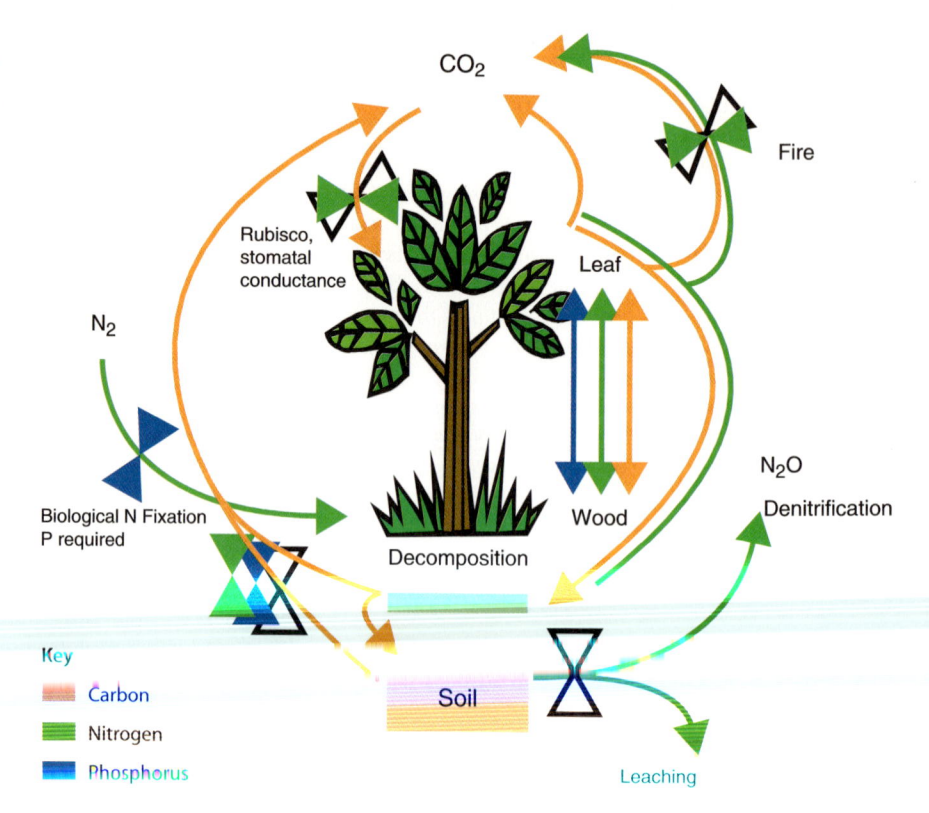

Fig. 13.1. The principal points at which the terrestrial carbon, water, nitrogen, and phosphorus cycles interact

controls the rate of carbon assimilation. The enzyme Rubisco, which often contains half of the nitrogen in the plant, controls the rate of photosynthesis, especially at high carbon dioxide levels. Without sufficient Rubisco, plant 'fertilisation' by carbon dioxide is limited. On the other hand, high-nitrogen tissues respire at a greater rate than low-N tissues. Secondly, nitrogen is an essential element in proteins, nucleic acids and other metabolic compounds, in well-defined proportions relative to carbon. Thus, the carbon and nitrogen cycle run in exact parallel with respect to these compounds, with strong and necessary stoichiometry. Thirdly, the rate at which carbon (and nitrogen) is released from organic bondage through the processes of decomposition is strongly influenced by the nitrogen content of the tissue (Enriquez et al. 1993). Low-N tissues decompose slowly, while high-N tissues decompose fast, all else being equal.

13.4 Do Land Ecosystems Retain Nitrogen Deposited From the Air?

Human activities such as high-temperature combustion, livestock raising and the manufacture and use of nitrogen fertilisers release nitrogenous gases to the atmosphere. The quantity is about 70 million t of N per year – equal to the total quantity cycled through the atmosphere naturally (Vitousek et al. 1997). These gases tend to be highly reactive and are deposited on the land surface within a few hundreds to thousands of kilometres of their sources, which are currently principally in the developed world. Since natural ecosystems are often nitrogen limited, it has been estimated that the additional carbon potentially stored through this mechanism, based on stoichiometric principles, may be around 0.6 ±0.3 Pg C yr^{-1} (Hudson et al. 1994). The most likely geographical location for such storage is consistent with the land carbon sink inferred from inverse modelling of the global distribution of atmospheric carbon: the Northern Hemisphere's Temperate Zone. Isotopic evidence, however, suggests that only a small fraction of the deposited N is actually incorporated into organic compounds (Nadelhoffer et al. 1999), particularly in ecosystems approaching 'nitrogen saturation'. The unincorporated nitrogen may remain in inorganic form in the soil, or leach out to the ground water, rivers or sea, where it is ultimately converted to N_2 or N_2O, a greenhouse gas. Estimated carbon uptake due to deliberate and inadvertent nitrogen fertilisation of the land remains uncertain, but is possibly smaller than the initial calculations suggest.

13.5 The Phosphorus-Nitrogen Link

Phosphorus is also an ingredient of key biological compounds. These compounds (rRNA, ATP, phospholipids), while essential to life, make up a tiny fraction of the biomass. The C : P ratio in land plants can therefore be very variable. A large extent of the tropical land surface is phosphorus-limited, and supports vegetation with adaptations to a low phosphorus supply. Frequently wet and frequently burned ecosystems both 'leak' nitrogen. The main way to replace it is through biological nitrogen fixation (BNF), for which phosphorus is a necessary cofactor. Current global vegetation models do not take direct or indirect phosphorus limitation into account. The result may be an overestimation of (*i*) the capacity of tropical ecosystems to increase their biomass in response to rising carbon dioxide, and of (*ii*) the potential for non-P adapted vegetation to establish in P-deficient areas under a future climate.

13.6 The Carbon Cycle During the Past 420 000 Years

The history of the carbon cycle in the recent Earth System history may provide some clues to the factors that control it and its possible future modes. The exquisitely detailed record of atmospheric carbon dioxide, as revealed by the bubbles trapped in Antarctic ice (Petit et al. 1999), exhibits the following major features (Fig. 13.2):

- It is periodic, with a main period of 110 000 years, corresponding to the eccentricity element of the orbital forcing (Berger 1978);
- It does not drop below a 'floor' of 180 ppm;
- Prior to the onset of the Industrial Age, around 1750, it had not risen above a 'ceiling' of 280 ppm for at least 420 000 years (though it may have been as high as 5 000 ppm hundreds of millions of years ago, before the rise of modern land plants);
- The rise from the floor to ceiling levels is abrupt (5 000 years), while the decline from ceiling to floor is gradual (50 000 years), and often interrupted by small spikes;
- There appears to be fine control around the floor level.

Palaeoecological data add the following observations:

- There is a rise in ^{13}C and ^{18}O in foraminifera (Imbrie et al. 1984), indicating a decrease in ice cover during the warm periods;
- The change in ice volume follows the increase in atmospheric CO_2 rather than precedes it (Shackleton 2000);
- There is an increase in airborne mineral dust during phases of the draw-down.

Many schemes have been advanced to explain features of this record, but no comprehensive mechanism is widely accepted to account for all aspects of it. Here follows an integrated hypothesis, drawn from several

Do land-ocean chemical links control changes in atmospheric carbon dioxide?

The hypothesis is as follows:

- The 'ground state' of the Earth System (meaning the one to which it is repeatedly drawn over the past million years) is the cold, dry, 'glacial', low-atmospheric CO_2 condition. The lower limit is maintained around 180 ppm by a 'biospheric compensation point', set by widespread failure of land ecosystems below this level. Such failure releases a small amount of CO_2 back into the atmosphere, allowing plant regrowth. In this state, there is a large quantity of organic C in ocean sediments and dissolved inorganic C in deep ocean waters, and an oceanic circulation with less ventilation than that which dominates during the warm, wet 'interglacials';
- A small, but sustained increase in radiative forcing during the high solar radiation phase of orbital eccentricity nudges the Earth System out of the narrow lower control band, by triggering a reorganisation of oceanic circulation. This releases deep ocean CO_2 to the atmosphere in greater quantities than can be taken up by the land or ocean surface ecosystems. The rising CO_2 amplifies the radiative forcing, which is further amplified by rising atmospheric water content, the retreat of the ice-and-snow-covered area, and other greenhouse gases such as methane;
- The upper limit is set by a new equilibrium established by ocean-surface solubility of carbon dioxide and the biological storage of carbon on land, given the quantity of CO_2 released from the deep ocean and the relatively fixed quantity of biospheric carbon at the millennial time-scale;
- This limit persists for several thousand years, while the newly rearranged terrestrial ecosystems accumulate carbon, nitrogen and phosphorus. The N and P eventually begin to leak to the oceans;
- The N and P flux from land to sea increases the efficiency of the 'oceanic biological pump', gradually transferring the stock of atmospheric CO_2 to the deep oceans. At some point (or several points) during the draw-down, reorganisation of the ocean circulation to the glacial state occurs. The newly-arid climate increases the transport of iron (and possibly silicon and other land-derived elements essential to phytoplankton and to oceanic BNF) to the oceans, sustaining the draw-down, until the biospheric compensation point is reached.

What are the implications of this model for the current and future times?

The prehistoric upper bound of atmospheric carbon dioxide (280 ppm) was easily breached during the Industrial Era, because the biospheric stock of C was altered through the burning of fossil carbon. Recovery towards this level could occur if human-caused CO_2 emissions were reduced to some very small value, but would require a period of several millennia. The new 'ceiling' would be somewhat higher than 280 ppm, and would eventually decline to around 180 ppm in the next glacial cycle. There is no known upper attractor above 280 ppm, but it is not impossible that one does exist.

What have we learned in the past decade?

Prior to the 1990s, the prevailing view of the global climate system was a physical one: it was controlled by in-

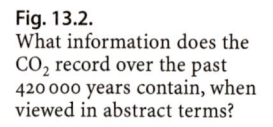

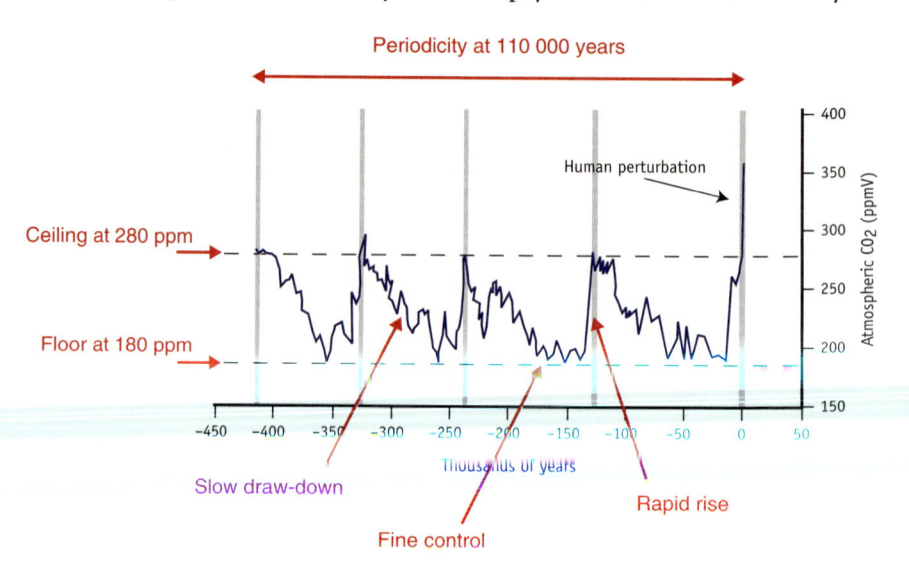

Fig. 13.2.
What information does the CO_2 record over the past 420 000 years contain, when viewed in abstract terms?

teractions between solar radiation, atmospheric absorption, ocean currents, clouds and ice bodies. The decade in which the IGBP has been active has brought considerable interdisciplinary synthesis and in particular the understanding that the biosphere is an active participant in the climate system and not a passive responder. The concept of Earth System physiology, largely mediated by biogeochemical transfers, has moved from the scientific fringes to the mainstream. The challenge for the next decade is to continue this integration beyond its current focus on the carbon cycle to include the other critical cycles and to link land, ocean and atmosphere in a fully dynamic way.

References

Berger A (1978) Long-term variations of daily insolation and Quarternary climate change. Journal of Atmospheric Science 35:2362–2367

Enriquez S, Duarte CM, Sand-Jensen K (1993) Patterns in decomposition rates among photosynthetic organisms: the importance of detritus C:N:P content. Oecologia 94:457–471

Falkowski P, Scholes RJ, Boyle E, Canadell J, Canfield D, Elser J, Gruber N, Hibbard K, Högberg P, Linder S, Mackenzie FT, Moore B III, Pedersen T, Rosenthal Y, Seitzinger S, Smetacek V, Steffen W (2000) The global carbon cycle: a test of our knowledge of Earth as a system. Science 290:291–296

Hudson RJM, Gherini SA, Goldenstein RA (1994) Modelling the global carbon cycle. Nitrogen fertilisation of the terrestrial biosphere and the 'missing' CO_2 sink. Global Biogeochemical Cycles 8:307–333

Imbrie JJ, Hays D, Matinson DG, MacIntyre A, Mix AC, Morley JJ, Pisias NG, Prell WL, Shackleton NJ (1984) The orbital theory of Pleistocene climate: support from a revised chronology of the marine $d^{18}O$ record. In: Berger AL, Imbrie JJ, Hays J, Kukla G, Saltzman B (eds) Milankovich and climate. Riedel, Dordrecht

Mooney HA, Canadell J, Chapin FS III, Ehleringer JR, Körner Ch, McMurtie RE, Parton WJ, Pitelka LF, Schulze E-D (1999) Ecosystem physiology. In: Walker BH, Steffen W, Canadell J, Ingram JI (eds) The terrestrial biosphere and global change. Cambridge, pp 141–189

Nadelhoffer KJ, Emmett BA, Gundersen P, Kjornaas OJ, Koopmans CJ, Schleppi P, Tietema A, Wright RF (1999) Nitrogen deposition makes a minor contribution to carbon sequestration in temperate forests. Nature 398:145–147

Prentice IC, Farquhar G, Fasham M, Goulden M, Heimann M, Jaramillo V, Kesghi H, Le Quere C, Scholes R, Wallace D (2001) The carbon cycle and atmospheric CO_2. In: Houghton J, Yihui D (eds) Climate change: The scientific basis. Intergovernmental Panel on Climate Change Third Assessment Report (Vol. 1, Chap. 3), Cambridge University Press

Petit JR, Jouzel J, Raynaud D, Barkov NI, Barnola JM, Basile I, Bender M, Chapellaz J, Davis M, Delaygue G, Delmotte M, Kotlyakov VM, Legrand M, Lippenkov VY, Lorius C, Pepin L, Ritz C, Saltzman E, Stievenard M (1999) Climate and atmospheric history over the past 420 000 years from the Vostok ice core, Antarctica. Nature 399:436–439

Rastetter EB, Shaver GR (1992) A model of multiple-element limitation for acclimating vegetation. Ecology 73:1157–1174

Schimel DS, House JI, Hibbard KA, Bousquet P, Ciais P, Peylin P, Braswell BH, Apps MJ, Baker D, Bondeau A, Canadell J, Churkina G, Cramer W, Denning AS, Field CB, Friedlingstein P, Goodale C, Heimann M, Houghton RA, Melillo JM, Moore B III, Murdiyarso D, Noble I, Pacala SW, Prentice IC, Raupach MR, Rayner PJ, Scholes RJ, Steffen WL, Wirth C (2001) Recent patterns and mechanisms of carbon exchange by terrestrial ecosystems. Nature 414:169–172

Scholes RJ, Schulze E-D, Pitelka LF, Hall DO (1999) Biogeochemistry of terrestrial ecosystems. In: Walker BH, Steffen W, Canadell J, Ingram JI (eds) The terrestrial biosphere and global change. Cambridge University Press

Shackleton NJ (2000) The 100 000-year ice-age cycle identified and found to lag temperature, carbon dioxide and orbital eccentricity. Science 289:1897–1902

Vitousek PM, Aber JD, Howarth RW, Likens GE, Matson PA, Schindler DW, Schlesinger WH, Tillman D (1997) Human alteration of the global nitrogen cycle:sources and consequences. Ecological Applications 7:737–750

Chapter 14

Can New Institutions Solve Atmospheric Problems? Confronting Acid Rain, Ozone Depletion and Climate Change

Oran R. Young

Some efforts to create international institutions to solve large-scale environmental problems are more successful than others. Although methodological problems plague efforts to arrive at precise measurements in this realm, it is not difficult to produce rough-and-ready assessments of relative success. By 1997, twelve years after the signing of the Vienna Convention for the Protection of the Ozone Layer and less ten years after the signing of the Montreal Protocol on Substances that Deplete the Ozone Layer, worldwide production of chlorofluorocarbons (CFCs) had dropped from a high of about 1.2 million t yr^{-1} to less than 160 000 t; it is probable that production of these chemicals will continue to decline during the foreseeable future (see Fig. 14.1).[1] During the six years following the signing of the UN Framework Convention on Climate Change (UNFCCC) in 1992, on the other hand, worldwide emissions of carbon dioxide, the principal greenhouse gas, rose between 7 and 8% (see Fig. 14.2). Nor has the signing of the 1997 Kyoto Protocol, the first measure committing participants to reducing greenhouse gas (GHG) emissions within well-defined timeframes, reversed this trend.

More generally, we can identify three types of trajectories that characterise the performance of international regulatory systems or regimes as they are commonly called (see Fig. 14.3). Some regimes get off to a prompt start that sets in motion a pattern of increased success with the passage of time. The performance of the ozone regime in the period following the signing of the Montreal Protocol exemplifies this pattern. Other regimes experience a slow start but then gather momentum and enjoy growing success with the passage of time. The accumulation of knowledge leading to the addition of new and more stringent protocols to the regime dealing with long-range transboundary air pollution in the Northern Hemisphere fits this pattern. Yet a third pattern occurs in cases where a slow start does not trigger developments that produce increased success with the passage of time. Although it is too early to pass any definitive judgement, the climate regime is currently on a trajectory of this sort. Specific regimes may jump from one of these trajectories to another over time. The climate regime may yet experience increased momentum, and the news is not entirely good with regard to phasing out the remaining uses of CFCs. But these basic patterns are sufficient to allow for first-order accounts of the performance of most international environmental regimes.

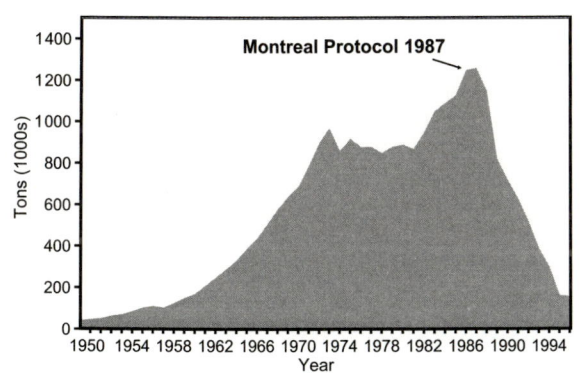

Fig. 14.1. World production of chlorofluorocarbons 1950–97 (reproduced with permission from the Worldwatch Institute)

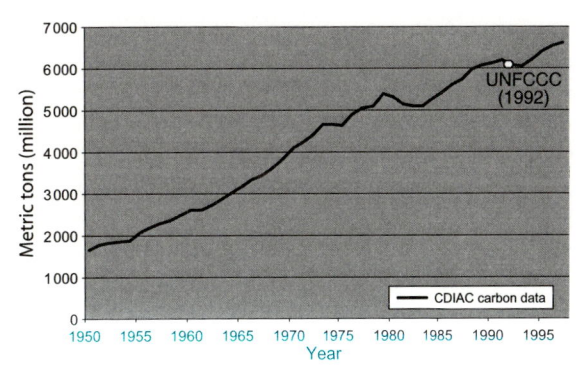

Fig. 14.2. World carbon emissions from fossil fuel burning 1950–1997 (data from the US Department of Energy through its Carbon Dioxide Information Analysis Center at Oak Ridge National Laboratory)

---

[1] One credible source has calculated that production would have been on course to reach 3 million tons per year in the absence of this regime (Porter et al. 2000).

Effectiveness

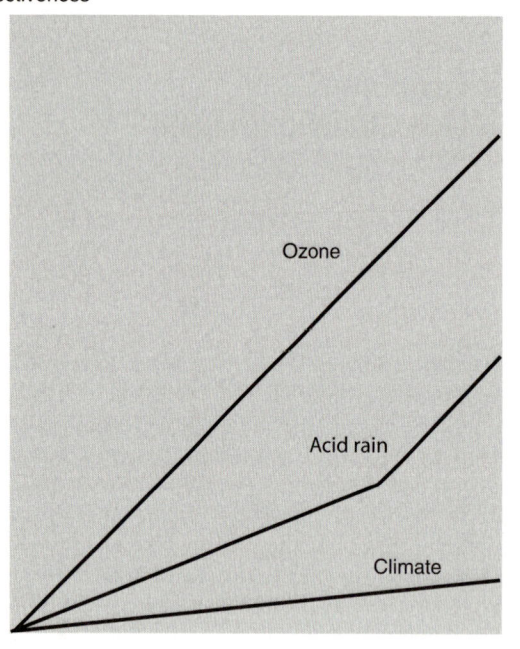

Fig. 14.3. Common institutional trajectories

How can we explain the patterns that emerge in specific cases? Why has the ozone regime performed well, whereas it has proven so difficult to put the climate regime on the road to success? Many analysts have sought to respond to questions of this sort by developing propositions stating necessary or (less often) sufficient conditions for the success of international regimes (Victor et al. 1998; Weiss and Jacobson 1998; Young 1999; Ostrom et al. 1999; Miles et al. 2001). For the most part, these efforts are rooted in broader perspectives on the nature of international society. Thus, neo-realists suppose that the active participation of a dominant actor or hegemon is necessary to the achievement of success (Baldwin 1993). Neo-liberals emphasis the importance of creating arrangements that increase transparency, reduce transaction costs, or lengthen the shadow of the future (Oye 1986). Social constructivists, by contrast, look to the emergence of consensual knowledge or the development of an epistemic community (i.e., an identifiable group of analysts and policy makers who share a common view of the nature of the problem to be solved and of the appropriate solution) as the key to success (Litfin 1994; Haas 1997). But these efforts invariably end in failure. For every proposal regarding a necessary condition, it is easy to find exceptions that undermine the claim. Some regimes succeed in the absence of a hegemon; transparency is not always essential to success; some arrangements are not based on consensual knowledge. There is little reason to suppose that a redoubling of effort along these lines will overcome this problem.

What is to be done? More specifically, how can we improve our understanding of the sources of institutional effectiveness, and in the process, develop knowledge that is usable in efforts to come to terms with issues like acid rain, ozone depletion and climate change? Here, I propose and begin to flesh out an alternative approach to this subject. In essence, I argue that all successful regimes must fulfil several key functions but that the specific ways in which successful regimes go about performing these functions can and ordinarily do vary from case to case. This means that we may never be able to formulate simple generalisations about the determinants of institutional effectiveness. Even so, it is reasonable to expect that we can identify the key functions associated with the performance of regimes and build up a repertoire of best practices regarding methods of fulfilling these functions under a variety of conditions (Honkanen et al. 1999). Among these functions, I argue, three stand out as matters requiring particular attention: (1) the development of *behavioural mechanisms* that can guide the actions of key actors toward desired ends, (2) the growth of *social practices* that can transform paper arrangements into growing concerns with loyal supporters, and (3) the creation of *steering systems* that can monitor progress toward agreed upon goals and adjust regulatory arrangements to move behavioural complexes toward fulfilling these goals.

14.1 Does the Regime Have Appropriate Behavioural Mechanisms?

No regime can succeed without finding ways to influence the behaviour of those actors – individuals, corporations, governments – whose actions give rise to the problems at hand (e.g., acid rain or ozone depletion). In every case, the resultant causal mechanism(s) must single out and redirect – to a greater or lesser degree – the actions of those who have leverage in the relevant situation.[2] The Antarctic Treaty regime, to start with a successful example, was able to break a jurisdictional impasse by targeting the minimum number of actors required to solve the problem, directing attention to the actions of national governments, and providing the governments of claimant states with a means of saving face even as their claims were being set aside in de facto terms (Peterson 1988). In most cases, those responsible for creating regimes face choices among alternative behavioural mechanisms, and the consequences of these choices are often far-reaching. The regime dealing with intentional oil pollution at sea, for example, proved ineffective so long as it relied on discharge standards re-

---

[2] Note that those possessing leverage in specific situations are not always the same actors as those whose activities give rise to the problems under consideration.

quiring compliance on the part of tanker captains, whose behaviour was difficult or impossible to monitor (Mitchell 1994). When the regime switched from discharge standards to equipment standards in the 1970s, however, the situation changed dramatically. Under the restructured system, classification societies cannot certify a new oil tanker for operation in the absence of proof that it has segregated ballast tanks (SBTs), and insurance companies cannot insure it in the absence of certification. There is no way to hide the absence of SBTs, and the resale value of tankers failing to meet these standards is low. The willingness of only a few key port states to enforce these standards is sufficient to ensure that tanker owners and operators cannot ignore them without facing financial ruin. This shift has not solved the overall problem of marine pollution; the regime contains no provisions covering land-based – in contrast to vessel-source – pollution. But the move from discharge standards to equipment standards offers a clear example of a transition from a relatively ineffective behavioural mechanism to one that has yielded improved results.

Among the three big atmospheric regimes, the ozone arrangement has the most effective behavioural mechanism. Although the ultimate concern is emissions of ozone-depleting substances (ODSs), the regime directs attention to the actions of a small number of large companies that produced most of the world's ODSs in the 1980s (DuPont alone accounted for about 25% of global production). Without exception, ODSs constituted a minor product line for these firms whose production processes were highly transparent. In addition, these actors had reason to anticipate that they would be well-positioned to dominate the market for any chemicals introduced as substitutes for CFCs and related chemicals. Contrast this comparatively advantageous situation with the problems facing the climate regime. Emissions of carbon dioxide and other greenhouse gases arise from many sectors of society, involve large numbers of distinct actors, and are associated with practices that are deeply embedded in industrial systems (e.g., the burning of fossil fuels as a source of energy). The Kyoto Protocol specifies overall targets and timetables for reductions in GHG emissions, but it does not offer any explicit guidance regarding the behavioural mechanisms needed to bring about such changes (Victor 2001). The effort to control acid rain constitutes an intermediate case in these terms. Dependent initially on the use of traditional and somewhat ineffective regulatory mechanisms, the campaign against transboundary air pollution has led over time both to new ways of thinking about air pollution (e.g., the development of the idea of critical loads) and to a range of experiments with incentive-compatible policy instruments (e.g., tradable permits) (Tietenberg 2002). Once accepted, these mechanisms become self-implementing in the sense that they do not require special sanctions to make them work.

14.2 Has the Regime Given Rise to a Robust Social Practice?

To make a difference in solving complex problems – especially social dilemmas like the tragedy of the commons (Dawes 1980) – regimes must progress from paper to practice, or in other words, trigger the development of recognised social practices that assign well-defined roles to key actors and engage the loyalties of these actors on some basis that transcends simple calculations of benefits and costs. Social practices in this sense can and normally do vary along a number of dimensions. The participants in social practices are individuals. But in specific instances, they include distinct mixes of government officials, corporate managers, representatives of nongovernmental organisations, and members of the research community. As a group, the participants in social practices may or may not adopt a common discourse or share consensual knowledge regarding the nature of the problems to be solved and the character of effective solutions (Litfin 1994). They may or may not form epistemic communities (Haas 1997). In the typical case, the activities of individuals participating in social practices reflect a variety of motives. As social practices evolve, however, the influence of what some observers call the logic of appropriateness relative to the impact of the utilitarian logic of consequences normally grows (March and Olsen 1998). As these observations suggest, the richness, or as some would put it, the thickness of social practices varies from one regime to another and from one time period to another with regard to individual regimes. But to the extent that they are effective, social practices turn the provisions of regimes into routinised activities or standard operating procedures, endow these procedures with a sense of legitimacy, and create constituencies ready, able and willing to defend regimes against external threats to their viability.

The regimes dealing with acid rain, ozone depletion, and climate change have all given rise to significant social practices. But the nature of these practices and their contributions to the track records of the three regimes vary substantially. The practice associated with acid rain centres on the operation of the Cooperative Programme for the Monitoring and Evaluation of the Long-Range Transmission of Air Pollutants in Europe (EMEP). Over time, EMEP has fostered both a sense of community among those working on the problem of acid rain and the development of new ways of thinking about the effects of air pollution which have slowly permeated not only the outlook of the scientific community but also the discourse of the policy community. In the case of ozone depletion, by contrast, the social practice features a conjunction of two distinct elements: a growing scientific consensus concerning the biophysical mechanisms involved that has been acknowledged by the corporate sector and a willingness on the part of leading industrialised countries to provide financial incentives (e.g., the Montreal Protocol Multilateral Fund) that have proven sufficient to persuade key developing countries (e.g., China, India) to become active participants in the regime. The result has been a steady tightening of the regime's regulatory provisions articulated in decisions of the Conference/Meeting of the Parties. Interestingly, the climate regime has given rise to a significant social practice of its own, involving the efforts of the scientific community (articulated in the work of the Intergovernmental Panel on Climate Change), the community of policy analysts (activated by an interest in debating the relative merits of different policy instruments), and the policy community (reflected in the efforts of the Conference of the Parties). As a number of observers have noted, the search for appropriate responses to climate change has become a growth industry. The problem is that there remain significant uncertainties regarding the science of climate change, major disagreements about the merits of different policy instruments and serious obstacles to the mobilisation of political will needed to underpin successful responses to the problem of climate change.

14.3 Does the Regime Have a Sensitive Steering System?

There is a running debate concerning the extent to which institutions, in contrast to factors like population, affluence and technology are significant drivers of the behaviour of coupled human and natural systems (Young 2002). For the most part, however, this debate misses the essential character of institutions. Institutions are best thought of in cybernetic terms. They are regulatory devices or guidance systems intended to monitor the dynamics of behavioural complexes and to steer them back on track when they start to produce results that fall outside an acceptable zone (Raustiala 2000).[3] Just as thermostats seek to control heating systems and automatic pilots seek to guide aircraft, environmental regimes are intended to steer the behavioural complexes giving rise to problems like acid rain, ozone depletion, and climate change. To succeed in endeavours of this sort, steering systems must meet three requirements. They must be equipped with well-specified and appropriate criteria regarding zones of acceptability. A thermostat may be set to maintain room temperature in the range of 67 to 70 °F, for instance, but what is the acceptable zone when it comes to concentrations of carbon dioxide in the Earth's atmosphere?[4] Steering mechanisms must also have the capacity to monitor complex systems to determine whether or not they are operating within zones of acceptability; early warning regarding impending departures from such zones is especially important in many cases. The recent focus on systems of implementation review (SIRs) in the literature on regimes is particularly relevant to this function (Victor et al. 1998). In addition, steering systems must have the capacity to effectuate changes in key elements (e.g., a car's accelerator, an aircraft's guidance system, or a factory's emission control system) needed to correct actions that produce deviations from zones of acceptability. In the case of environmental regimes, regular meetings of the conferences of the Parties are intended to perform this function.

The three big atmospheric regimes differ substantially in these terms (Social Learning Group 2001). Among the sources of the success of the ozone regime as a steering system are the existence of a high level of consensus on the need to phase out the production and consumption of several families of chemicals entirely, the ability of the Conference of the Parties to accelerate phase-out schedules for designated families of chemicals without requiring ratification on the part of individual parties, and the presence of a financial mechanism under its own control. Contrast this situation with the climate regime in which there is little consensus regarding acceptable levels of GHGs resident in the Earth's atmosphere, a Conference of the Parties that often produces gridlock, and a separate funding mechanism (the Global Environment Facility) that has an agenda of its own. Here, too, the acid rain regime constitutes an intermediate case. The success of EMEP has given the regime a considerable capacity to monitor patterns of acid rain, and what some have called "tote board diplomacy" allows the regime to respond in a differentiated fashion to trends regarding specific pollutants (e.g., sulphur dioxide, nitrogen oxides, volatile organic compounds, heavy metals and persistent organic pollutants) (Levy 1993). But there are continuing disagreements regarding the ecological effects of acid rain, and the regime's capacity to respond to changing conditions is often limited by the vagaries of high-level bargaining among member countries. Needless to say, these circumstances may change over time. The greatest challenge regarding steering systems during the foreseeable future may well be to define zones of acceptability for atmospheric concentrations of GHGs and to equip the climate regime with effective control mechanisms that can be activated when trends in emissions threaten to exceed these zones.

---

[3] For an early but still significant effort to think of governance more generally in cybernetic terms, see Deutsch (1963).

[4] The UNFCCC calls for the "stabilization of greenhouse gas concentrations in the atmosphere at a level that would prevent dangerous anthropogenic interference with the climate system" (Art. 2). Many scientists take a doubling of atmospheric concentrations of carbon dioxide from preindustrial levels as a benchmark in this context. But there is no hard evidence to use in identifying a zone of acceptability in this case.

14.4 Implications for the Climate Regime

Perhaps the most pressing atmospheric problem of our time centres on the looming threat of climate change and variability. The effort to create an international regime to control ozone-depleting substances has proven successful; the performance of the regime dealing with long-range transboundary air pollution is improving. But the climate regime is in crisis, with the collapse of the Kyoto Protocol following hard on the heels of the latest and strongest conclusions of the Intergovernmental Panel on Climate Change regarding the probable consequences of rising atmospheric concentrations of greenhouse gases. Yet crises can be periods of opportunity as well as harbingers of the onset of times of troubles. What does the argument of this paper tell us that may prove helpful in taking advantage of the current crisis to reconstruct the climate regime in such a way as to improve its prospects for success?

The emergence of the problem of climate change and the growth of the climate regime more specifically have triggered the development of a robust social practice. That is the good news. But the climate regime in its current form lacks effective behavioural mechanisms. Relying on general "targets and timetables" that do not impose obligations on identifiable actors will not suffice. Espoused by some as a way to promote efficiency, this approach lacks teeth as a means of changing the behaviour of specific actors. What is needed at this juncture is a suite of explicit directives aimed at governments, nonstate actors and private individuals, calling on key actors to change their current behaviour in significant and easily verifiable ways. Such directives should include – but not be limited to – imposing higher fuel economy standards on auto manufacturers, requiring municipal governments to act now to improve public transportation, mandating national governments to strengthen incentives for the development and adoption of alternative fuels and energy conservation measures, and creating an international financial mechanism analogous to the Montreal Protocol Multilateral Fund to address the problem of integrating developing countries into an effective climate regime.

There is a need as well for better steering systems to allow the climate regime to adapt easily and quickly to changes in our understanding of the nature of the problem of climate change and to make regular adjustments or mid-course corrections based on ongoing assessments of trends in atmospheric concentrations of GHGs. A system that gives rise to ratification battles every time there is a need to revise timetables for reductions in emissions of GHGs will not suffice. Climate change is a far harder problem to solve than the depletion of stratospheric ozone; there is no guarantee that international society will succeed in managing – much less solving – this problem over the course of next few decades. Yet we are already in a position to identify the functional requirements for success in this realm with a relatively high degree of precision. What is needed now is political will and the leadership to use it to good advantage (Young 1992).

References

Baldwin DA (ed) (1993) Neorealism and neoliberalism: The contemporary debate. Columbia University Press, New York
Dawes R (1980) Social dilemmas. Annual Review of Psychology 31:169–193
Deutsch KW (1963) The nerves of government: Models of political communication and control. Free Press, Glencoe IL
Haas PM (ed) (1997) Knowledge, power, and international policy coordination. University of South Carolina Press, Columbia
Honkanen ML, von Moltke K, Hisschemöller M (1999) Report of the concerted action on the effectiveness of international environmental agreements. Noordwijk Workshop, October 15–18, 1998. Institute for Environmental Studies, Vrije Universiteit of Amsterdam, Report No. R-99/05
Levy MA (1993) European acid rain: The power of Tote-Board Diplomacy. In: Haas PM, Keohane RO, Levy MA (eds) Institutions for the Earth: Sources of effective international environmental protection. MIT Press, Cambridge, pp 75–132
Litfin KT (1994) Ozone discourses: Science and politics in global environmental cooperation. Columbia University Press, New York
March JG, Olsen JP (1998) The institutional dynamics of international political order. International Organization 52:943–969
Miles EL, Underdal A, Andresen S, Wettestad J, Skodvin T, Carlinet EM (2001) Environmental regime effectiveness: confronting theory with evidence. MIT Press, Cambridge
Mitchell RB (1994) Intentional oil pollution at sea: Environmental policy and treaty compliance. MIT Press, Cambridge
Ostrom E, Burger J, Field CB, Norgaard RB, Policansky D (1999) Revisiting the commons: Local lessons, global challenges. Science 284:278–282
Oye KA (ed) (1986) Cooperation under anarchy. Princeton University Press, Princeton
Peterson MJ (1988) Managing the frozen south: The creation and evolution of the Antarctic Treaty System. University of California Press, Berkeley
Porter G, Welsh Brown J, Chasek PS (2000) Global environmental politics, 3rd ed. Westview Press, Boulder
Raustiala K (2000) Compliance and effectiveness in international regulatory cooperation. Case Western Reserve Journal of International Law 32:387–440
Social Learning Group (2001) Learning to manage global environmental risks, 2 vols. MIT Press, Cambridge
Tietenberg T (2002) The tradable permits approach to protecting the commons: What have we learned? In: Ostrom O et al. (eds) The drama of the commons. National Academy Press, Washington DC, pp 197–232
Victor DG (2001) The collapse of the Kyoto Protocol and the struggle to slow global warming. Princeton University Press, Princeton
Victor DG, Raustiala K, Skolnikoff EB (eds) (1998) The implementation and effectiveness of international environmental commitments. MIT Press, Cambridge
Weiss EB, Jacobson HK (eds) (1998) Engaging countries: Strengthening compliance with international environmental accords. MIT Press, Cambridge
Young OR (1992) Political leadership and regime formation: On the development of institutions in international society. International Organization 45:281–308
Young OR (ed) (1999) The effectiveness of international environmental regimes: Causal connections and behavioral mechanisms. MIT Press, Cambridge
Young OR (2002) Are institutions intervening variables or basic causal forces? Causal clusters vs. causal chains in international society. In: Brecher M, Harvey F (eds) Reflections on international studies at the dawn of the new millennium. University of Michigan Press, Ann Arbor

Part IIIb

**Land-Ocean Interactions:
Regional-Global Linkages**

Chapter 15

Emissions from the Oceans to the Atmosphere, Deposition from the Atmosphere to the Oceans and the Interactions Between Them

T. Jickells

The atmosphere and the oceans are intimately linked in a complex web of physical and chemical cycles that play an important role in regulating the biogeochemistry and climatology of Planet Earth. In this short paper, I will show first that atmospheric inputs are a major route by which some key nutrients for phytoplankton growth reach the ocean, and second that exchanges of a variety of gases and sea salt itself between the ocean and the atmosphere are of major importance in regulating climate and the properties of the atmosphere. These exchanges are regulated by biological and physical processes that are themselves sensitive to the inputs of nutrients, and to changes in physical forcing arising from climate change. We are beginning to understand the nature and regulation of some of these air-sea exchange processes, but the challenge for the next decade is to try to quantitatively understand the interactions and feedbacks between these systems. Only then can these processes be incorporated into Earth System models and contribute to improved predictions of global change.

15.1 Atmospheric Inputs to the Oceans

15.1.1 Nitrogen

Primary productivity of the oceans is limited by the supply of several key nutrients, particularly nitrogen. It is now clear that inputs of nitrogen to the oceans via the atmosphere are comparable in size to other inputs (Table 15.1). By contrast, atmospheric inputs of the other key phytoplankton macronutrients, phosphorus and silicon, are small. River and inorganic atmospheric nitrogen inputs have been approximately doubled as a result of human activity, primarily from combustion processes releasing nitrogen oxides into the atmosphere and ammonia emissions from intensive agricultural practices. Organic nitrogen (representing relatively high molecular weight material deposited with aerosols and in rain water) makes up an important but very uncertain (between a third and three quarters) fraction of the total atmospheric nitrogen deposition. Because the source of this organic matter is not known, it is not yet possible to quantify the scale of human impact on nitrogen inputs to the oceans. The characterisation of the organic component of aerosols and its bioavailability, along with understanding its role in cloud-water processes and impact on aerosol light scattering represents a major challenge in the coming years.

Assessing the impact of atmospheric deposition of nitrogen is not straightforward. While rivers enter the oceans at a defined point and discharge continuously (albeit with some temporal variability), atmospheric deposition is highly episodic and dispersed unevenly in both space and time over the whole ocean. Most primary production in the oceans appears to be supported by mixing of nitrate from deep waters to the surface sunlit region so the immediate effects of increased atmospheric deposition are usually small. However, high atmospheric deposition events, particularly in summer under stratified and nutrient-depleted conditions, could have a significant effect on phytoplankton growth. On longer time scales the loss of organic carbon (and hence

Table 15.1. Nitrogen inputs to the oceans

Source	Flux 10^{12} mol yr^{-1}	Source
Biological N$_2$ fixation	5.7	Capone et al. 1997
Lightning	0.6	Price and Penner 1997
Rivers	5.4	Galloway et al. 1995
Atmosphere	3.1–7.9	Galloway et al. 1995; Cornell et al. 1995

Note: N$_2$ gas can only be utilised by specialist nitrogen-fixing organisms (biological N$_2$ fixation), and hence the air-sea exchange of N$_2$ is not considered as nitrogen input. Lightning estimate is scaled from a global figure and assumes 2/3 falls into the ocean; this will be an overestimate, since most lightning is generated over land. Lightning and biological fixation inputs are essentially unperturbed by human activity. There is a significant emission of ammonia from the oceans but this is assumed to redeposit in the oceans.

CO_2) and nitrogen from the oceans to the deep sea must be balanced by external inputs of nitrogen from the atmosphere and rivers, so in this context, atmospheric inputs are of major importance.

Over coastal waters, atmospheric deposition processes for nitrogen are particularly complex due to a number of factors. Interactions in the atmosphere between nitrogen species from land and sea salt increase nitrogen deposition rates. This enhances atmospheric nitrogen deposition in coastal seas that are already under pressure from waste discharges, over-fishing and other activities. Quantification of these atmospheric inputs is difficult because of complex physical and chemical processes occurring throughout the coastal region. In addition, coastal seas, and ocean waters can emit ammonia to the atmosphere, if its concentrations in water are high and atmospheric concentrations are low, and this flux needs to be evaluated in nitrogen budgets.

As an example of the importance of atmospheric inputs to coastal waters, it is estimated that more than 25% of the total nitrogen input to the North Sea from land arrives through the atmosphere. Atmospheric inputs to some coastal regions are much higher than those to the open oceans. For example, nitrogen deposition rates to the southern North Sea are four times those to the central North Atlantic gyre and ten times those to the Southern Indian Ocean. The gradients for primary productivity are much less steep, suggesting that atmospheric inputs will contribute proportionately more of the nitrogen to fuel production in coastal waters compared to offshore, though a comparison based on "new" production (i.e., excluding that based on internal cycling) might be more appropriate. Extreme deposition events in these coastal environments can be quantitatively important in comparison to primary productivity, and several research projects are under way to try and quantify such effects and identify possible emission management strategies to mitigate them.

15.1.2 Iron

Dust transport can significantly influence radiative forcing of climate directly. Over the last decade it has become clear that dust can also play a critical role in ocean biogeochemistry, and hence climate regulation, as a source of iron, which is now recognised to be a key micronutrient for phytoplankton growth. Iron is extremely insoluble in river water and so is readily removed to the sediments in estuarine and coastal areas. Hence, the main external supply of iron to open ocean waters is via the atmosphere as wind blown dust, even though the solubility of iron in this material is also very low.

Most dust is derived from the major Northern Hemisphere deserts of North Africa, Arabia and Asia. Being furthest from these sources, the Southern Ocean is the area with the lowest atmospheric dust supply, and recent studies have clearly demonstrated that primary productivity in this region is limited by the supply of iron (Fig. 15.1).

A second issue related to iron supply links the nitrogen and iron inputs to the oceans. Marine bacteria also require iron for nitrogen fixation, so it has been suggested that in addition to direct effects on photosynthesis, iron supply can modify surface water productivity by increasing nitrogen fixation rates. This impact is most likely to be seen in tropical and subtropical oceans, because these conditions best suit *Trichodesmium*, the main nitrogen fixing cyanobacteria in the ocean.

Large-scale increases in dust production occurred during the ice ages and may have contributed to increased productivity in the Southern Ocean, leading to lower atmospheric CO_2 concentrations and a colder climate. Land use and climate changes are currently significantly altering dust production and hence iron supply to the oceans. We do not know what effect this may be having on biogeochemical processes in the oceans.

15.2 Emissions from the Ocean to the Atmosphere

Biological processes in the oceans result in the emissions of a wide range of important gases to the atmosphere including the sulphur-containing species dimethylsulphide (DMS) and carbonyl sulphide (COS), ammonia, and halogen gases including methyl iodide, methyl bromide and bromoform. Although the production of all these gases is linked to some extent to biological processes in the surface waters of the ocean (e.g., Fig. 15.1),

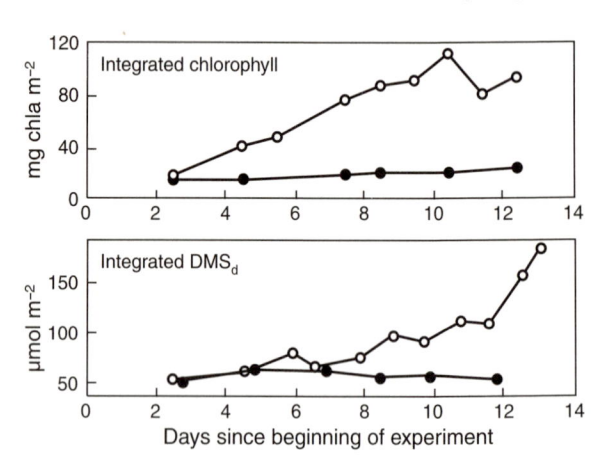

Fig. 15.1. Graph from the SOIREE experiment in the Southern Ocean, showing the increases in phytoplankton biomass (expressed as chlorophyll concentrations) and dimethylsulphide (DMS_d) in a patch of water labelled with a tracer and fertilised with iron on day 0 (*open circles*) compared to water outside the patch (*dark circles*) over the 14 days of the experiment. Results are integrated over the upper 65 m of the water column (based on Boyd et al. (2000). A mesoscale phytoplankton bloom in the polar Southern Ocean stimulated by iron fertilisation. Nature 407:695–702, © 2000 MacMillan Magazines, with permission)

different gases appear to be produced by different biological and/or photochemical mechanisms. Hence it is difficult to extrapolate from the production of one gas to another. An improved knowledge of the production and cycling processes of these gases is therefore needed before fluxes can be predicted from models of marine biogeochemical processes.

The exchange of these gases between the ocean and the atmosphere depends on the concentration gradient across the air-sea interface. In most cases the oceans are supersaturated, and hence are sources of these gases to the atmosphere. However, in the cases of CO_2 and ammonia, the direction of air-sea exchange varies considerably in space and time, depending particularly on temperature and water column biological activity. Rates of transfer of supersaturated gases across the sea-air interface increase rapidly with increasing wind speed and hence are very sensitive to climate change. Our quantitative understanding of this process is inadequate to allow accurate predictions of fluxes and future trends.

Marine biogenic gas exchange plays a number of important roles in global biogeochemical cycles and climate regulation:

- Air-sea exchange regulates the very large ocean sink for anthropogenic CO_2, though we cannot predict with confidence how this may change in the future;
- Sulphate particles dominate the fine mode aerosol in the atmosphere, and the main natural source for these particles are marine biogenic sulphur gases, emissions of which still dominate over anthropogenic sulphur emissions in the Southern Hemisphere. These gases are oxidised in the atmosphere to sulphuric acid, and this can then react with ammonia to form ammonium sulphate aerosols. These aerosols have a significant but very uncertain effect on radiative forcing;
- The emissions of methyl iodide and other organoiodine gases represent a major route by which iodine reaches the terrestrial environment, where it is an essential nutrient for humans. Halogen gases also break down in the atmosphere to various radicals that play an important role in oxidation reactions and ozone cycling.

Coastal waters appear to be particularly important sources of some gases including COS, N_2O and organoiodine gases. The latter arises in part because of the extreme bioaccumulation of iodine by seaweeds which are found predominantly in coastal waters. A consequence is that sampling from coastal sites may not be representative of the marine atmosphere, and that ship-based sampling is required.

Another important emission process from the oceans is that of sea salt itself. This process is physically controlled and increases strongly with wind speed, so is again very sensitive to climate change. Sea salt aerosol plays a direct role in climate regulation by scattering light and producing cloud condensation nuclei. Sea spray is also an important source of alkalinity in aerosols, halogens radicals and a site of oxidation for DMS. Indeed, these processes are linked, since halogen production and DMS oxidation mechanisms are pH-dependent. Sea spray also influences nitrogen deposition to the oceans as discussed earlier.

15.3 Interacting Cycles – the Challenge for the Future

Research over the last decade or so has revealed the importance and nature of the air-sea exchange processes. There have also been a number of suggestions about how these processes may be linked into climate-regulating processes.

In one of the earliest of these studies it was suggested that emissions of DMS could influence radiation balance via the formation of sulphate aerosols, which act as cloud condensation nuclei. The resultant changes in radiation balance could in turn influence primary productivity and hence DMS emissions. While plausible, research to date has not been able to confirm if such a system would create a positive or a negative feedback system. Subsequently more complex interactions have been proposed, including roles for iron and ammonia in these air-sea cycles. Overlain on these cycles are several more that may be important, such as the interactions between nitric acid and sea salt and the role of DMS oxidation on sea salt which can release halogens and hence influence ozone cycling and atmospheric oxidation rates. These will all vary with sea salt production rates and hence wind speed. In addition, changes in ocean productivity and wind speed can affect air-sea CO_2 exchange and nitrogen and iron supply. Although they have been described as separate entities, it is clear that all these potential feedback systems will interact with each other.

The potential for an extremely complex and important climate regulation system is evident, as is also the scale of the challenge ahead. We need to develop our understanding of the current systems, the controls on them and their interactions so we can predict how they will respond to and influence global change. These are the challenges before the new IGBP project called SOLAS (Surface Ocean Lower Atmosphere Study). Meeting this challenge will require a new improved collaboration between marine and atmospheric scientists. The time is opportune because of the array of new techniques that are now available to study the oceans, particularly the capacity to conduct real experiments on marine ecosystems, as illustrated in Fig. 15.1, and the power of new satellite sensors to describe the physical and chemical properties of the atmosphere and oceans.

References

Boyd PW, Watson AJ, Law CS, Abraham ER, Trull T, Murdoch R, Bakker DCE, Bowie AR, Buesseler KO, Chang H, Charette M, Croot P, Downing K, Frew R, Gall M, Hadfield M, Hall J, Harvey M, Jameson G, LaRoche J, Liddicoat M, Ling R, Maldonado MT, McKay RM, Nodder S, Pickmere S, Pridmore R, Rintoul S, Safi K, Sutton P, Strzepek R, Tanneberger K, Turner S, Waite A, Zeldis J (2000) A mesoscale phytoplankton bloom in the polar Southern Ocean stimulated by iron fertilization. Nature 407:695–702

Capone DG, Zehr JP, Paerl HW, Bergman B, Carpenter EJ (1997) *Trichodesmium*: a globally significant marine cyanobacterium. Science 276:1221–1229

Cornell S, Rendell A, Jickells T (1995) Atmospheric inputs of dissolved organic nitrogen to the oceans. Nature 376:243–246

Galloway JN, Schlesinger WH, Levy II H, Michaels A, Schnoor JL (1995) Nitrogen fixation: Anthropogenic enhancement-environmental response. Global Biogeochem Cycl 9:235–252

Price C, Penner J (1997) NO_x from lightning Global distribution based on lightning physics. J Geophys Res 102:5929–5941

Bibliography

No attempt is made here to create a comprehensive reference list. Readers requiring a comprehensive summary of literature in this area are referred to the SOLAS Science Plan, which is available on *www.solas-int.org*.

In addition to the references associated with Table 15.1 and Fig. 15.1, the following reviews may be valuable further reading.

Cornell S, Mace K, Coeppicus S, Duce R, Huebert B, Jickells T, Zhuang L-Z (2001) Organic nitrogen in Hawaiian rain and aerosol. J Geophys Res 106:7973–7983

Duce RA (1995) Sources, distributions and fluxes of mineral aerosols and their relation to climate. In: Charlson RJ, Heintzenberg J (eds) Aerosol forcing of climate. Wiley

Falkowski PG, Barber RT, Smetacek V (1998) Biogeochemical controls and feedbacks on ocean productivity. Science 281:200–206

Jickells TD (1998) Nutrient biogeochemistry of the coastal zone. Science 281:200–206

IPCC (2001) Climate Change 2001: The Scientific Basis. Contribution of Working Group I to the Third Assessment Report of the Intergovernmental Panel on Climate Change (Houghton JT, Ding Y, Griggs DJ, Noguer M, van der Linden PJ, Dai X, Maskell K, Johnson CA (eds)). Cambridge University Press, Cambridge, U.K. and New York, NY, USA, 881pp

Vogt R, Sander R, von Glasow R, Crutzen PJ (1999) Iodine chemistry and its role in halogen activation and ozone loss in the marine boundary layer. J Atmos Chem 32:375–395

Watson AJ, Liss PS (1998) Marine biological controls on climate via carbon and sulphur geochemical cycles. Phil Trans Roy Soc Lond B 353:41–51

Chapter 16

The Impact of Dams on Fisheries: Case of the Three Gorges Dam

Chen-Tung Arthur Chen

Rivers are the major channels for the passage of water, nutrients, organic material and particulate matter from land to sea. Yet humans have being continuously modifying river inputs into the ocean with their frequent unscrupulous use of land. In many areas, land is used for agriculture, forestry and mining, not to mention infrastructure and commercial development, and this has drastically increased the rate of terrestrial denudation (Fig. 16.1). Levels in excess of ten times the natural rate of erosion are not uncommon. The effect of increased denudation is, however, not necessarily reflected in an immediate increase in the flux, or flow, of particulate matter to the ocean. Rather, the eroded material from mountainous areas may be stored for centuries or even millennia in upland areas before it finds its way to the ocean. For example, there was rapid erosion in the hilly Piedmont areas of the central and southern Atlantic coast of the United States, starting in 1900. This continued until soil conservation methods were introduced during the first half of the 1900s. Even so, most of the eroded material still lies stored on the hill slopes and on the floors of valley streams, but not in the oceans.

16.1 The Effect of Dams on Deltas and Estuaries

River basin development, notably from the construction of dams, has an immediate, profound impact on river inputs to the oceans. As far back as 2800 BC, the Egyptians built the Sadd el-Kafara Dam just south of Cairo (Newson 1997). Now the 35 000 dams worldwide trap sediments and regulate river flow, which can result in a sharp increase in the amount of consumption of the river water. To illustrate this point, it has been well documented that the completion of the large Aswan Dam on the Nile River in Egypt, among others, has drastically reduced fresh water outflow and that this has been central to the substantial reductions in fish stocks in the connecting estuaries. Briefly put, dams block the downstream transport of small particles, known as particulate matter, which replenish a delta and are an important source of nutrients and food for aquatic biota, or living plants and animals. Reduced sediment discharge contributes to the erosion of the delta, the transformation and disruption of the entire ecosystem, as well as to the

Fig. 16.1.
Severe denudation upstream of the Three Gorges Dam (taken by the author in 1998)

starvation of fishes. The more subtle effects, however, go far beyond the deltas and the estuaries.

16.2 The Case of the Three Gorges Dam

Taking the case of the Three Gorges Dam (Fig. 16.2) as an example, evidence is provided to show that, despite a large riverine input of nutrients to the East China Sea (ECS), only a small fraction (7% for phosphorus and 33% for nitrogen) of the total external nutrient supply that is required to support the new production of phytoplankton actually comes from that input. It is now clear that the major nutrient supply stems from the onshore advection, or transfer by horizontal currents, of nutrient-rich, subsurface Kuroshio waters (Chen 1996; Chen and Wang 1999).

Since there is not so much nutrient outflow from the rivers in the first place, any disruption of flow from the dam would not have a great effect on the riverine nutrient flux to the ECS. It is argued, however, that the completion of the Three Gorges Dam on the Yangtze River is likely to result in diminished biological productivity in the ECS, home of the largest fishing grounds in the world. This is because the decreased water flow will reduce cross-shelf upwelling, or upward flow of subsurface water, which is the major source of nutrients to the shelf. As a result, the disruption of water flow caused by the Yangtze River Dam will indeed reduce the important flux of nutrients from upwelling.

Because of the ability of filaments and eddies to move water and other material on and off the continental shelf, investigations of buoyancy-forced flow must be conducted on spatial scales large enough to separate this effect (Henrichs et al. 2000). One method, using the so-called Box model, is frequently used for such an investigation (LOICZ 1997). For instance, cutting back the Yangtze River outflow by a mere 10% will reduce the cross-shelf water exchange by about 9% because of a reduced buoyancy effect, and at the same time, it will cut the onshore nutrient supply by nearly the same amount. It can therefore be expected that primary production and fish catch in the ECS will decrease proportionately (Chen 2000).

16.3 Threat to Other Shelves

Similar situations probably exist in the case of other shelves with large freshwater inputs. For instance, the Gulf of Alaska and the shelves off Oregon are high in biological productivity, but the rivers draining the Alaskan coast and the Columbia River have low nutrient concentrations and are probably not a significant source. Instead, upwelling is the most important driving force behind productivity (Henrichs et al. 2000).

From a global perspective, 60% of the largest 227 rivers are strongly fragmented by dams, diversions and canals. At least one large dam modifies 46% of the world's 106 primary watersheds (World Commission on Dams 2000). The number of large dams has increased sevenfold since 1950, and by the early part of the 1990s more than 13% of the global river flow to the sea had already been dammed or diverted. This figure may exceed 20% within a few decades (Nilsson and Berggren 2000; Revenga et al. 2000). In fact, Vorosmarty et al. (1997) estimated that more than 40% of global river discharge is already intercepted by 663 of the world's largest reservoirs. The list of the major dams in Asia, as an example, is given in Table 16.1. Most affected will probably be wide shelves with large riverine inputs. World-

Fig. 16.2.
Qutang Xia Gorge, one of the Three Gorges (taken by the author in 1992)

Table 16.1. Major Asian Dams (Source: United Nations, Guidebook to Water Resources, Use and Management in Asia and the Pacific 1995). Only dams with capacities of 10^9 m³ or more are included

Name	Country	Year completed	Capacity (10^6 m³)
Baishan	China	1984	4 967
Dongjiang	China	1989	8 120
Gezhouba	China	1992	1 580
Liujiaxia	China	1968	5 700
Longyangsia	China	1986	24 700
Wuijiangdu	China	1981	2 140
Yuccheng	China	1970	1 220
Balimela	India	1977	3 610
Bhakra	India	1963	9 621
Hirakud	India	1957	8 105
Idukki	India	1974	1 996
Pung Beas	India	1974	8 570
Ukai	India	1972	8 511
Mangla	Pakistan	1967	7 252
Tarbela	Pakistan	1976	13 690
Thac Ba	Vietnam	1971	3 600
Hoa Binh	Vietnam	1991	9 450
Tri An	Vietnam	1985	1 056

wide, approximately 40% of the fresh water and particulate matter entering the oceans is transported by the ten largest rivers, and this is in the form of buoyant plume formation on the open shelves. Globally, over 80% of modern carbon burial occurs on deltaic shelves, which are associated with buoyant plumes (Berner 1982; Milliman 1992). These shelves face diminished fish production when damming reduces fresh-water outflow and the buoyancy effect. Further, silicate retention in the reservoirs behind the dams affects the ecosystem structure, and certainly disturbs the biogeochemistry of the coastal seas (Ittekkot et al. 2000).

In isolated basins such as the Black Sea, another threat cannot be overlooked. If all the major rivers leading to the Black Sea, namely the Danube, Dnieper, Don, Rioni and Sakarya, were completely dammed, for example, the decreased fresh-water flux to the Black Sea could reduce the brackish or salty surface mixed layer, which could lead to a shoaling of deeper waters rich in highly toxic hydrogen sulphide. This would mean the deeper toxic waters would move into shallower zones. The consequence of this would undoubtedly have major environmental repercussions on the sea and on all of the bordering countries.

References

Berner RA (1982) Burial of organic carbon and pyrite sulfur in the modern ocean: Its geochemical and environmental significance. American Journal of Science 282:451–473

Chen CTA (1996) The Kuroshio intermediate water is the major source of nutrients on the East China Sea continental shelf. Oceanologica Acta 19:523–527

Chen CTA, Wang SL (1999) Carbon, alkalinity and nutrient budgets on the East China Sea continental shelf. J Geophys Res 104:20675–20686

Chen CTA (2000) The Three Gorges Dam: reducing the upwelling and thus productivity of the East China Sea. Geophysical Research Letters 27:381–383

Henrichs S, Bond N, Garvine R, Kineke G, Lohrenz S (2000) Coastal ocean processes (CoOP): Transport and transformation processes over continental shelves with substantial freshwater inflows. Report on the CoOP Buoyancy-Driven Transport Processes Workshop, Oct. 6–8, 1998, Salt Lake City, CoOP Report No. 7, 131 pp

Ittekkot V, Humborg C, Schafer P (2000) Hydrological alterations and marine biogeochemistry: A silicate issue? BioScience 50:776–782

LOICZ (1997) JGOFS/LOICZ Workshop on Non-Conservation Fluxes in the Continental Margins. LOICZ Meeting Report No. 25, JGOFS Report No. 25, 25 pp

Milliman JD (1992) Flux and fate of fluvial sediment and water in coastal seas: In: Mantoura R, Martin JM, Wollast R (eds)Ocean margin processes in global change. John Wiley & Sons, New York, pp 69–90

Newson M (1997) Land, water and development. Routledge, London, 423 pp

Nilsson C, BerggrenK (2000) Alterations of riparian ecosystems caused by river regulation. BioScience 50:783–792

Revenga C, Brunner J, Henninger N, Kassem K, Payne R (2000) Pilot analysis of global ecosystems: freshwater systems. World Resources Institute, 65 pp

Vorosmarty CJ, Meybeck M, Fekete B, Sharma K (1997) The potential impact of neo-Castorization on sediment transport by the global network of rivers. In: Human impact on erosion and sedimentation. Proceedings of the Rabat Symposium, IAHS Pub. No. 245:261–273

World Commission on Dams (2000) Dams and development. Earthscan Pub. Ltd., London, 404 pp

Chapter 17

Global Change in the Coastal Zone: The Case of Southeast Asia

Liana Talaue-McManus

17.1 Where is the Coastal Zone and Why Southeast Asia?

Southeast Asia highlights the human-natural interactions and feedbacks between changing climate regimes, a rapidly increasing population and an extremely rich though seriously threatened living resource base within the coastal domain. The current state of coastal science in the region has begun to unravel the local-global dynamics of these interactions. What is known to date provides compelling reasons to make the knowledge base transparent to the policy process so that local action throughout the world can maintain the life support functions of the global coastal zone.

The coastal zone represents the interphase domain between the atmosphere, land and sea. It includes the coastal plains, estuaries and embayments and extends to the edge of continental and island shelves (Pernetta and Milliman 1995). On a global scale, the coastal zone occupies about 20% of the Earth's surface, but accounts for 90% of the world's fisheries. The rich ecosystems in shallow waters, including coral reefs, seagrasses and mangroves, provide homes to some of the most diverse groups of living organisms as well as areas where nutrients cycle among their dissolved and particulate forms.

Because of their proximity to land, the living resources in the coastal zone are heavily exploited, often beyond rates at which these can regenerate. In addition, the coastal ecosystems continue to deteriorate with heavy harvesting or are further altered for other uses such as aquaculture. The coastal zone receives waste generated by land-based activities including sewage, sediments and industrial effluents. Agricultural chemicals, notably fertilisers and pesticides, degrade coastal waters worldwide. The continuing assault on the coastal zone through extraction of goods and through habitat modification has profound impacts on food and environmental security for humankind.

For this paper, I have chosen Southeast Asia as a focus to highlight the changes in the global coastal zone for a number of reasons. This region showcases the most diverse assemblage of marine life inhabiting shallow waters. Twenty of 50 known seagrass species, 45 of 51 mangrove species, and 50 of 70 hard coral genera are found in the region. The countries of Southeast Asia have among the highest population growth rates in the world and economies that heavily rely on living resources from land and sea. The archipelagos and low-lying areas of islands and continental shelves are most vulnerable to changes in sea level rise, and to flooding and storm surges that result from a changing monsoonal climate. Thus, the interaction between human societies and the environment within the domain of the coastal zone is most dramatic in Southeast Asia. Indeed, the coastal zone of this region is a global environmental hot spot.

17.2 Pressures on the Coastal Zone: Population and a Resource-Dependent Economy

Southeast Asia is home to 500 million people. Ninety percent of them live within 100 km of a coast (Burke et al. 2000). About 85 million, who constitute 7% of the world's poor, subsist on US$1.00 per person per day or less (ESCAP 2001). The developing economies of the region are largely defined by the trade of natural resources to earn foreign exchange as evidenced by the significant contribution of agriculture to the gross domestic product (GDP) of each country (Table 17.1). The fluxes of materials (gaseous emissions, sediments, nutrients, solid and liquid waste) resulting from this trade determine, along with other factors, the current state of the coastal zone in the region. With high annual population growth rates and a high degree of dependence on natural resources, the people of Southeast Asia are among those most vulnerable to global environmental change.

Vertical interactions: air pollution and climatic changes.
Because 90% of Southeast Asians live near the coast, anthropogenic gas emissions in the region mostly emanate from the coastal domain. In a global context, Eastern and Southeast Asia release 17–24% of total gas emissions worldwide (Lelieveld et al. 2000) (Fig. 17.1). Lelieveld and co-workers have found that the collective composition of these gases reduces the oxidising power of the atmosphere, allowing methane gas to remain longer in the air, hence increasing the latter's impact as a greenhouse gas. In a number of studies, Paerl (1997) and Paerl et al. (1999)

Table 17.1. Socioeconomic indicators for Southeast Asian countries

Country	Coastal length[a] (km)	Area of continental shelf[b] (×1 000 km^2)	Population in 2000[b] (×1 000)	annual growth rate[c] (%)	with 100 km from coast[a] (%)	Per caput fish food supply[d] (kg P^{-1} yr^{-1})	Per caput GDP[c] (US$)[e]	GDP from agriculture[c] (%)[e]
Cambodia	1 127	34.6	12 212	2.4	24	(1995) 9.0	309	51
Indonesia	95 181	1 847.7	224 784	1.3	96	(1998) 16.3	1 018	24
Malaysia	9 323	335.9	21 793	2.0	98	(1999) 69.0	4 523	12
Myanmar	14 708	216.4	41 735	1.8	49	(1994) 16.6	220	38
Philippines	33 900	244.5	81 160	2.0	100	(1998) 25.9	644	17
Thailand	7 066	185.4	61 231	1.0	39	(1998) 23.6	1 970	11
Vietnam	11 409	352.4	78 774	1.4	83	(1996) 11.5	267	40
Cf Japan	29 020	304.2	126 550	0.2	96	(1997) 64.0	24 070[f]	2

[a] Burke et al. 2000. [b] The World Almanac 2001. [c] ESCAP and ADB 2000. [d] Fishery Country Profile from www.fao.org/fi/fcp.
[e] Same year as food supply. [f] In 1998.

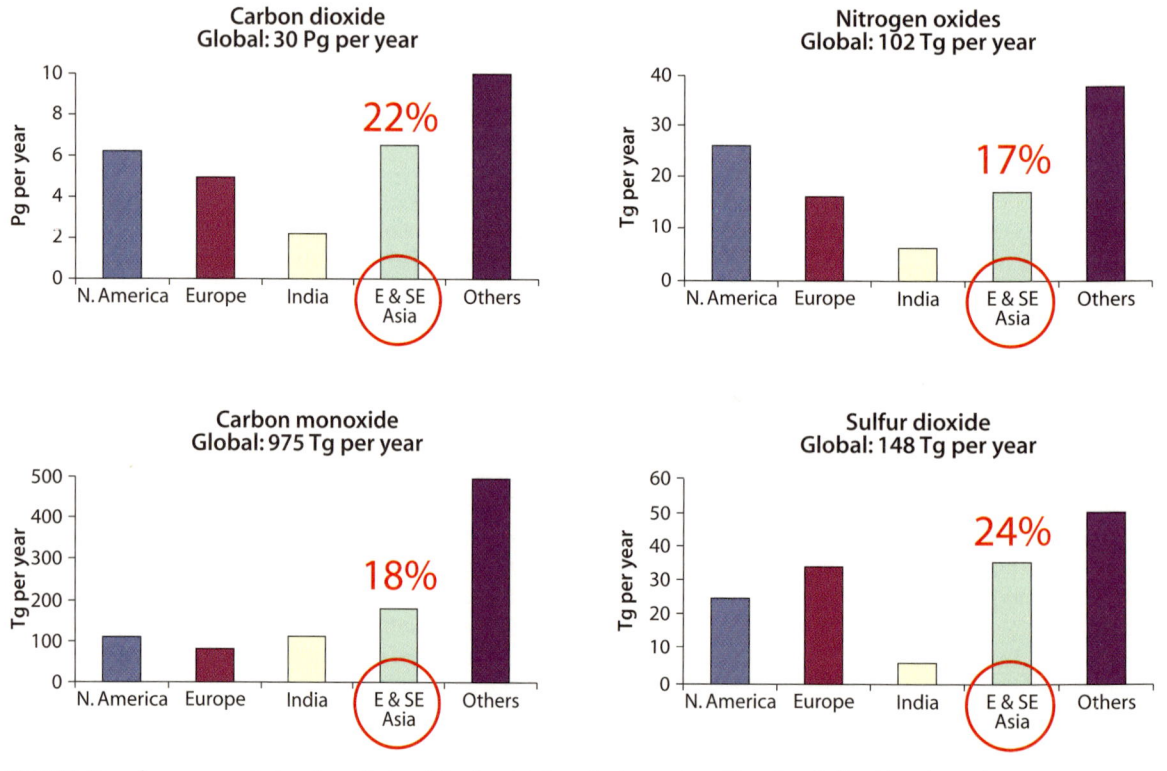

Fig. 17.1. Greenhouse gas emissions from East and Southeast Asia and other regions (data from Lelieveld et al. 2000)

warn that the increasing amount of nitrogen in the atmosphere, such as in the form of nitrous oxides, can lead to an enhanced direct deposition of nitrogen through rain. Nitrogen loading through precipitation can increase the flux of new nitrogen to the coastal zone and potentially exacerbate eutrophication.

Over a 37-year period, Manton et al. (2001) have observed profound climatic changes in Southeast Asia, which directly influence material fluxes, and the temperature regime at which chemical transformations occur on land, in air and sea. They have noted that the number of hot days and warm nights per year has increased. In addition, the contribution of extreme events like La Niña to the annual rainfall in the region has increased. These regional changes interact with larger-scale increases in the heat content of the ocean, which Barnett et al. (2001) have shown to be occurring over the last 45 years. The monsoonal shifts in the climate patterns have begun to have dramatic effects on the natural resource-based economies of Southeast Asia and on the increasing vulnerability of low-lying areas to increasing frequency of typhoons and flooding events.

Horizontal interactions: sediments and nutrients. The vertical interactions between the coastal zone and the atmosphere represent one complex subset of changes. The horizontal interactions of the coast with the land and the sea represent another. From deforested land and poorly managed tillage, large quantities of soil end up as mud in tropical estuaries with serious environmental consequences (Wolanski and Spagnol 2000). Asian rivers account for approximately 40% of total annual sediment discharge from land to sea or about $3000\ t\ km^{-2}\ yr^{-1}$ (Milliman and Syvitsky 1992). Burial of filter-feeding animals including corals (Wesseling et al. 1999) and benthic plants (causing a decrease in their biodiversity and productivity) (Terrados et al. 1998) decreased water transparency for phytoplankton and other autotrophs, and economic losses from degraded tourist attractions are among the major impacts of increased sedimentation. Biogeochemically, the sediments contain a significant amount of organic carbon that is an important component of the global carbon cycle (Schlünz and Schneider 2000).

Nutrients from minimally treated domestic waste of a rapidly increasing population and from fertiliser applications in a widening expanse of tillage account for eutrophication being the most pressing pollution problem in coastal waters (Tilman et al. 2001). Asia currently uses 50% of annual global fertiliser production or about 70 million t. In three study sites in Southeast Asia, dissolved inorganic nitrogen discharged to coastal basins has been found to represent anywhere from 10 to 50% of waste generated by land-based activities, notably agriculture and the household sectors (Talaue-McManus in prep.; Table 17.2).

Nutrients reaching near-shore waters via rivers or direct loading lead to profound changes in ecosystem structure and function. These can include toxic algal blooms (Tilman et al. 2001), shifts from coral- to algal-dominated coral reef communities in synergism with the overharvest of herbivorous fish (McManus et al. 2000), and the occurrence of hypoxic zones that produce nitrous oxide and methane, both potent greenhouse gases (Naqvi et al. 2000; Purvaja and Ramesh 2000). Hypoxic zones probably commonly occur in the shallow waters of Southeast Asia given intense rainfall, large river runoff, and high nutrient and organic matter loading. The frequency and extent of their occurrence remain to be established and quantified.

Coastal ecosystems: Corals and mangroves. The changing climatic patterns, the extent of material delivery, as well as harvest and habitat modification determine the state of the coastal ecosystems of Southeast Asia. This section highlights the status of these life support systems in the region.

Southeast Asia contains 25% of the world's charted reefs, with Indonesia and the Philippines accounting for 80% of this (i.e., 20% of known global area). The most recent risk assessment of corals is being conducted under the aegis of the World Resources Institute with partners from the region (see *www.wri.org*). Their most recent findings as of October 2000 indicate that 86% of all reefs in the region are at medium or higher anthropogenic threat. Overfishing affects about 60% of reefs, destructive fishing 50%, while coastal development and sedimentation each impacts about 20% of reefs in Southeast Asia. Preliminary analyses further show that only less than 1% of the reefs are in well-managed marine protected areas.

Superimposed on the human-induced threats to coral reefs are anomalous occurrences of prolonged high sea surface temperatures, as witnessed during the 1997/1998 El Niño event. Mortality because of bleaching as documented during this extreme event was unprecedented in the past 3000 years (Aronson et al. 2000). In addition to elevated sea surface temperatures, the increasing concentration of carbon dioxide in the air has been shown to decrease the extent to which corals can produce chalk or calcium carbonate (Kleypas et al. 1999). Leclerq et al. (2000) predict that the calcification rate of reef-dominated communities including corals, calcareous algae, crustaceans, gastropods and echinoderms may decrease by as much as 21% from the preindustrial period (1880) to the time when CO_2 is expected to double its concentration in 2065. Although Baker (2001) reports results to indicate that coral bleaching has adaptive value that allows corals to expel suboptimal symbionts and to acquire healthy ones, the question remains to what extent corals under siege by

Table 17.2. Nutrient fluxes in 3 Southeast Asian sites (Talaue-McManus et al. in prep). *DIN* (dissolved inorganic nitrogen) is a major component of sewage and fertiliser that cause eutrophication in coastal waters

Parameter	Red River Delta, Vietnam	Lingayen Gulf, Philippines	Merbok Estuary, Malaysia
Coastal population (×1 000)	19 870	2 600	300
Drainage area (km^2)	117 700	8 810	550
Coastal basin area (km^2)	1 510 (mudflats)	2 100	10 (waterways) 45 (mangroves)
Anthropogenic generated *DIN* (mmol km^{-2} basin area yr^{-1})	4 410	800	2 480
DIN discharged to coastal basin (mmol km^{-2} basin area yr^{-1})	405	420	600

anthropogenic threats can survive adverse climate change.

Like corals, mangroves in Southeast Asia represent the most heavily altered coastal ecosystem in the reduction of its area to about a third of the cover estimated for the early 1900s. Estimated loss rates range from 1 to 4% of total area per year. At these present rates, the region will lose its mangrove forests by about 2030 (Talaue-McManus 2000). The dominant pressure, as always, is economic. Mangrove swamps continue to be converted to short-lived shrimp production ponds to earn foreign exchange, among other uses. Thus, four countries in the region accounted for 50% of global shrimp trade in 1984, increasing to 70% in 1994 (FAO 1996). Naylor et al. (2000) estimate a reduction of fish biomass of about 434 g for every kg of farmed shrimp because of habitat conversion alone. In addition, the loss of major ecological functions such as sediment trapping and shoreline protection serves to underscore the unsustainability of mangrove conversion as a means to increase food production.

Regionalising global change science for policy contexts. The interactions and feedbacks between the natural and anthropogenic components of the Earth System within the coastal domain are complex and require a multiplicity and synergy of actions from many disciplines and stakeholders. Coastal zone science and management are daunting tasks, indeed. In the Southeast Asian context, good science and sound management are matters of survival. So what can we do?

We need to put Earth System science in regional contexts so that the variability and magnitude of environmental change can be made transparent and accessible to the policy-making process. Global models do not sufficiently provide the finer-grain nuances of the causes and impacts of environmental change that policy needs in formulating economic and legal instruments within jurisdictional limits. At the same time, action plans from various states will need to be harmonised so that they can effectively address the transboundary features of coastal issues and problems. Thus, regional models should aim to evoke local action that addresses issues of global significance.

A prudent strategy to pursue might be to amplify existing scientific and management support at the local scale with the goal of providing the experience for regional collective action. The Land Use and Cover Change (LUCC) and the Land-Ocean Interactions in the Coastal Zone (LOICZ) Projects of the International Geosphere-Biosphere Programme have designed regional projects in Southeast Asia that address the complexity of natural and human interactions in the coastal zone and associated catchments. They are good templates for much needed integrative and synthetic research at local scales and are excellent platforms for site comparisons that can allow researchers to evolve regional scenarios of change. One hopes that the knowledge base these initiatives provide will stimulate a progressive policy climate for a holistic and functional management of the coastal domain at the local and national levels. At the regional scale, regional conventions may be appropriately designed to address transboundary interactions.

References

Aronson RB, Precht WF, Macintyre IG, Murdoch TJT (2000) Coral bleach-out in Belize. Nature 405:36

Baker A (2001) Reef corals bleach to survive change. Nature 411:765–766

Burke L, Kura Y, Kassem K, Revenga C, Spalding M, McAllister D (2000) Pilot analysis of global ecosystems: Coastal ecosystems. World Resources Institute, Washington, DC 93 p

Economic and Social Commission for Asia and the Pacific (2001) Economic and social survey of Asia and the Pacific. See http://www.unescap.org

Economic and Social Commission for Asia and the Pacific, Asian Development Bank (2000) State of the environment in Asia and the Pacific. United Nations, New York, see http://www.unescap.org

Food and Agricultural Organization (1997) The state of world fisheries and aquaculture 1996. FAO, Rome, 125 p

Kleypas JA, Buddemeier RW, Archer D, Gattuso J-P, Langdon C, Opdyke B (1999) Geochemical consequences of increased atmospheric CO_2 on coral reefs. Science 284:118–120

Leclercq N, Gattuso J-P, Jaubert J (2000) CO_2 partial pressure controls the calcification rate of a coral community. Global Change Biology 6:329–334

Lelieveld J, Crutzen PJ, Ramanathan V, Andreae MO, Brenninkmeijer CAM, Campos T, Cass GR, Dickerson RR, Fischer H, de Gouw JA, Hansel A, Jefferson A, Kley D, de Laat ATJ, Lal S, Lawrence MG, Lobert JM, Mayol-Bracero OL, Mitra AP, Novakov T, Oltmans SJ, Prather KA, Reiner T, Rodhe H, Scheeren HA, Sikka D, Williams J (2000) The Indian Ocean experiment: Widespread air pollution from South and Southeast Asia. Science 291:1031–1036

Manton MJ, Della-Marta PM, Haylock MR, Hennessy KJ, Nicholls N, Chambers LE, Collins DA, Daw G, Finet A, Gunawan D, Inape K, Isobe H, Kestin TS, Lefale P, Leyu CH, Lwin T, Maitrepierre L, Ouprasitwong N, Page CM, Pahalad J, Plummer N, Salinger MJ, Suppiah R, Tran VL, Trewin B, Tibig I, Yee D (2001) Trends in extreme daily rainfall and temperature in Southeast Asia and the South Pacific: 1961–1998. International Journal of Climatology 21:269–284

McManus JL, Meñez AB, Reyes K, Vergara S, Ablan M (2000) Coral reef fishing and coral-algal phase shifts: implications for global reef status. ICES Journal of Marine Science 57(3):572–578

Milliman J, Syvitsky J (1992) Geomorphic/tectonic control of sediment discharge to the ocean: The importance of small mountainous rivers. Journal of Geology 91:1–21

Naqvi SWA, Jayakumar DA, Narvekar PV, Naik H, Sarma VVSS, D'Souza W, Joseph S, George MD (2000) Increased marine production of N_2O due to intensifying anoxia on the Indian continental shelf. Nature 408:346–349

Naylor R, Goldburg R, Primavera J, Kautsky N, Beveridge M, Clay J, Folke C, Lubchenco J, Mooney H, Troell M (2000) Effect of aquaculture on world fish supplies. Nature 405:1017–1024

Paerl HW (1997) Coastal eutrophication and harmful algal blooms: Importance of atmospheric deposition and groundwater as "new" nitrogen and other nutrient sources. Limnology and Oceanography 42 (5 part 2):1154–1165

Paerl HW, Willey JD, Go M, Peierls BL, Pinckney JL, Fogel ML (1999) Rainfall stimulation of primary production in western Atlantic Ocean waters: Roles of different nitrogen sources and co-limiting nutrients. Marine Ecology Progress Series 176:205–214

Pernetta JC, Milliman JD (eds) (1995) Land-ocean interactions in the coastal zone. Implementation plan. IGBP Report No. 33., International Geosphere-Biosphere Programme, Stockholm, 215 p

Purvaja R, Ramesh R (2000) Human impacts on methane emission from mangrove ecosystems in India. Regional Environmental Change 1(2):86–97

Schlünz B, Schneider RR (2000) Transport of terrestrial organic carbon to the oceans by rivers: re-estimating flux and burial rates. International Journal of Earth Sciences 88:599–606

Talaue-McManus L (2000) Transboundary diagnostic analysis for the South China Sea. EAS/RCU Technical Report Series No. 14. UNEP, Bangkok

Terrados, Duarte JCM, Fortes MD, Borum J, Agawin NSR, Bach S, Thampanya U, Kamp-Nielsen L, Kenworthy WJ, Geertz-Hansen O, Vermaat J (1998) Changes in community structure and biomass of seagrass communities along gradients of siltation in SE Asia. Estuarine, Coastal and Shelf Science 46:757–768

Tilman D, Fargione J, Wolff B, D'Antonio C, Dobson A, Howarth R, Schindler D, Schlesinger WH, Simberloff D, Swackhamer D (2001) Forecasting agriculturally driven global environmental change. Science 292:281–284

Wesseling I, Uychiaoco AJ, Aliño PM, Aurin T, Vermaat JE (1999) Damage and recovery of four Philippine corals from short-term sediment burial. Marine Ecology Progress Series 176:11–15

Wolanski E, Spagnol S (2000) Environmental degradation by mud in tropical estuaries. Regional Environmental Change 1:152–162

Part IIIc

**The Climate System:
Prediction, Change and Variability**

Chapter 18

Climate Change Fore and Aft: Where on Earth Are We Going?

Thomas F. Pedersen

Recent results from a variety of palaeoarchives show that rapid and profound shifts in Earth's climate occurred repeatedly during the last half million years. Rapid variations in the concentrations in the atmosphere of the greenhouse gases CO_2, CH_4 and N_2O occurred in lockstep with temperature, implying a causative relationship. Abrupt changes in the character of thermohaline circulation in the sea, particularly in the North Atlantic, generated swings in temperature across the Northern Hemisphere. Shifts in hydrologic balances in continental interiors were marked, producing severe drought in some regions.

In contrast, the climate of the Holocene has been thought to have been relatively stable, and this has probably contributed to the flourishing of human societies. But as we continue to improve our understanding of the past and struggle to predict the future, the intimate interconnectedness of the climate system is becoming increasingly obvious. Many of the past variations in climate that have recently been discovered can be described as having occurred abruptly; that is within decades or less. The Holocene has not been immune to such swings; abrupt changes have occurred within this epoch, albeit with amplitude swings smaller than in the Pleistocene.

The palaeodata suggest that the sensitivity of Earth's climate system to forcings is well within the scale of anticipated anthropogenic perturbations (Fig. 18.1). Thus, the historical records offer a warning: should we continue along the climate-modulating trajectory that we have established, we should anticipate sudden changes. These may (or will?) include: (a) potentially major variations in regional precipitation patterns; (b) abrupt changes in average regional temperatures; (c) sudden increases in the rate of addition of methane to the atmosphere from marine sedimentary (clathrate) sources; and (d) rapid shifts in the surface and deep circulation of the oceans.

There can be little doubt that the social and economic consequences of such variations will be negative. Furthermore, human societies are constantly making themselves more vulnerable to climate shifts. Consider episodic drought. In central Africa, borders now prevent the past migrations of populations that occurred when shifts to drier conditions rendered some regions inhospitable. In northern Texas, groundwater resources have been mined from the High Plains Aquifer to the point where pumping from wells will soon no longer be able to provide a buffer against the dry spells that have occurred in that area for millennia. In Gujarat, northwestern India, the rapidly growing population is increasingly at the mercy of the monsoon, for groundwater has been all but exhausted in the region. The state is now extremely vulnerable to even the smallest negative shifts in net precipitation.

Consider sea level rise. The burgeoning populations of megacities on coastal margins are increasingly at risk of inundation. For some, there is double jeopardy. Sea level is rising globally by about 2 mm per year as a result of warming of the ocean and continued melting of

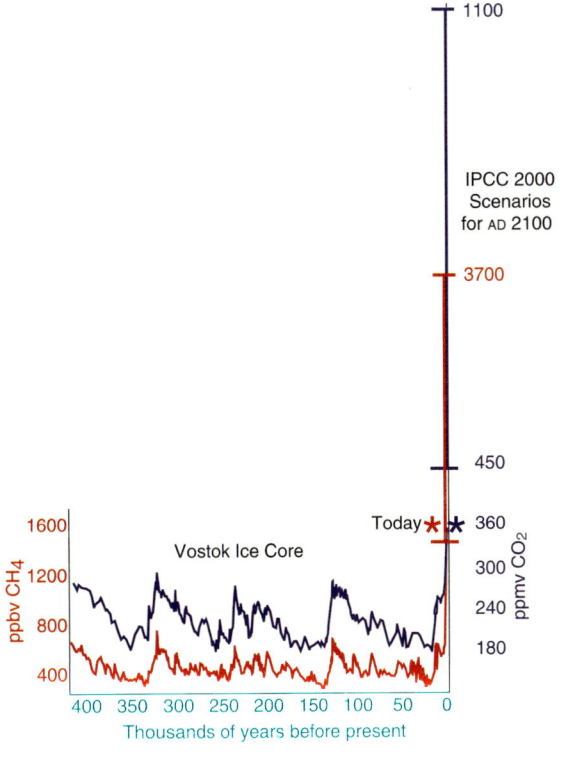

Fig. 18.1. Atmospheric concentrations of the greenhouse gases CO_2 and CH_4 over the last four glacial-interglacial cycles from the Vostok ice core record. The present-day values and estimates for the year 2100 are also shown (data from the World Data Center for Paleoclimatology and IPCC)

alpine glaciers, but on top of that, groundwater extraction is causing subsidence. Tianjin, at 10 million the third most populous city in China, sits on the coastal plain that rings the Gulf of Bohai. The surface elevation of the city fell by about 2.7 m over an area of 8 000 km^2 between 1959 and 1987 as a direct consequence of groundwater extraction. Parts of the city are now below sea level, and exceptionally at risk to flooding. In Shanghai to the south, the recent rate of subsidence has reached 10 mm yr^{-1}, five times the rate of global sea level rise. Groundwater extraction is now severely constrained, and the city is considering building a 315 km long dike to protect against future storm surges and rising sea level. Overall, some 75 million people on the Chinese coastal plain are today directly threatened by the relative rise in the level of the sea.

Consider deforestation. The extraordinary floods in the Yangtze River Valley in 1998 displaced more than 200 million people. But historical records show that the flooding was not the result of record rainfall. Rather, it was the direct consequence of severe deforestation and the associated inability of the soils in the Yangtze watershed to act as a buffer during heavy rains.

Consider insect ecology. A combination of forest-fire suppression and consecutive recent warm winters in western Canada has led to a massive infestation of pine forests by the mountain pine-bark beetle. Mature trees covering some 5 million ha of prime forest, an area larger than Holland, are dying rapidly. The tinder-dry dead trees represent an enormous direct economic loss and present the prospect of a forest fire of possibly unprecedented scale. Continued warming can only exacerbate this risk.

We know now that in geological terms, abrupt natural changes are common. We should fully expect them in the future. Given additional, and geologically rapid, human-induced forcing of climate (Fig. 18.1), we can reasonably expect enhanced societal distress as climate changes are superimposed on increasingly vulnerable populations and ecosystems. The case for mitigative actions to limit such forcings is thus compelling. Political fiddling while Rome burns is not an acceptable option.

Chapter 19

Climate Change – Past, Present and Future: A Personal Perspective

Raymond S. Bradley

The record of instrumental temperature measurements clearly documents a systematic rise in global temperatures since the mid-19th century. However, there are insufficient data to say if this warming is part of a longer-term trend, a quasi-periodic oscillation or even if it is quite unusual. In order to answer such questions, we must place the relatively short instrumental record in a longer-term perspective, and to do that we must rely on palaeoclimatic information. Palaeoclimatic archives are natural phenomena that have in some way recorded in their structure a record of past climate. They are a treasure trove of the climatic and environmental history of the planet.

Important palaeoclimatic archives include tree rings, ice cores, lake and ocean sediments, cave deposits (speleothems) and banded corals. The climatic signal in these archives is often embedded in a lot of non-climatic "noise", and the task of the palaeoclimatologist is to extract a meaningful climatic history from the material. This may involve detailed geochemical studies (for example, examining isotopes in corals as an index of past sea-surface temperatures) or taking measurements of specific physical characteristics (such as tree ring widths, or wood density changes). These measurements are calibrated with overlapping instrumental records to develop equations that can then be applied to periods for which we have no instrumental data. In this way, a picture of past climate variability (temperature, rainfall conditions, etc.) can be built up.

Some palaeoclimatic archives can also provide information about factors that may have directly influenced climate variations in the past ("forcing factors"). For example, ice cores record variations in atmospheric chemistry that track the past history of explosive volcanic eruptions, many of which had significant impacts on global temperatures. Ice cores have also provided records of past variations in the cosmogenic isotope ^{10}Be, which varies in response to changes in solar activity. Studies of ^{10}Be show that it tracks sunspot variations and thus provides an index of solar irradiance changes over time. Ice cores may also yield samples of the Earth's atmosphere from the past, trapped in bubbles within the ice. These have provided remarkable insights into variations of important greenhouse gases (such as carbon dioxide, methane and nitrous oxide) and reveal how far beyond the limits of natural variability atmospheric conditions now are as a result of recent anthropogenic activity.

Palaeoclimatic records thus tell us about conditions in the past before we had instrumental records, enabling us to build up a much longer perspective on the climatic and environmental history of the Earth. They unlock a world that humans experienced but no longer remember. And they provide a salutary warning that our recent experiences are but a small sample of the full range of natural variability. They also help us to understand the mechanisms that cause climate to change, and provide a framework for us to assess the nature of future changes that we are likely to experience as greenhouse gases continue to accumulate in the atmosphere.

Ironically, many of the archives that are necessary to achieve these goals are currently under threat from human activities and the very climatic changes that we need to better understand. It is alarming to see so many of these unique records disappearing at a dramatically increasing rate, before we have had the opportunity to sample them and so preserve a record of the past. Throughout the world we see old growth trees being cut, corals being dynamited for marine developments, or eroding away following bleaching episodes associated with record high ocean temperatures, tropical glaciers and ice caps melting. There is an urgent need to establish a programme to recover these archives, and to study the records they contain before they are lost forever. They are part of our history, just as much as the Pyramids, the Sistine Chapel in Rome or the terracotta soldiers in the Tomb of Emperor HuangTi in Xi'an. Would we seriously contemplate such treasures of our past being removed by purposeful acts, or even by benign neglect? Such destruction of our cultural archives would be accompanied by an international outcry, and we must recognise that the loss of the Earth's natural archives is no less of a tragedy, requiring international attention. There is an urgent need for an international programme to recover, calibrate and analyse these unique and rapidly disappearing archives. Such an effort should be carried out in parallel with the implementation of the Global Climate Observing System (GCOS), the Global Ocean Observing System (GOOS) and the Global Terrestrial Observing System

(GTOS). GCOS, GOOS and GTOS will shed light on how climate changes in the future, but without the perspective of the past, they will have limited value.

19.1 Lessons from the Past

Drought reconstruction from networks of trees in the USA shows that periods of drought occurred in the past that were far beyond recent experience. In the 16th century, droughts in the southwestern United States were longer, more extreme and geographically more extensive than anything registered in the instrumental period. In fact, for part of the time, such a drought spread across the USA, and was in part responsible for the demise of the early colonial settlement at Jamestown, Virginia. If similar droughts occurred today across this important agricultural region in an area of fast-growing population, the results would be devastating. Interestingly, the drought pattern is similar to that we often see associated with La Niña conditions. Could the prolonged droughts in the 16th century be a reflection of a different mode of El Niño activity in the past? Unfortunately, our understanding of sea surface temperature variations associated with the quasi-periodic shift from El Niño to La Niña conditions is very short – less than a century of spotty observations. But isotopic records from banded corals in the Pacific can provide a longer perspective on these variations. For example, spectral analysis of ^{18}O (a proxy for sea surface temperatures) in the carbonate of corals from Maiana Atoll in the central equatorial Pacific shows that in the mid to late 19th century, sea surface temperatures varied on longer time scales than have been observed during the period of instrumental records. Thus, the palaeoclimatic evidence suggests that the ENSO system may indeed have oscillated at longer periods than our recent experience suggests, leading to more persistent anomalies than anything experienced in the 20th century.

Other evidence also points to conditions in the recent past that were quite different from modern conditions. For example, studies of lake sediments in the upper Midwest of the United States reveal that the relatively moist conditions that this region experiences today may be atypical of what prevailed before AD 1200. This information is obtained from variations in the type of diatoms found in the sediments. These indicate changes in water chemistry that reveal the past history of water balance in the region. In the first millennium, much drier conditions prevailed. If these conditions were to return, it would have an enormous impact on agricultural productivity of the region, and the resulting economic disruption would ricochet through the region and far beyond.

Some of the most remarkable palaeoclimatic records have come from ice cores, taken in both north and south polar regions, as well as from high elevation tropical ice caps. High resolution ice core records can reveal important insights into past climate variations that (if they occurred again today) would have catastrophic consequences. One such record, recovered from over 7 000 m at the crest of the Himalayas, registers monsoon precipitation variations, and clearly reveals a sequence of monsoon failures in the late 18th century, as recorded by dust and chloride levels in the ice. Nothing like this has occurred in the last 180 years of instrumental records from this region, but the lesson from the ice core is that it happened not long before instrumental records began – and it could happen again. We need to understand what caused these anomalies, because in 1792, at least 600 000 people died in just one region of northern India from the droughts associated with this event. Considering that the population in this region today is hundreds of times greater than it was in the late 1700s, a recurrence of this kind of anomaly would be nothing short of catastrophic.

19.2 Global Temperature Change

Networks of palaeo-data can also be used to reconstruct large-scale temperature changes. By combining the best temperature-sensitive records from around the globe, several studies have used palaeoclimate proxies to reconstruct past temperature variations. Figure 19.1a shows one example, a record of mean annual temperatures, averaged over the Northern Hemisphere, from Mann et al. (1999). The record, expressed as anomalies from the 1901–1980 average, is shown in *black*, with low-frequency changes in *purple* and estimated statistical uncertainties in *yellow*. The long-term perspective provided by this study indicates that temperatures slowly declined over the last millennium, but this trend was abruptly terminated at the start of the 20th century. Superimposed on the long-term millennial-scale trend are century to multi-decadal-scale fluctuations and high frequency (inter-annual to multi-year) variations. Modelling studies using realistic estimates of various forcing factors over the last millennium suggest that much of the variability of the record can be explained by past changes in solar irradiance and by large-scale explosive volcanic eruptions. But the temperature change in the 20th century can not be explained by such factors, and it appears that it is largely the result of accumulating greenhouse gases in the atmosphere, as a consequence of human activities.

There are now several other studies that generally confirm these findings (Fig. 19.1b). Differences relate to the fact that some studies have reconstructed a particular season of the year, rather than mean annual temperature, or the precise geographical domain of the study in question is less than hemispheric in scale. Nevertheless, they all show a cooling trend from the early part of

Fig. 19.1.
Top: Temperature variability of the Northern Hemisphere over the last 1 000 years (from Mann et al. 1999) (*yellow shading* indicates statistical uncertainty; *black line* shows interannual variability with low frequency changes superimposed (*purple line*); *red line* is calibration period, 1901–1980, followed by the instrumental record from 1980–1998) (© American Geophysical Union).
Bottom: Estimates of long-term temperature changes from various sources (only low frequency variations shown). *Blue line* is the low frequency record from the upper figure (from Briffa et al. 2001) (© American Geophysical Union)

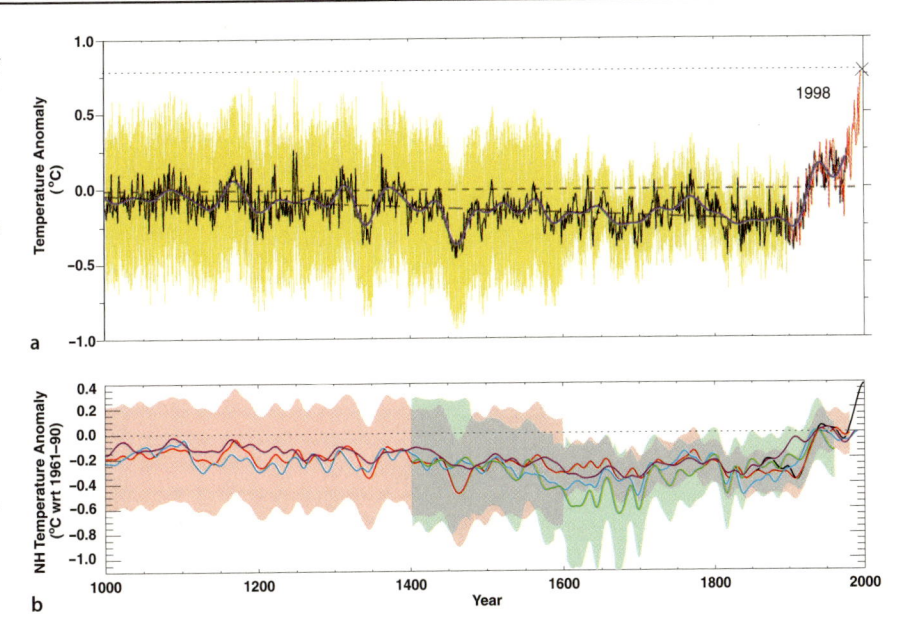

the millennium, to lowest temperatures in the 16th to 19th centuries – the so-called "Little Ice Age" – followed by the dramatic warming of the 20th century, which is unprecedented in rate and amplitude in the entire one thousand-year record.

19.3 The Future in Perspective of the Past

It is instructive to compare the variability reconstructed for the last millennium, with that expected in the future, as projected by IPCC under different energy use and population growth scenarios (Fig. 19.2). All of them are far outside the range of natural variability on the time scale of the last 1 000 years. But will a change of a few degrees make much of a difference? It is important to recognise that the changes shown here are *hemispheric* or *global* averages. One should superimpose on these projections higher frequency variability of the sort that is shown for the last millennium in Fig. 19.1a. As an example, the interannual range for the last thousand years amounted to ~3 times the variability over 50 year periods. So individual years over the next century will be both considerably warmer and colder than these averages suggest. Furthermore, general circulation models, running for periods of ~1 000 years, show that regional-scale changes are larger than hemispheric changes by a factor of 3–5. This means that areas about the size of Montana, or Poland, (i.e., a typical GCM grid box) could expect to experience anomalies on a scale that could significantly amplify the overall hemispheric changes. Of course, regional impacts related to changes in rainfall patterns may be more important economically and ecologically than temperature changes, but such changes are extremely difficult to predict. The bottom line is that the climate variability that all regions have experienced in the last millennium is unlikely to capture the variability we can expect in the future, which should give us cause for concern. These changes will be a direct consequence of the unprecedented increase in greenhouse gas levels in the atmosphere, together with associated feedbacks.

Ice cores reveal that CO_2 levels have generally varied over glacial to interglacial time scales from ~180 ppmv to ~275 ppmv. CO_2 levels today at ~370 ppmv have far exceeded anything that the Earth has experienced for millions of years, and the growth rate has been unprecedented. But such long-term views are sometimes hard to grasp. What is the time scale relevant for society today? Perhaps it is more instructive to consider just the period over which civilisation has developed, from the time when the first cities were established around 6 000 years ago. If we consider this in terms of the minute hand on a clock, with time ticking away as we approach the present, each minute represents a century (Fig. 19.3). 6 000 years ago, CO_2 levels were around 270 ppmv. When the first writing was developed around 3 800 years ago (20 minutes to the hour in terms of the "civilisation clock") CO_2 levels were still close to 280 ppmv … and they remained at this level throughout most of our history – throughout Egyptian and Chinese, Incan and Mayan dynasties, all through the Crusades, the Inquisition, the Renaissance and the period of European settlement of the New World. But things began to change significantly following James Watt's invention of the steam piston engine 250 years ago, and Karl Benz and Gustav Daimler's introduction of the first petrol-driven internal combustion engines in 1886 – a little over 100 years ago and only 2 or 3 minutes ago on our clock of civilisation. As we began this process of industrialisation, CO_2 levels were 275 ppmv … but today they have reached 360 ppmv and are rising rapidly. Thus, on the time scale of this clock of civilisation, it's the last

Fig. 19.2.
An estimate of the temperature variability of the Northern Hemisphere over the last 1 000 years (from Mann et al. 1999), and of the changes in mean *global* temperature that might occur in the next millennium, according to several different general circulation model experiments, used by the IPCC in their 2001 assessment, assuming a range of different energy consumption scenarios

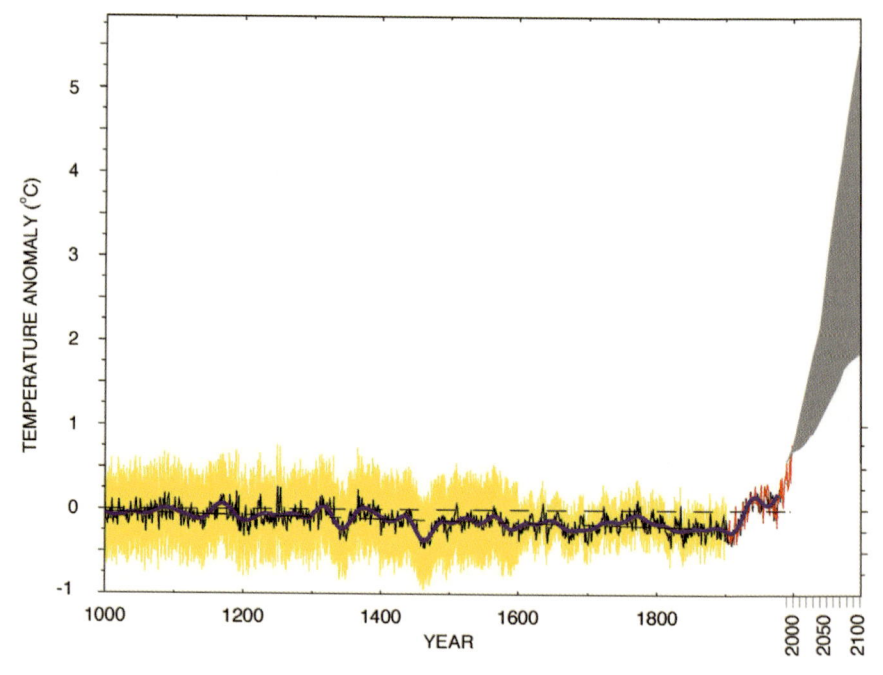

2 minutes where things have gone badly wrong on a global scale ... and it's in the "next minute" – that is, within this century – that we must fix the problem. We cannot mindlessly hand it off to our children and to future generations as yet unborn, with the hope that something will turn up – some technological fix, some magic solution that may relieve us of this burden. We cannot ruthlessly pin our hopes in unsupported speculation that feedbacks will bring these unprecedented and incredibly rapid changes into balance. And we cannot rely on alternative energy systems which produce waste products that need to be isolated from living things for a hundred thousand years – 15 times longer than the entire history of our civilisation. What an appalling legacy that would be. What right do we have to burden future generations with this responsibility?

This is not merely an issue of scientific importance – it's an ethical and moral issue. We who are the inheritors of all that civilisation has provided, of all the wisdom accumulated in our literature and science, all the beauty in art and music handed down for generations – *we* have to fix this problem. And we must act quickly, within the next "one minute" on the time scale of our civilisation's development.

Civilisation implies civility – the development of a society that is caring and respectful of its citizens *and* its environment. Sadly, we have diverted from such a trajectory. The abuses our civilisation have heaped on the world in "the last couple of minutes" will require a major effort to resolve, particularly given the expected further growth in world population. There will be 50% more people on Earth within this century. Morally, ethically

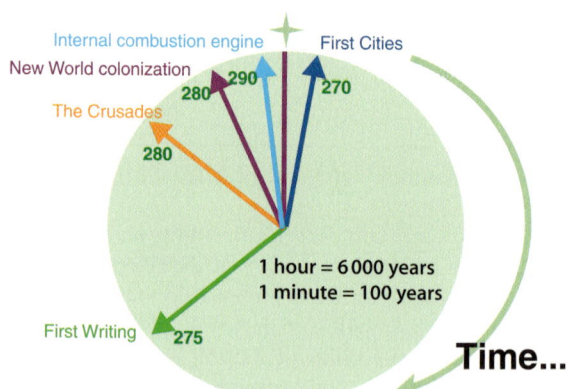

Fig. 19.3. The "clock of civilization"; assuming the first cities (~5 700 years ago) mark the start of modern civilization, each minute represents 100 years. CO_2 levels at various times are shown in *black* (based on ice core measurements)

and scientifically we have no choice but to act boldly, and quickly, *as a global community*, to resolve these global-scale problems – our heritage from civilisations in the past, and our obligations to civilisations in the future require that we deal with these issues now.

References

Briffa KR, Osborn TJ, Schweingruber FH, Harris IC, Jones PD, Shiyatov SG, Vaganov EA (2001) Low-frequency temperature variations from a northern tree ring density network. J Geophys Res 106D: 2929–2941

Mann ME, Bradley RS, Hughes MK (1999) Northern Hemisphere temperatures during the past millennium: inferences, uncertainties, and limitations. Geophysical Research Letters 26:759–762

Chapter 20

The Changing Cryosphere: Impacts of Global Warming in the High Latitudes

Oleg Anisimov

Many recent studies conclude that anthropogenic climate change will develop more rapidly at high latitudes than in other parts of the world. Climate in the Arctic will be markedly warmer; in the following few decades air temperature may rise by several degrees C. Polar regions incorporate important environmental thresholds associated with phase changes of water. Sustained warming across the freezing point will have discernible impact on natural systems and phenomena that contain snow and ice, i.e., on the cryosphere.

20.1 What Is the Cryosphere and How Does It Contribute to Global Climate Change?

The cryosphere is the part of the climate system that includes snow, all forms of ice and frozen ground. Snow and permafrost have the largest areal extent of the cryospheric components. Glaciers, ice caps, lake and river ice, though individually small, are important regulators of regional climate through their influence on water balance and runoff. The role of the cryosphere in the climate system is three-fold.

Most importantly, the cryosphere stores nearly 80% of all fresh water. If all ice on the planet were to melt, the sea level would rise almost a hundred metres. Less than one percent of the planetary ice melts and refreezes annually and is thus involved in the water balance. Snowmelt in early spring causes a dramatic increase of runoff and seasonal floods in the river systems that can lead to losses of human life and property. Global warming enhances melting of snow and ice ultimately leading to sea level rise. Many studies agree that in the 21st century, sea level rise is not likely to exceed 0.8 m, which is enough to cause serious environmental problems in many coastal zones.

Secondly, the cryosphere has a powerful control over the energy balance and surface temperature due to the latent heat involved in phase changes of ice/water, and the insulating effect of snow cover on land and floating ice on water. In spring, the surface temperature does not rise above freezing point as long as the snow or ice cover persists, even if the near-surface air temperature is well above zero.

And lastly, the cryosphere contains several important drivers of global climate change; two of them are associated with snow and ice, and one with permafrost. Snow and ice reflect much of the incoming solar radiation back to space. Anthropogenic warming will shrink the cryosphere, increasing the absorbed radiation and surface heating, primarily at high latitudes. A second driver acts through ocean thermohaline circulation, which works like a pump transferring heat with water currents. The thermohaline circulation depends on the salinity of surface waters, and with increased fresh water, input from thawing ice may slow down or even collapse, causing regional changes of climate in many parts of the world. Thawing permafrost has a potential to facilitate further climate change through the release of trace gases to the atmosphere, which enhances the "greenhouse" effect.

20.2 What Is the Empirical Evidence of Climate Change in the High Latitudes?

Meteorological records contain a high degree of variability, which complicates the detection of climate change. However, a significant warming dating from the beginning of the 20th century was detected in the Arctic and confirmed by several indirect indicators, including sea ice, snow cover, permafrost and vegetation. Cumulatively, these data provide strong empirical evidence of climate change in the high latitudes (see also Fig. 20.1).

- During the 20th century, climate has become up to 5 °C warmer over extensive land areas in the Arctic, with areas of cooling in Eastern Canada, the North Atlantic and Greenland;
- Extent of ice in the Nordic Seas has decreased by 30% over the last 130 years; sea ice thickness exhibits pronounced variability over the last three decades and has thinned at some locations;
- Snow cover extent in the Northern Hemisphere has decreased by 10% since 1970, while precipitation has increased during all seasons, except winter;
- Reduction of the near-surface permafrost area, warmer ground temperatures and deeper seasonal thawing have been observed in many areas;
- Altered composition of plant species was documented in the Arctic tundra, and the tree line moved northward.

Fig. 20.1.
Air temperature changes (°C) on Alaska (*1*) and Siberia (*2*); distribution of permafrost over the continents of the Northern Hemisphere and shelf of the Nordic seas (see map legend); (*3*) changes in the extent of ice in the Nordic seas (10^6 km^2): *top:* total area, *middle:* eastern part, *bottom:* western part

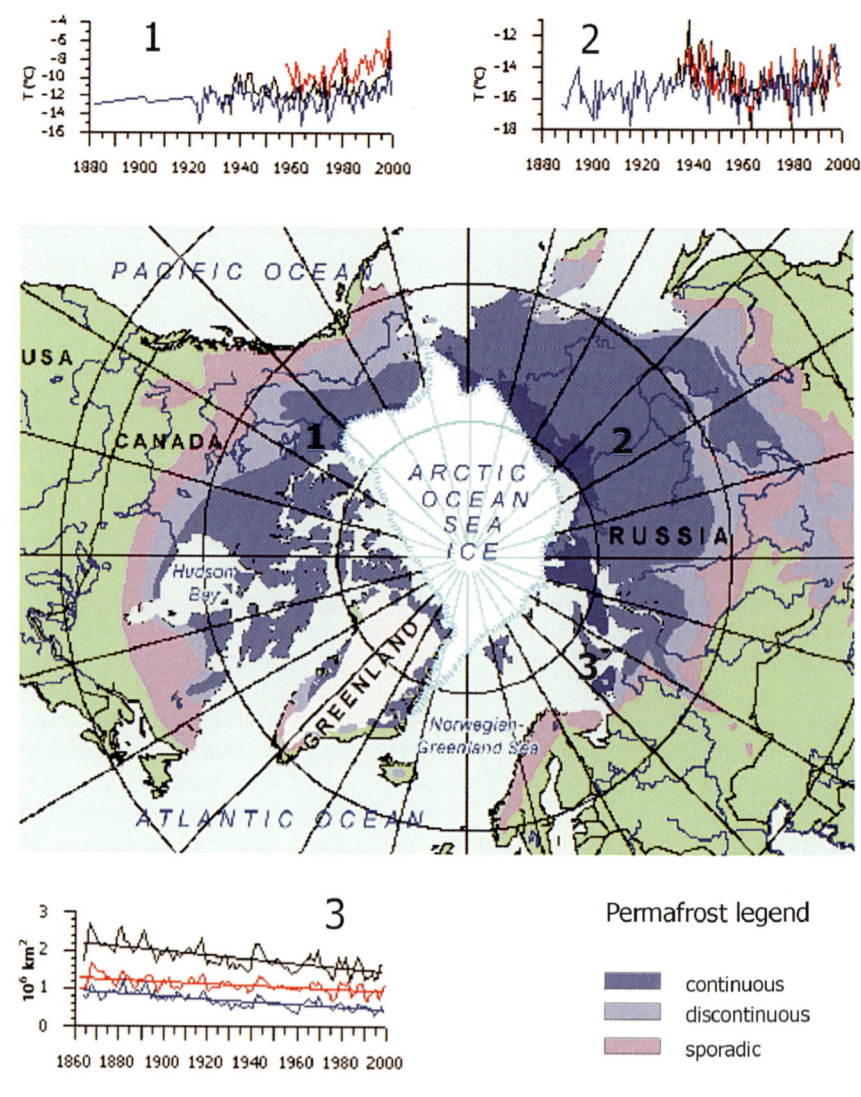

20.3 Impacts of Global Warming in the High Latitudes

Many scientific studies have been focused on the impacts that global climate change may have on natural systems in the Arctic. The results suggest that major environmental impacts will include changes in the distribution and thinning of the sea ice, lengthening of the snow-free period, changes of the temperature, distribution, and depth of seasonal thawing of permafrost, and changes in the plant cover and biodiversity. Such changes in the natural systems will interfere with the human environment and have direct and immediate implications on land use, the economy and human life in the Arctic.

Climate models predict that summer sea ice in the Arctic could shrink by 60% under conditions of CO_2 doubling, opening new sea routes for northern navigation, fisheries, and ecotourism. With more open water, there will be a moderation of temperatures and increase in precipitation in Arctic lands. Lengthening of the ice-free period is likely to cause increased coastal erosion through wave action in the Nordic seas. In the Antarctic, changes will be less pronounced due to the severe cold climate, however, sea ice volume is predicted to decrease by 25% with its boundary retreating about two degrees latitude. Projected retreat of ice shelves on the Antarctic Peninsula will expose more bare ground, which may cause changes in terrestrial biology through introduction of exotic plants and animals.

Climate change will have discernible impact on Arctic hydrology, acting through the increase of precipitation, longer snow-free and ice-free periods, and higher air temperature. Reduction of the near-surface permafrost area and deeper summer thawing increase water storage capacity of the ground. Earlier transition from winter to spring leads to protraction of the snowmelt period. Interplay of these factors will decrease the runoff extremes in spring, while summer base flow will increase. Thus, there will be less seasonal fluctuation in runoff through the year. Some of the Arctic wetlands along the southern limit of permafrost may disappear because of the better drainage conditions, increased evaporation and transpiration. In contrast, in the zone of ice-rich permafrost, the surface warming could lead to ground depression and formation of ponds and new wetlands.

Lengthening of the snow-free and ice-free period and general warming of climate will cause changes in Arctic biota, which will ultimately become more productive. However, animals such as seals, walrus and polar bears, which are dependent on sea ice, will be disadvantaged. Poleward shifts in species assemblages, changes of migration and geographic ranges of animals will have an impact on indigenous peoples following traditional lifestyles.

Discernible impacts on permafrost are expected; it is likely to shrink and become warmer in response to climatic change. Currently, permafrost underlies nearly one quarter of the Earth's land area and plays an important role in the economy and human life in the Arctic supporting constructions built upon it. Warming, thawing, and disappearance of permafrost have accelerated in recent decades, damaging structures and raising public concerns. By the middle of the 21st century, anthropogenic climate change may cause 2–3 °C warming of the frozen ground, 12–16% reduction of the near-surface permafrost area, and 15–30% deeper penetration of seasonal thawing. Such changes may have serious implications for land use, economy and engineered works in the northern lands, and affect the people's lifestyle.

Nordic countries have a well-developed infrastructure, which includes industrial and residential buildings, extracting facilities in oil fields with an associated network of pipelines and pump stations, roads, bridges, and airport runways built upon permafrost. The mechanical properties of frozen ground decrease with warming, resulting in the possible failure of pilings for buildings and pipelines, and road beds. In the zone of discontinuous permafrost, a 2.0 °C rise in air temperature may decrease the bearing capacity of frozen ground under buildings by more than a half. Where permafrost is ice-rich, thawing may cause uneven ground settlement and formation of specific landscape known as thermokarst terrain. Infrastructure is particularly susceptible to thermokarst, which may eventually lead to severe distortions of constructions, the collapse of buildings, and the loss of human lives and property. The problem is particularly severe in Russia, where a large number of residential buildings constructed in the permanent permafrost zone between 1950 and 1990 are already weakened or damaged, probably as a result of climate change.

Many natural and human systems in high latitudes are very vulnerable to climate change and have only limited ability to adapt. Ecosystems may adapt in limited range through migration and changing composition of species and eventually are likely to increase the overall productivity. Impacts on human systems in the Arctic and subarctic are mostly adverse, although there are more options to adapt. Detrimental environmental changes are not necessarily abrupt; they may instead evolve gradually and can be monitored and predicted. Assessments of climate-induced risks to human infrastructure provide important information to assist decisionmaking in engineering and land use planning in permafrost regions, allowing for the mitigation of the adverse impacts of warming.

Chapter 21

The Coupled Climate System: Variability and Predictability

Antonio J. Busalacchi

21.1 Introduction

Research on climate variability and predictability spans time scales from months to centuries and ranges from studies of natural variability to the response of the climate system to anthropogenic forcing. Progress on understanding the coupled climate system requires that we describe and understand the physical processes responsible for climate variability and predictability on seasonal, interannual, decadal, and centennial time scales, through the collection and analysis of observations and the development and application of models of the coupled climate system. Process studies, empirical studies and modelling studies require the record of climate variability be extended over the time scales of interest through the assembly of quality-controlled instrumental and palaeoclimatic data sets. One of the short-term benefits of this research is the extension of the range and accuracy of seasonal to interannual climate prediction through the development of global coupled predictive models. On longer time scales, we seek to understand and predict the response of the climate system to increases of radiatively active gases and aerosols and to compare these predictions to the observed climate record in order to detect the anthropogenic modification of the natural climate signal.

During the 1980s and 1990s, as outgrowths of the World Climate Research Programme's (WCRP) Tropical Ocean Global Atmosphere programme and the World Ocean Circulation Experiment, considerable progress was made in understanding, monitoring, and predicting the El Niño-Southern Oscillation (ENSO) phenomenon and discerning the role of the oceans in climate, respectively. This line of research on how the ocean, atmosphere and land interact as a coupled system is continuing and expanding under the auspices of the WCRP's Climate Variability and Predictability (CLIVAR) study. Research under CLIVAR seeks to describe and understand the physical processes responsible for climate variability and predictability on seasonal, interannual, decadal, and centennial time scales. Such a research agenda requires the collection and analysis of climate observations and the development and application of numerical models of the coupled climate system. This paper will discuss not only the present status of our understanding of ENSO, but also other potential modes of the coupled climate system such as the North Atlantic Oscillation, tropical Atlantic variability, the Pacific Decadal Oscillation, and monsoon circulation in Asia, the Americas and Africa. Efforts to extend the range and accuracy of seasonal to interannual climate prediction, to extend the record of climate variability through the use of instrumental and palaeoclimate data sets, and to understand the possible anthropogenic modification of natural climate variability will be highlighted.

21.2 Seasonal to Interannual Climate Variability

The largest interannual signal of the coupled climate system is the El Niño or El Niño/Southern Oscillation (ENSO) phenomenon. A necessary prerequisite to any understanding of the anthropogenic effect on climate requires that we be able to distinguish between the natural variability of the coupled climate system such as ENSO and that due to external forcing of global warming. El Niño has its origin in the equatorial Pacific Ocean, but its effects have worldwide implications. El Niño appears quasi-periodically every 4–7 years as a result of coupled interactions between the tropical Pacific Ocean and the atmosphere above it. Simply put, anomalous changes in the equatorial Pacific trade winds cause changes in the equatorial Pacific Ocean circulation. The resultant perturbations in the ocean give rise to changes in sea surface temperatures (SST). The fluctuations in SST induce changes in the surface wind field, and this cyclical atmosphere-ocean interaction begins anew. A common indicator of El Niño is when the warmest water of the global ocean shifts from the dateline in the Pacific Ocean eastward by about 5 000 km, inducing SST anomalies of 2–4 °C. This eastward migration of a critical heat source to the atmosphere changes global weather patterns (i.e., precipitation and temperature) far beyond the equatorial Pacific. In 1982–1983, what was referred to at the time as the "El Niño event of the century" occurred with global economic consequences totalling more than $13 billion. The recently concluded 1997–1998 El Niño was the *second* "El Niño event of the century" in the last 15 years. Economic losses of this major climatic event were estimated to be upward of $89 billion.

The strongest and most direct of ENSO interannual climate variations are in the tropical belt and are usually represented as shifts in the normal precipitation patterns. Along the coast of Ecuador and Peru, El Niño brings torrential rains to a region that is normally semi-arid. In other areas such as Australia, Indonesia, Northeast Brazil, and South Africa, an El Niño event implies the coming of drought conditions. For example, El Niño often results in disastrous stress in the wheat and sheep sectors in the State of Queensland, Australia. In Northeast Brazil, the socioeconomic and political sectors have learned to implement mitigation strategies based on ENSO forecasts. Crop planting decisions and water resource management are guided by El Niño climate forecasts. In Zimbabwe, drought and resultant famine have been found to be strongly influenced by El Niño.

Since the 1982–1983 the El Niño event, the international research community has implemented observational and modelling studies that have culminated in an ability to predict El Niño events at least 6 months in advance. The key to advance warnings of El Niño events lies in the ocean and the manner in which ocean temperature and currents evolve on much slower time scales than the atmosphere. Figure 21.1 shows the six-month lead forecasts for the 1997–1998 El Niño performed by the European Centre for Medium-Range Weather Forecasts.

Seasonal to interannual climate variability does not begin and end merely with El Niño. The Asian-Australian (AA) monsoon is another key component of the Earth's climate system and affects the livelihood of more than 60% of humanity, primarily in developing nations. Better predictions of the monsoon will greatly benefit the social and economic well-being of this large segment of the world's population. The prediction of the monsoon is limited by the high frequency variability arising from the internal dynamics of atmospheric and surface hydrologic processes within the monsoon region. There are predictable elements of the Asian-Australian monsoon system, which when fully exploited, will be extremely beneficial to society. For instance, there is statistical evidence that, e.g., the ENSO phenomenon, snow cover of Asia and the 30–60 day waves influence the onset and strength of the Asian-Australian monsoon.

21.3 Decadal Climate Variability

As is the case for the El Niño phenomenon, it is also clear that the ocean holds the primary memory of the climate system on longer time scales. The wind driven circulation carries deep heat and freshwater anomalies around the ocean basins, which present a slowly changing SST field to the winter atmosphere. The thermohaline circulation, involving high latitude sinking and distributed upwelling and mixing at lower latitudes, is heavily involved in the longer time scales of climate variability. Tropical SST changes have displayed decadal trends and strong correlations with long-term rain and drought cycles as well as tropical storm frequency. However, we remain largely ignorant of the processes controlling such variability as the North Atlantic Oscillation (NAO), tropical Atlantic dipole mode, Atlantic thermohaline circulation, or the Pacific Decadal Oscillation.

The North Atlantic Oscillation (NAO) is a large-scale alternation of atmospheric mass with centres of action

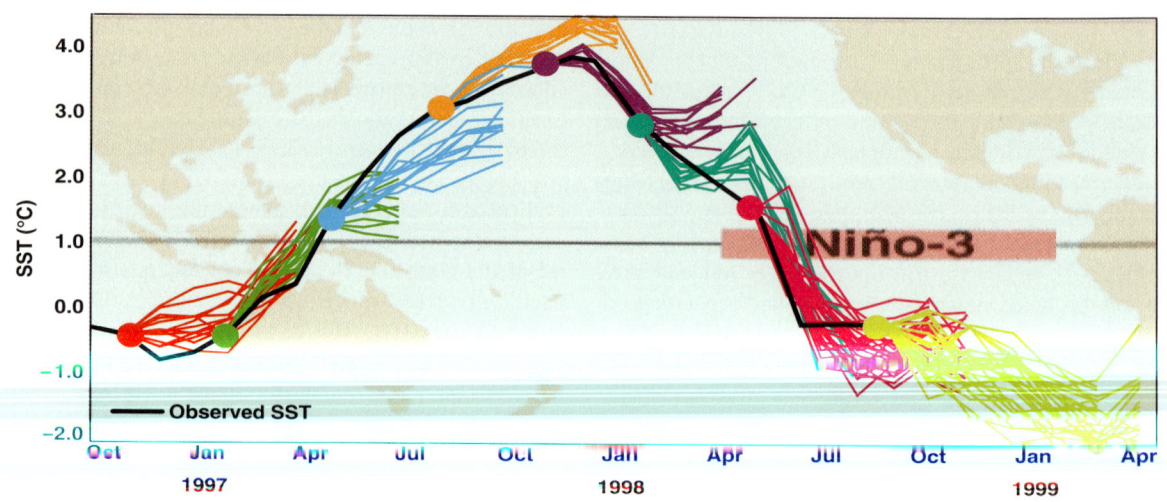

Fig. 21.1. Six month lead forecasts of sea surface temperature (*SST*) versus observations for the 1997–1998 El Niño as performed at the European Centre for Medium-Range Weather Forecasts

near the Icelandic low-pressure region and the Azores high-pressure region. It is the dominant mode of atmospheric behaviour in the North Atlantic sector (Fig. 21.2). The NAO is most pronounced in winter but detectable as a characteristic pattern in all months. The winter NAO exhibits variability on time scales from a few months through several decades in the instrumental record, including an evolution from quasi-biennial predominance late in the last century to increasing decadal and interdecadal prominence through this century. The winter NAO pattern contributes the largest fraction of the Northern Hemisphere temperature variability of any mid-latitude or tropical mode of fluctuation. A pattern involving both SST fluctuations and surface air temperature seesaws between northern Eurasia and eastern Canada and Greenland, and between Mediterranean Eurasia/Africa and the eastern United States. NAO fluctuations are also found in the patterns of precipitation in these same areas as well as storminess over the ocean and adjacent land areas. Records are too short for this to be viewed as conclusive, but a major transition of the NAO between decadal periods of extreme states (low to high) in the early 1970s coincided with shifts in the intensities and pathways of severe hurricanes. Also concurrent with this transition were the shifts in rainfall patterns in the western Sahel region of Africa and Northeast Brazil. All these changes point to a link between the variability in the NAO and decadal variability in the tropics.

Tantalising evidence for predictability is accumulating in new data on deep-rooted oceanic temperature anomalies, which appear to propagate around the Atlantic Basin on decadal and longer time scales with clear potential to affect air-sea interaction processes. The large heat capacity of the ocean is the likely memory of this coupled ocean-atmosphere system. That memory survives the seasonal cycle of upper ocean forcing though the sequestering action of deep winter mixed layers. Winter thermal anomalies are stored in these subsurface layers during the summer and transition seasons and reexposed to the atmosphere in subsequent winters, with progressive displacements through the agent of oceanic advection. The NAO is also associated with much of the recent increase in Northern Hemisphere temperature and has a pattern quite similar to the expected "fingerprint" of global warming.

At low latitudes, strong SST variability in the tropical Atlantic, often appearing as a "dipole" pattern across the equator, is a second signal of decadal climate variability. This much debated form of variability, seemingly unique to the tropical Atlantic Ocean, is often referred to as the Atlantic SST dipole. This feature involves a low-frequency oscillation of the SST gradient across the equator, which has spatially coherent SST patterns in the subtropics of either hemisphere. Although sea surface temperature anomalies in the tropical Atlantic are weaker than those associated with the Pacific El Niño, they can lead to shifts in climatic patterns over the Americas and Africa that can have major and sometimes disastrous environmental and socioeconomic impacts. The well-known drought cycle of Northeast Brazil, for example, is closely related to the variability of sea surface temperature in the tropical Atlantic.

Lastly, the thermohaline circulation, with its strong conveyor belt circulation, is another major factor in climate, as it carries much of the poleward heat flux, and has been characterised as having dramatic variability in the palaeoclimatic record. A major focus of the ther-

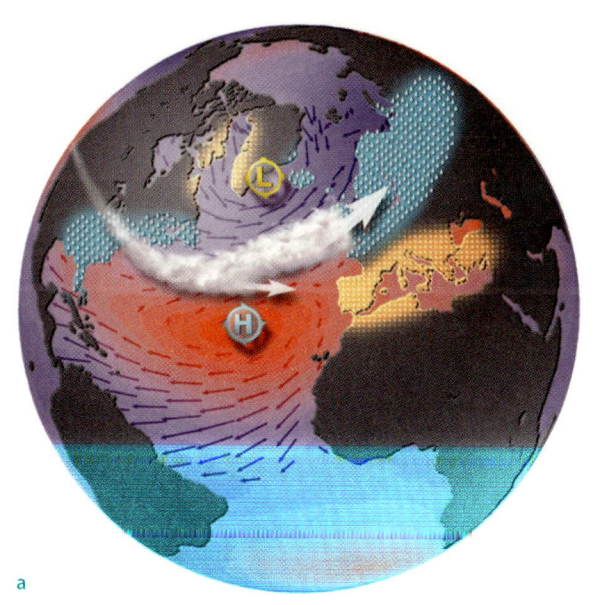

Fig. 21.2. The two phases of the NAO (courtesy of USCLIVAR and M. Visbeck/LDEO). **a** Positive phase leading to mild and wet winters over Northern Europe and dry conditions in the Mediterranean. **b** Opposite conditions during the negative phase

mohaline circulation is in the Atlantic, but the thermohaline circulation in the Southern Hemisphere is of interest as well. Also, the role of the Southern Ocean in the communication of heat and fresh-water anomalies and water masses between basins, and their influence on climate are of interest. It is especially important to develop a monitoring strategy for this vast, infrequently sampled stretch of ocean. Similarly, sampling and monitoring strategies must be designed for the Indian and Pacific Oceans.

For example, the SST in the tropical Pacific and at mid-latitudes influence the climate around the Pacific basin. At decadal time scales (peaking in the 20–30 year range), the SST pattern is symmetric about the equator with the tropics out of phase with mid-latitudes; the amplitude at mid-latitudes is approximately equal to the tropical amplitude. This pattern is roughly similar to that of the strong ENSO variability at seasonal to interannual time scales, with the exception that the amplitude of ENSO is much higher in the tropics than at mid-latitudes. This 20–30 year Pacific Decadal Oscillation modulates the more rapid ENSO signal with sometimes significant and marked effects. The decadal pattern of SST correlates strongly with atmospheric pressure patterns over North America on decadal time scales; it is likely to be part of a much larger pattern, which also includes the Indian Ocean and possibly the western North Atlantic. The consequences of the extremes of this 20–30 year mode are well-known in the North Pacific, where it is associated with the variations of the strength of the Aleutian Low; impacts include those on temperature and water resources, and on a wide range of environmentally-related variables such as fish stocks.

21.4 Detection and Attribution of Climate Change

The studies of the natural variability of the coupled climate system, as discussed above on seasonal, interannual and decadal time scales, provide the fundamental basis for distinguishing between natural and anthropogenic climate change. Detection of climate change is the process of demonstrating that an observed variation in climate is highly unusual in a statistical sense. Detection of climate change requires demonstrating that the observed change is larger than would be expected to occur by natural internal fluctuations. Attribution of change to human activity requires showing that the observed change cannot be explained by natural causes, forced or unforced. This is the process of establishing cause and effect relations, including the testing of competing hypotheses.

Hence, studies of anthropogenic climate change require an understanding, modelling and predictive capability of the response of the climate system to the anthropogenic increases in radiatively active gases and changes in aerosols. Many of the same modelling tools used to study natural climate fluctuations are invoked and modified to study global change. In addition, study of climate change requires the identification of patterns of anthropogenic modification to the mean state and to the variability of the climate system. In this manner, advances in our understanding of natural climate variability on interannual to decadal time scales serve as a baseline for detecting the trends and signatures associated with increases in greenhouse gases and the effects of other anthropogenic changes.

Part IIId

Hot Spots of Land-Use Change and the Climate System: A Regional or Global Concern?

Chapter 22

Hot Spots of Land-Use Change and the Climate System: A Regional or Global Concern?

Pavel Kabat

Traditionally, Earth's vegetation cover has been considered as a more or less passive component of the climate system. Alexander von Humboldt (1849), for example, imagined that sometime in the 'dark past' a strong subtropical Atlantic storm flooded the Sahara and washed away the vegetation and fertile soil which led to the desertification of North Africa. Koppen (1936) described vegetation as 'crystallised, visible climate', thereby considering vegetation as being an entirely passive, climate-determined system, which itself had no significant impact on atmospheric or oceanic processes. Not too long ago (up to the middle of the 1990s), coupled atmosphere-ocean models were regarded as state-of-the-art climate models in which vegetation patterns were kept as simple as possible in space and constant in time.

On the other hand, many studies, most of which have been carried out during the last five to ten years, point clearly towards the key role that land surface and biospheric processes may play in weather and climate at local, regional and global scales. The role of the land surface ranges from purely physical influences, for example, the aerodynamic drag on the atmosphere or the role of soil characteristics in controlling soil moisture and runoff, to some major biological influences, such as the response of leaf stomata to environmental changes and biogeochemical cycling. Hence, a more general definition of a climate system that encompasses not only the abiotic world (atmosphere, hydrosphere, cryosphere, and pedosphere) but also the living world, the biosphere, is now becoming more common.

In this section of the book, we focus on Earth's most important 'hot spots' of past and current land-cover and land-use changes, and systematically investigate the possible role of these changes for the past, present and future climate. A large amount of observational, analytical and modelling evidence has gradually been acquired for these regions, all of which point towards a significant role for land surface and biogeochemical processes and feedbacks in determining our present and future climate. This evidence comes from a wide range of spatial scales, from point and local measurements up to global scale multi-century modelling.

From theoretical considerations about the Earth's climate system, large-scale regional land-cover changes, particularly in the tropics, should be expected to have

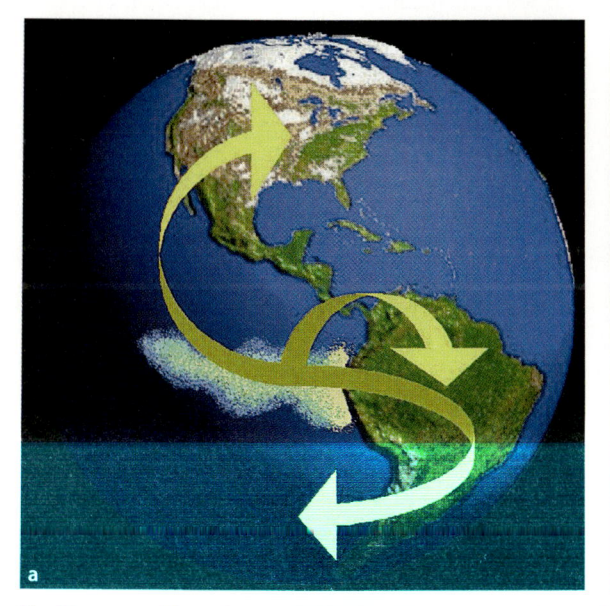

Fig. 22.1. A model simulation of the potential impact of widespread deforestation of Amazonia on regional and global circulation patterns. Such changes could even lead to a decrease in precipitation in Southeast Asia (courtesy of R. Avissar, unpublished)

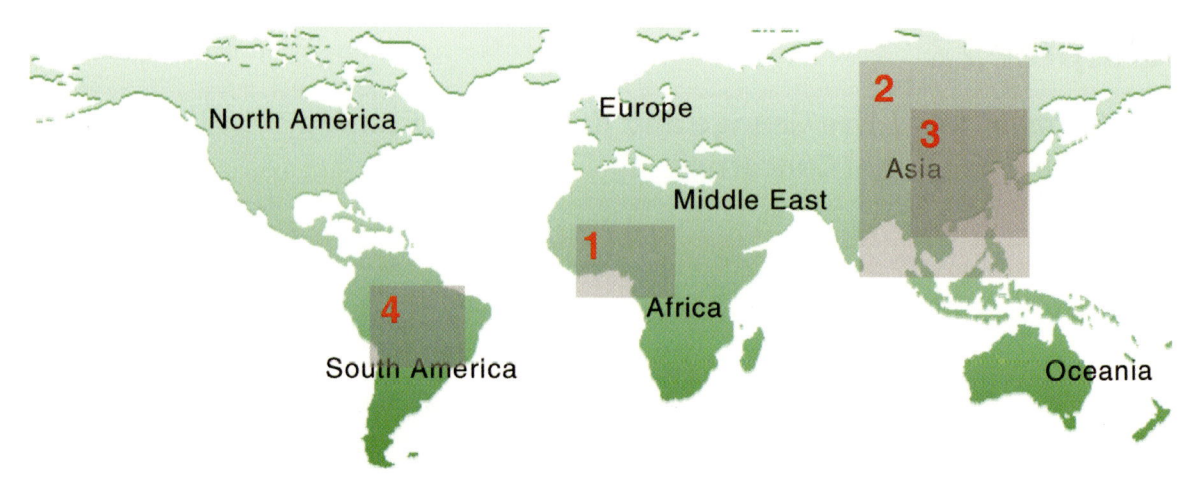

Fig. 22.2. The locations of the four case studies presented in this part

remote climatic effects. Therefore, we make an attempt to put our hot-spot regions into a global Earth System perspective, and try to answer some questions about possible global feedbacks and implications. Given that the three major tropical convective heating centres are associated with the land surfaces in Africa, Amazonia and the maritime region of Southeast Asia, changes in vegetation cover in these regions could affect the structure, strength and positioning of convective storms. Even small changes in the magnitude and spatial pattern of tropical convection may then alter the magnitude and pattern of upper-level tropical outflow, which feeds the higher altitude zonal jet stream, thereby affecting regions far beyond the actual hot spots of land-use change (Fig. 22.1).

Apart from affecting the mean zonal flow, alteration of tropical convection may also force anomalous Rossby waves that can propagate to higher latitudes. Therefore, land-cover changes, which result in changes in tropical convection, may affect weather and climate remotely both in the tropics and the high latitudes, analogously to well-documented remote effects attributed to the opposing phases of the El Niño – Southern Oscillation. In addition, changes in tropical and mid-latitude vegetation cover appear to play a significant role in tropical monsoon circulations, which have global effects and interactions extending far from the tropics.

Even though more complex global implications and feedbacks of regional land-use changes are just starting to be explored more systematically, for example, their implications for the terrestrial carbon cycle, there is increasingly strong evidence pointing towards global implications of large-scale regional land-use changes as a part of the Earth's physical and biogeochemical climate system. Adding the human and socioeconomical dimension as a part of an already advanced globalisation process will only exacerbate the importance of regional to global land-use change – climate feedbacks.

In this part, we present four case studies that illustrate various aspects of the influence of land surface and biospheric processes on atmosphere and climate (Fig. 22.2). The studies represent a journey through time, starting with an example from North Africa of biosphere-climate coupling in the mid-Holocene, continuing with examples of the impacts of contemporary land-cover change in Southeast and East Asia on the Asian Monsoon system, and concluding with initial results on the effects of deforestation in the Amazon Basin on the atmosphere and the carbon cycle.

References

Koppen W (1936) Das Geographische System der Klimate. In: Koppen W, Geiger R (eds) Handbuch der Klimatologie, Bd.1, Teil C. Berlin

von Humboldt A (1849) Ansichten der Natur, 3rd ed. J.G. Cotta, Stuttgart Tübingen, Reprint 1969. P. Reclam, Stuttgart

Chapter 23

Africa: Greening of the Sahara

M. Claussen · V. Brovkin · A. Ganopolski

23.1 Africa: A Hot Spot of Nonlinear Atmosphere-Vegetation Interaction

In many classical considerations about climate, its interaction with the biosphere played a dominant role. For example, Köppen (1936) described vegetation as "crystallised, visible climate" and referred to it as an indicator of climate much more accurate than our instruments. However until a few decades ago, many climate researchers doubted that vegetation could have a strong and significant impact on global climate changes besides those impacts related to modification of carbon storage and hence atmospheric CO_2 concentration. Today, we are convinced that in some regions, which we call "hot spots", vegetation also affects continent-scale atmospheric motion – North Africa is an excellent example of such a hot spot.

Charney (1975) formulated a theory of biogeophysical interaction in subtropical deserts. He argued that the high albedo of sand deserts tends to stabilise deserts by changing the radiative budget above deserts, thereby shifting large-scale atmospheric flow. Later on, numerical models and theory (Brovkin et al. 1998) indicate that the interaction between atmosphere and land cover in North Africa is highly nonlinear. When initialised with different land surface conditions, an atmosphere-vegetation model can yield different solutions (Claussen 1997). With present-day conditions or with deserts covering all continents, the model predicts the present-day climate. But when continents are covered with vegetation initially, the model simulates a more humid climate with a Sahara greener than today. The qualitative feature of multiple solutions emerging mainly in the subtropical North Africa has been corroborated with completely different atmosphere-vegetation models (Wang and Eltahir 2000; Zeng and Neelin 2000). Moreover, the existence of multiple equilibria has been used to interpret decadal rainfall variability in the Sahel (Wang and Eltahir 2000).

Fig. 23.1.
Equilibrium solutions of an atmosphere-vegetation model for present-day ocean temperatures and insolation (Claussen 1997) shown in terms of global biome patterns

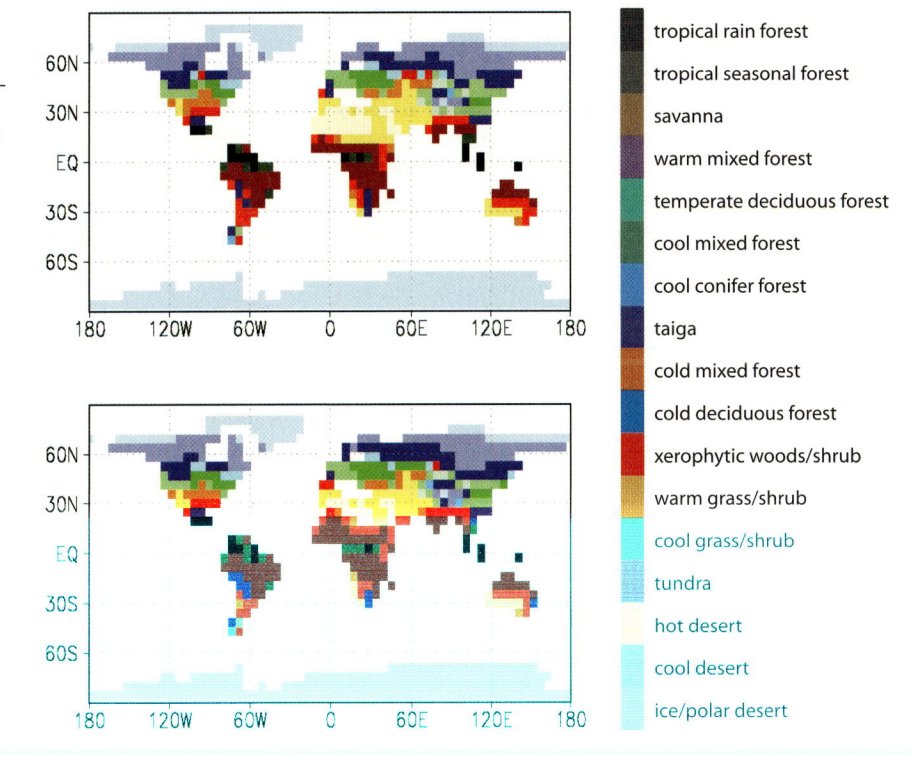

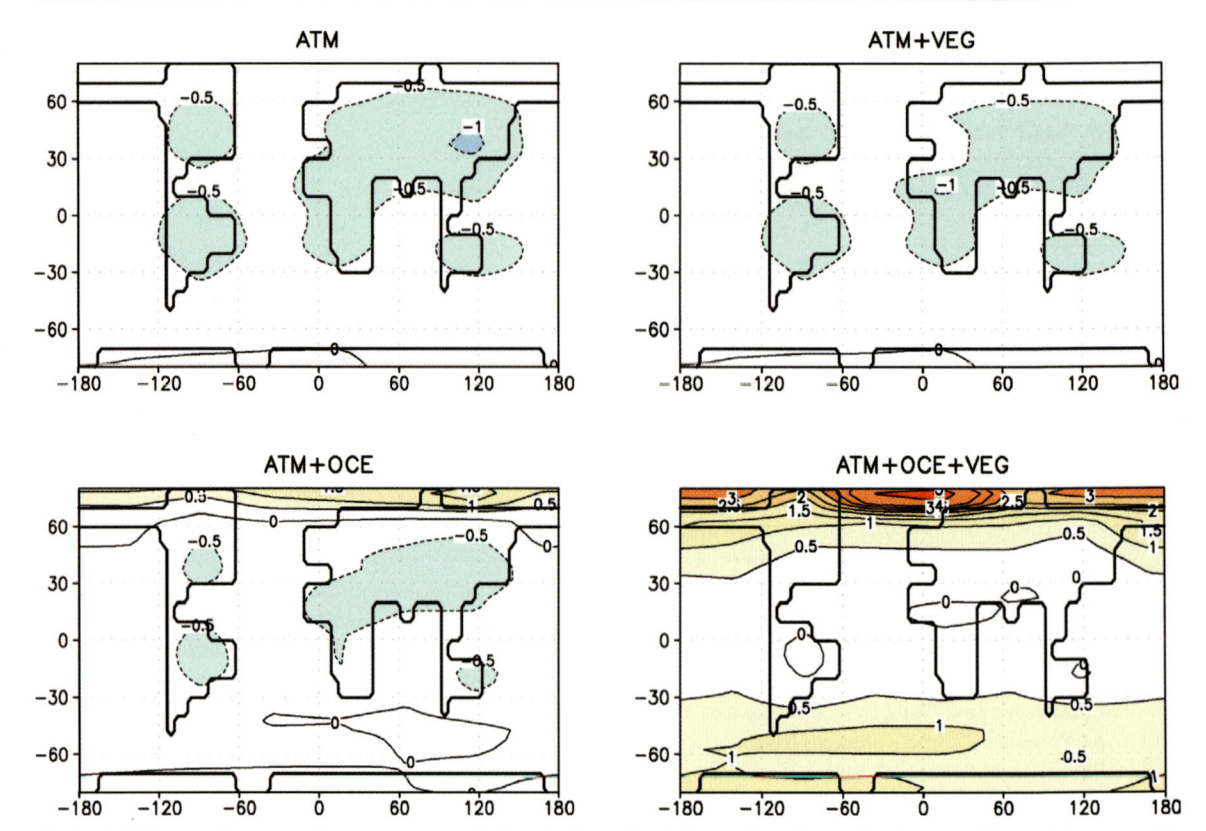

Fig. 23.2. Differences in near-surface temperatures during Northern Hemisphere winter (December, January, February) between mid-Holocene climate and present-day climate simulated by Ganopolski et al. (1998). The authors used different model configurations: the atmosphere-only model (labelled *ATM*), the atmosphere-ocean model (*ATM + OCE*), the atmosphere-vegetation model (*ATM + VEG*) and the fully coupled model (*ATM + OCE + VEG*). In *ATM*, *ATM + OCE*, and *ATM + VEG*, present-day land-surface and ocean-surface conditions – depending on the model configuration – are used (reprinted with modifications from Wasson and Claussen (2002) Quat Sci Rev © 2002, with permission from Elsevier Science)

23.2 The African Wet Period

A "green Sahara" is not just a model artefact. Palaeoclimatic reconstructions indicate that during the so-called Holocene climatic optimum some 9–6 thousand years ago, the summer in the Northern Hemisphere was warmer than today. The North African monsoon was stronger than today according to lake level reconstructions (Yu and Harrison 1996) and estimates of aeolian dust fluxes (deMenocal et al. 2000), for example. Moreover, palaeo-botanic data (Jolly et al. 1998) reveal that the Sahel reached at least as far north as 23° N. (The present boundary extends up to 18° N.) Hence, there is an overall consensus that during the Holocene optimum, the Sahara was much greener than today (e.g., Prentice et al. 2000).

The greening was tentatively attributed to changes in orbital forcing (e.g., Kutzbach and Guetter 1986) that amplified the North African summer monsoon. However, orbital forcing of the palaeo-monsoon alone did not seem to be strong enough to explain widespread greening (Joussaume et al. 1999). Models which include atmosphere-vegetation interaction yielded a stronger greening (Claussen and Gayler 1997; Doherty et al. 2000). It has also been suggested that the changes in the ocean could have contributed to a stronger summer monsoon (Braconnot et al. 1999), but a complete factor separation analysis undertaken by Ganopolski et al. (1998) demonstrates the overwhelming role of interactive dynamic vegetation.

23.3 Abrupt Changes in North Africa

Until some 6 000 years ago, the Sahara was much greener than today; how did the Sahara become what it looks like today? Alexander von Humboldt (1849) imagined the aridification of North Africa to be caused by an oceanic impact. He argued that somewhere in the "dark past", the subtropical Atlantic gyre was much stronger and flooded the Sahara, thereby washing away vegetation and fertile soil. By contrast, Claussen et al. (1999) suggested that vegetation itself strongly affected the change from a green North Africa to present-day Sahara. They have analysed the transient structures in the global vegetation pattern and atmospheric characteristics using the coupled atmosphere-ocean-vegetation model of Ganopolski et al. (1998). Their simulations show that subtle changes in orbital forcing triggered changes in North African cli-

Fig. 23.3.
Simulation of transient development of precipitation (**b**) and vegetation fraction in the Sahara (**c**) as response to changes in summer insolation on the Northern Hemisphere (**a**). These results obtained by Claussen et al. (1999) are compared with data of terrigenous material and estimated flux of material in North Atlantic cores off the North African coast (**d**) (reprinted with modifications from deMenocal et al. (2000) Quat Sci Rev 19:347–361, © 2000, with permission from Elsevier Science)

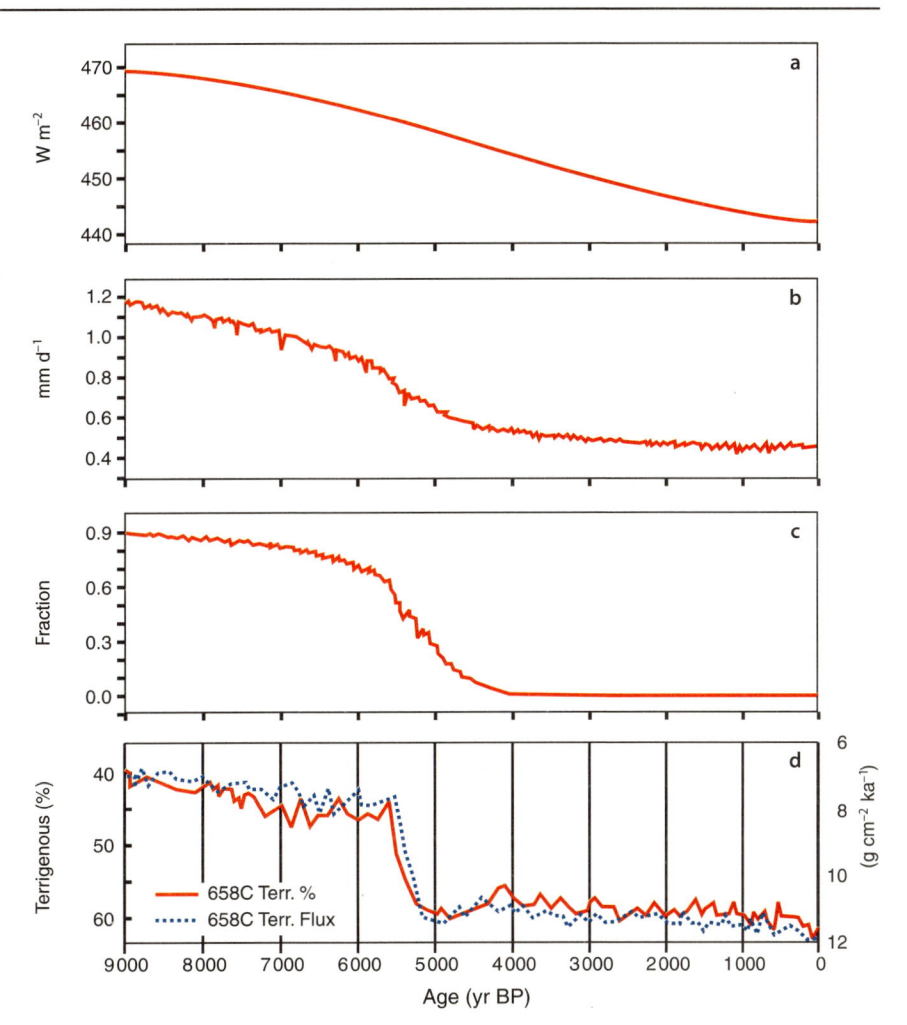

mate, which were then strongly amplified by biogeophysical feedbacks in this region. The abrupt aridification – abrupt in comparison with the subtle change in orbital forcing – is a regional effect caused by the nonlinearity of the atmosphere-vegetation feedback. The timing of it turned out to depend, however, on global processes such as continent-scale meridional temperature gradients that are governed, besides other processes, also by changes in boreal vegetation and Arctic sea ice.

23.4 Will North Africa Become Green Again in the Near Future?

Petit-Maire (1990) asked whether the greenhouse effect will green the Sahara, thereby hypothesising that an increase in atmospheric CO_2 concentrations owing to fossil fuel burning would lead to a warmer climate, which in some respect could resemble the Holocene climate optimum with its greener Sahara.

Most climate models reveal an increase of precipitation in the African tropics and the Sudan and a decrease in the subtropics, if atmospheric CO_2 concentration is prescribed to continuously increase. For today's Sahara, however, there seems to be no unequivocal result. Some models show an increase in precipitation, others a decrease. On the other hand, if precipitation were to increase over today's Sahara, then we can expect that it will be amplified by the atmosphere-vegetation feedback. Indeed, sensitivity tests in progress (Claussen et al. submitted) suggest that some expansion of vegetation into today's Sahara is theoretically possible, if atmospheric CO_2 concentration increases well above preindustrial values and if vegetation growth is not disturbed by grazing, for example. Depending on the rate of changes in atmospheric CO_2 concentrations, the rate of greening can be fast, up to one tenth of the Saharan area per decade. Such a greening, on the other hand, does not imply that the mid-Holocene climate optimum with its strong reduction of North African deserts can be considered a direct analogue for future greenhouse-gas induced climate change. According to the model study, not only the global pattern of climate change between the mid-Holocene model experiments and the greenhouse gas sensitivity experiments differ, but also the relative roles of mechanisms which lead to a greener Sahara. Moreover, the amplitude of

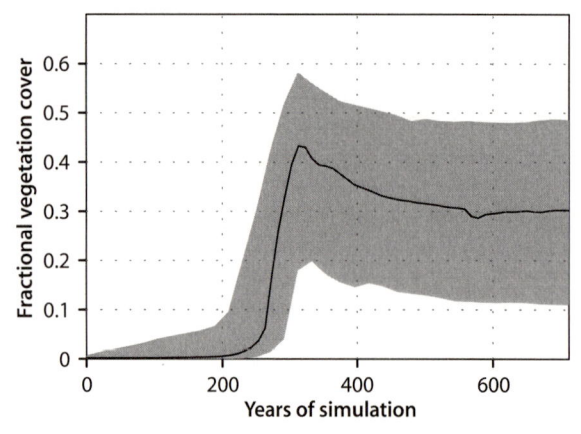

Fig. 23.4. Increase in fractional vegetation cover in the Sahara as a response to an increase in atmospheric CO_2 concentration simulated by an Earth System model of intermediate complexity. During the first 200 years of simulation, CO_2 changes are taken from reconstruction and measurements for the period of 1800 until 2000. Thereafter, CO_2 is prescribed to increase by 1% per year until a value of 1000 ppmv is reached, then the CO_2 concentration is kept constant at 1000 ppmv. The *full line* indicates a best guess. The *shaded area* indicates the uncertainty range owing to changes in the parameterisation of vegetation processes in the model

simulated vegetation cover changes in North Africa is less than estimated for mid-Holocene climate.

23.5 Outlook

North Africa is a fascinating example of a hot spot of atmosphere-vegetation interaction. Theory and model experiments suggest that this interaction is highly non-linear. It presumably affects Sahelian rainfall variability. Moreover, palaeoclimatic changes cannot be fully understood, when atmosphere-vegetation interaction is ignored. Whether North Africa will become greener again, as our stone-age ancestors have witnessed, cannot be forecasted because of model uncertainty and because socioeconomic boundary conditions are not known for the next centuries. There is some theoretical evidence that the Sahara could become greener, but direct anthropogenic land-cover change in North Africa is likely to be much more important than greenhouse-gas induced climate change in this respect.

Acknowledgements

We would like to acknowledge the team spirit of our CLIMBER-2 (Climate and Biosphere) model group at the Potsdam Institute for Climate Impact Research, which stimulates strong cooperation and fruitful discussion. Our colleagues are: Eva Bauer, Rainer Calov, Claudia Kubatzki, Vladimir Petoukhov, and Stefan Rahmstorf.

References

Braconnot P, Joussaime S, Marti O, deNoblet-Ducoudre N (1999) Synergistic feedbacks from ocean and vegetation on the African monsoon response to mid-Holocene insolation. Geophys Res Lett 26, No. 16:2481–2484
Brovkin V, Claussen M, Petoukhov V, Ganopolski A (1998) On the stability of the atmosphere-vegetation system in the Sahara/Sahel region. J G R 103 (D24):31613–31624
Charney JG (1975) Dynamics of deserts and droughts in the Sahel. Quart JR Met Soc 101:193–202
Claussen M (1997) Modelling biogeophysical feedback in the African and Indian Monsoon region. Climate Dyn 13:247–257
Claussen M, Gayler V (1997) The greening of Sahara during the mid-Holocene: results of an interactive atmosphere-biome model. Global Ecology and Biogeography Letters 6:369–377
Claussen M, Kubatzki C, Brovkin V, Ganopolski A, Hoelzmann P, Pachur HJ (1999): Simulation of an abrupt change in Saharan vegetation at the end of the mid-Holocene. Geophys Res Lett 24 (14)2037–2040
Doherty R, Kutzbach J, Foley J, Pollard D (2000) Fully coupled climate/dynamical vegetation model simulations over Northern Africa during the mid-Holocene. Climate Dyn 16:561–573
Ganopolski A, Kubatzki C, Claussen M, Brovkin V, Petoukhov V (1998) The influence of vegetation-atmosphere-ocean interaction on climate during the mid-Holocene. Science 280:1916–1919
Humboldt A von (1849) Ansichten der Natur. 3rd ed. J. G. Cotta, Stuttgart Tübingen, Reprint 1969. Stuttgart: P. Reclam, 20–21
Jolly D, Harrison SP, Damnati B, Bonnefille R (1998) Simulated climate and biomes of Africa during the late Quaternary: comparison with pollen and lake status data. Quat Sci Rev 17(6–7):629–657
Joussaume S, Taylor KE, Braconnot P, Mitchell JFB., Kutzbach JE, Harrison SP, Prentice IC, Broccoli AJ, Abe-Ouchi A, Bartlein PJ, Bonfiels C, Dong B, Guiot J, Herterich K, Hewit CD, Jolly D, Kim JW, Kislov A, Kitoh A, Loutre MF, Masson V, McAvaney B, McFarlane N, de Noblet N, Peltier WR, Peterschmitt JY, Pollard D, Rind D, Royer JF, Schlesinger ME, Syktus J, Thompson S, Valdes P, Vettoretti G, Webb RS, Wyputta U (1999) Monsoon changes for 6 000 years ago: Results of 18 simulations from the Paleoclimate Modeling Intercomparison Project (PMIP). Geophys Res Lett 26 (7):859–862
Köppen W (1936) Das geographische System der Klimate. In: Köppen W, Geiger R (eds) Handbuch der Klimatologie, Band 5, Teil C. Gebrüder Borntraeger, Berlin, 44 pp
Kutzbach JE, Guetter PJ (1986) The influence of changing orbital parameters and surface boundary conditions on climate simulations for the past 18 000 years. J Atmos Sci 43:1726–1759
deMenocal PB, Ortiz J, Guilderson T, Adkins J, Sarnthein M, Baker L, Yarusinski M (2000) Abrupt onset and termination of the African Humid Period: Rapid climate response to gradual insolation forcing. Quat Sci Rev 19:347–361
Petit-Maire N (1990) Will greenhouse green the Sahara? Episodes 13 (2):103–107
Prentice IC, Jolly D, BIOME 6000 members (2000) Mid-Holocene and glacial-maximum vegetation geography of the northern continents and Africa. J Biogeography 27:507–519
Wang G, Eltahir EAB (2000) Biosphere-atmosphere interactions over West Africa. 2. Multiple Equilibira. Q J R Meteorol Soc 126:1261–1280
Wasson RJ, Claussen M (2002) Earth system models: a test using the mid-Holocene in the Southern Hemisphere. Quat Science Rev, in press
Yu G, Harrison SP (1996) An evaluation of the simulated water balance of Eurasia and northern Africa at 6 000 y BP using lake status data. Climate Dyn 12:723–735
Zeng N, Neelin JD (2000) The role of vegetation-climate interaction and interannual variability in shaping the African savanna. J Climate, J Climate 13:2665–2670

Chapter 24

The Role of Large-Scale Vegetation and Land Use in the Water Cycle and Climate in Monsoon Asia

Tetsuzo Yasunari

24.1 The Asian Monsoon As a Huge Water Cycling System

The Asian monsoon is well-known as a land-atmosphere-ocean system. The land-ocean heating contrast between the Tibetan Plateau and the Indian Ocean produces strong southwesterly monsoon flow across the equator, which transports huge amounts of water vapour over south, southeast and east Asia. Another water vapour channel is located from the tropical Pacific Ocean toward East and Southeast Asia. Water vapour transport from the Indian and Pacific Oceans through these two main channels plays an essential role in maintaining large-scale cumulus convection and precipitation over humid Asia, or "monsoon Asia", which reinforces the monsoon circulation system through latent heat release. As result, a humid climate and green world dominate the eastern half of the Eurasian continent. This humid monsoon climate maintains a meridionally-oriented dense vegetation zone called the "green belt", from tropical Southeast Asia to sub-polar Siberia, which is one of the centres of biodiversity on the Earth, and a centre of world population.

24.2 Atmospheric Water Cycle over Monsoon Asia and the Eurasian Continent

Precipitation is the most important factor for life on land, i.e., for vegetation as well as a water resource for human beings. What, then, contributes to the origin of precipitation (P) in monsoon Asia and the interior of the Eurasian continent? In tropical monsoon Asia, water vapour transport (with its convergence) (C) should be a major moisture source for precipitation. Evapotranspiration (E) is another source for precipitation. The analysis of atmospheric water balance over various regions by using the objectively-analysed global meteorological data for 15 years (1979–1993) from the ECMWF (European Centre for Medium-range Weather Forecasts) revealed time-space characteristics of the P-C-E relationship. In the humid tropics, both C and E contribute nearly equally to P, or C is preferably a main source for P. In fact, over the Indian subcontinent and the Indo-China peninsula, the contribution of C is larger than that of E. In the mid to high latitudes of the continent, in contrast, E is a major source for P, though the amount of P (and E) itself is far smaller than that in the tropics. The interannual variability of P is highly correlated with C both in the tropical monsoon and the higher latitudes, associated with the variability of large-scale monsoon and atmospheric circulation. The large contribution of E to P in the mid and high latitudes strongly suggests the essential role of vegetation in the surface water (as well as CO_2) balance through transpiration. In the tropics, although C plays an important role in precipitation, E from the vegetated surface is equally important if we consider its larger amount compared to that in the mid/high latitudes.

24.3 Is Monsoon Rainfall Decreasing? The Impact of Deforestation on the Water Cycle in Thailand

The rainfall in the whole of Thailand in the late monsoon season (September) shows a remarkable decreasing trend since the 1950s. This decrease in rainfall may at least partly be attributed to the change of surface water balance induced by a change in the surface vegetation condition, i.e., the deforestation over the whole of the country as shown in Fig. 24.1. A recent model study (Kanae et al. 2001) has suggested that the decreasing trend of rainfall in September may be related at least partly to the recent deforestation in Thailand. That is, the deforestation over a wide region in the plain area of Thailand is likely to induce reduction of surface roughness and evaporation efficiency, and increase of albedo, which result in an increase in sensible heating and decrease in evaporation. This change of surface heat balance in turn affects atmospheric stability to decrease cumulus convection and rainfall. However, this effect of deforestation decreasing rainfall is noticeable only in September, when the southwest monsoon current has become weak. In July and August, when the monsoon current is still strong, the change of land surface condition by deforestation tends to change the distribution of rainfall rather than decrease the overall rainfall amount over the deforested area.

Fig. 24.1.
Deforestation across Thailand (shown here as proportional reduction in forested area) may be responsible for the sharp decrease in monsoon rainfall observed since the 1950s (source: Kanae et al. 2001)

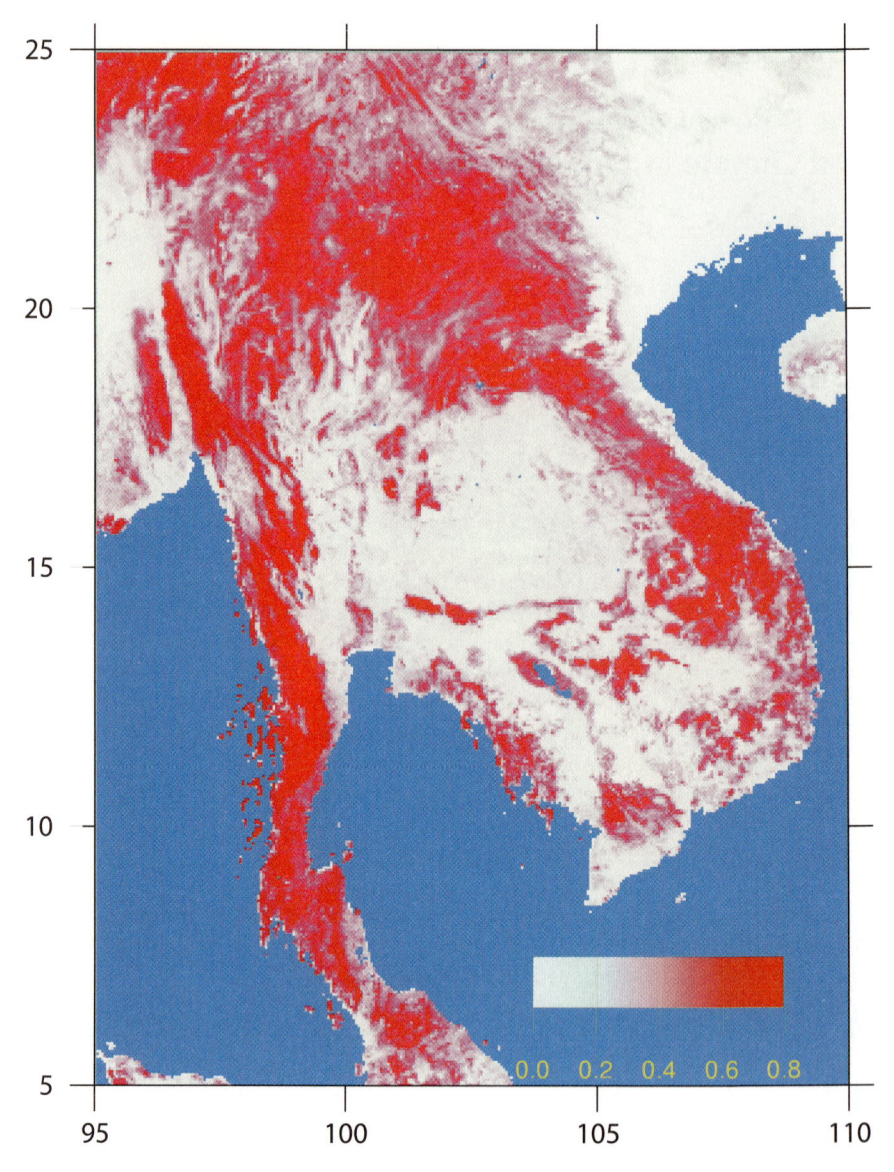

24.4 Do Water-Fed Rice Paddy Fields Increase Rainfall in Monsoon Asia?

Monsoon Asia is characterised by a traditional rice-cultivation society, which actually has enabled this region to support more than half of the population of the world. Rice paddy fields occupy a huge area in Southeast and East Asia. In Thailand, we have suggested that deforestation and widespread mixed shrub/paddy fields may reduce rainfall by making the land surface smoother and decreasing evaporation efficiency compared to the forest zone. In this case, the Bowen ratio (ratio of sensible heat to latent heat) at the surface has increased with deforestation, and the moist static stability of the atmosphere increased, thus reducing rainfall. On the other hand, water-fed paddy fields may function to increase evaporation compared to drier farmland and fields. This may be the case in the southern part of the China plain. In June and July, the China plain is strongly affected by the Meiyu (or Baiu in Japanese) frontal activity, which is a major part of the Asian monsoon system in East Asia. The southwest monsoon current from South and Southeast Asia is a main moisture source for rainfall in the Meiyu frontal zone.

In addition, most (more than 80%) of the target area is occupied by water-fed paddy fields in this rainy season. A recent model study based on the observational evidence of surface energy and water fluxes has shown an interesting result that the wide area of water-fed rice paddy fields in the southern and central China plain may play a key role in enhancing deep cumulus convection and rainfall there (Shinoda and Uyeda 2001). Based upon the data of GAME-HUBEX meso-scale and surface experiments, they conducted numerical simulations of the cloud and precipitation system using a cloud-resolving

Fig. 24.2.
Land use affects rainfall: simulations by Shinoda and Uyeda (2001) show that where rice paddy fields exist (*upper graph*), the atmospheric boundary layer becomes wetter than that for the farmland (*lower graph*), and deep convection easily develops to produce strong rainfall

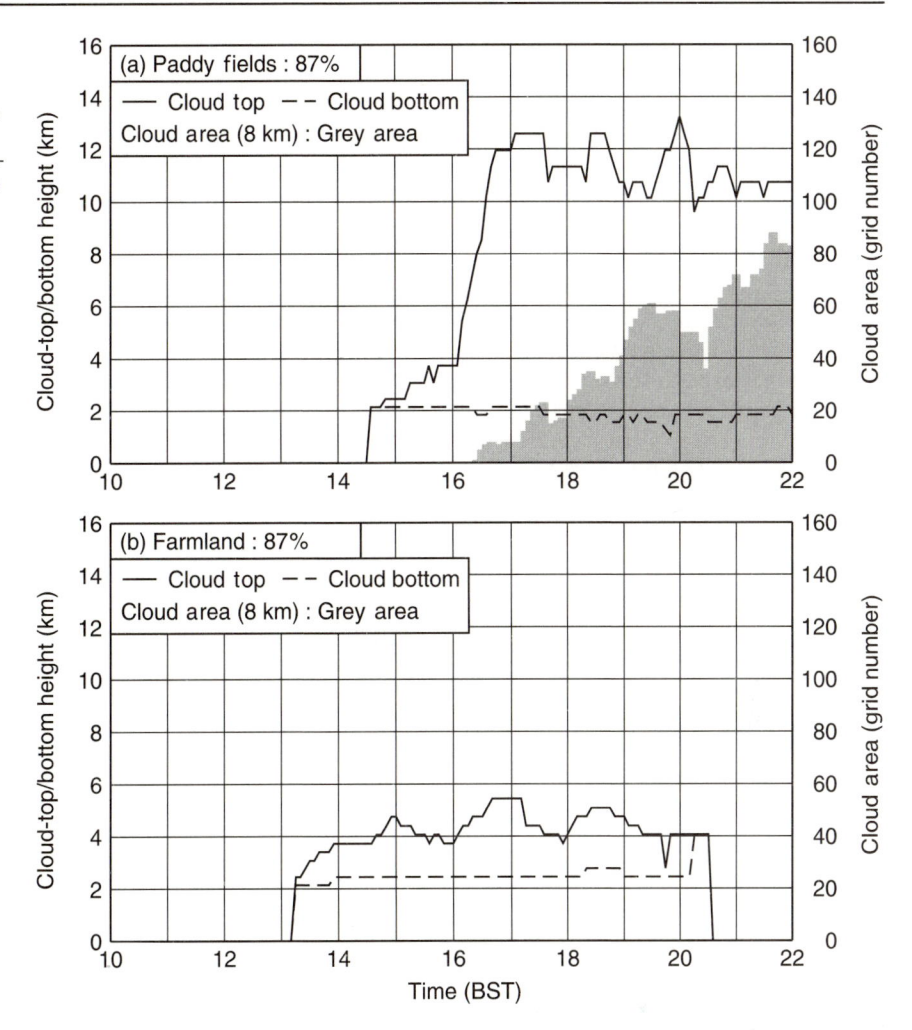

meso-scale model with two different boundary conditions: one with water-fed paddy fields, where evaporation (latent heat flux) dominates the sensible heat flux, and another with farmland conditions where sensible heat flux dominates. The experiments showed that in the paddy rice field case, the atmospheric boundary layer becomes wetter than that for the farmland, and deep convection easily develops to produce strong rainfall, as shown in Fig. 24.2. This result suggests that the rain-fed rice paddy fields in East and Eoutheast Asia provide a landscape well-harmonised with the monsoon climate through the positive hydrological feedback between the surface and the atmosphere.

24.5 A Hydro-Climate Memory Effect of the Taiga-Permafrost System in Siberia

It is well-known that one climatic zone corresponds well with one dominant vegetation type (e.g., of Koppen's). However, we should also note that the vegetation over a broad region of continents may, in turn, affect the atmosphere and surface climate. This implies that the vegetation and climate, at least on a large spatial-scale, should be considered as an interactive system rather than a one-way relation from one to the other.

Eastern Siberia is a broad permafrost zone, and correspondingly, the sub-polar boreal forest, Taiga, is distributed on it. The photosynthetic activity and associated evapotranspiration rate of this Taiga zone are very high in the summer season. Long-term energy and water flux measurements in the typical Taiga forest of the Lena River basin have revealed some interesting features in the seasonal change of this eco-climate system. In spring, snowmelt occurs in April to May, but the sensible heat flux was dominant and the latent heat flux was very small or negligible for a while even after the disappearance of snowmelt. The latent heat flux suddenly increased in June when the foliation of trees started. Why did the evapotranspiration start so suddenly? The perfect answer to this question has not been given yet, but the melting process of permafrost and the root depth of the trees are likely to be closely related to this problem; i.e., when the depth of the melted layer reached about 20 cm, where a large proportion of larch roots are distributed in the soil, the foliation and transpiration from the new leaves may suddenly have started. Most of the evapotranspiration occurred as the transpiration from the Taiga forest.

On the other hand, we already noted that most water vapour for precipitation in summer is occupied by evapotranspiration from the surface of this area, implying that most of the water is recycling between the vegetation and the atmosphere in summer. We also found a significant positive correlation between evapotranspiration in summer and precipitation in summer two years previously (Yasunari and Yatagai 2001). This strongly suggests that the permafrost has a hydro-climatic memory effect, so that an anomalous soil moisture memory in summer, produced as a result of the water cycle of the Taiga ecosystem, is retained through freezing/melting processes in winter/spring, and conveyed to the succeeding summers. We are considering whether this symbiotic system might be seriously affected by the recent remarkable warming in winter and spring over this particular region due to changes in the seasonal timing of permafrost melting, foliation etc.

These hydro-meteorological studies in Siberia have strongly suggested a symbiotic system of climate and vegetation through the hydrological cycle, where Taiga maintains itself by recycling water between the atmosphere and the vegetated surface of the permafrost layer, while the permafrost is also maintained by the Taiga through suppressing sensible heating in summer. The recent rapid warming in Siberia, which may be inducing melting of permafrost, may affect this "Taiga-permafrost symbiotic system".

References

Kanae S, Oki T, Musiake K (2001) Impact of deforestation on regional precipitation over the Indochina Peninsula. J Hydrometeor. 2:51–70

Shinoda T, Uyeda H (2001) Generation of deep convection over Eastern China during the summer monsoon. Submitted to J Meteor Soc Japan

Yasunari T, Yatagai A (2001) A hydro-climatic memory effect of Taiga-Permafrost ecosystem in Eastern Siberia. Submitted to *Nature*

Chapter 25

Can Human-Induced Land-Cover Change Modify the Monsoon System?

Congbin Fu

Two aspects of the climate-land cover interaction in the Asian monsoon region are particularly strong. In terms of natural processes, the highly variable monsoon climate, especially precipitation and temperature, forces changes in the function and structure of terrestrial ecosystems on various time scales through changing their physiological processes (e.g., Fu and Wen 1999). The long-term monsoon climate changes can even alter the bio-geographic distributions of ecosystems (An et al. 1990). These changes in the meantime bring about feedback effects on the monsoon climate. In addition, more than half the world's population lives in the Asian monsoon region. The long history of civilisation has caused significant changes of land cover across Asia. It is likely that anthropogenic modification of the monsoon system will occur through changing the surface fluxes of energy, water and greenhouse gases under the different land-use patterns.

25.1 History of Land-Cover and Land-Use Changes over East Asia

Owing to increasing population, industrialisation and urbanisation, natural ecosystems in Asia such as forests, grasslands and wetlands have been encroached upon by farmland and other man-made ecosystems on a large scale. For example, there has been significant expansion of farmland in China since the 11th Century BC (Deng et al. 1983). In the early days, all the country was covered by various kinds of natural ecosystems, except for a very narrow band of farmland in the lower reaches of the Yellow River basin. The agricultural area was expanded gradually both southward and northward and to the upper reaches of the major river basins. Around the late 19th century, the land-use pattern was generally similar to the land-use pattern today, although land cover has continuously changed until recently. Now man-made ecosystems cover nearly 80% of the total terrestrial area. Such human-induced land-cover changes have been as great in Asia as in other areas of the world, if not greater. To what extent has the monsoon system been modified by such changes? In this paper, numerical experiments with a high-resolution regional climate model have been performed to examine the most likely response of the Asian monsoon system to human-induced land cover change.

25.2 Design of the Numerical Experiments

The natural vegetation has been so altered in East Asia over millennia that its reconstruction other than by modelling is rather difficult. However, it is feasible to specify the equilibrium climax vegetation that may be

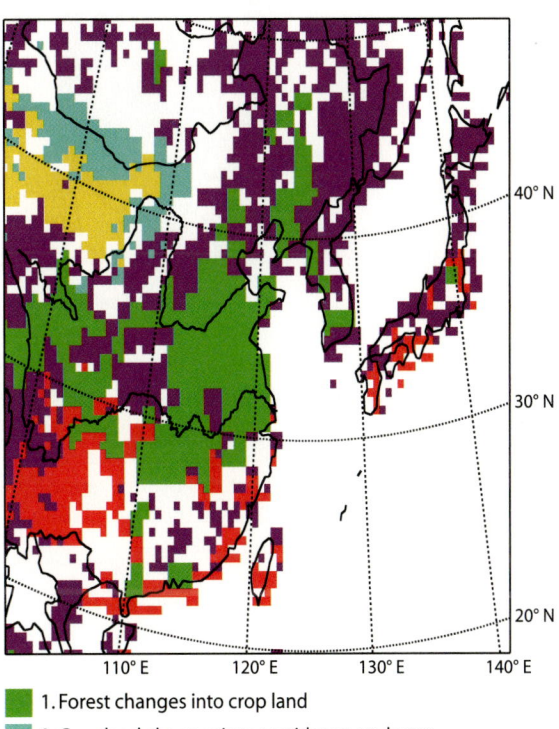

Fig. 25.1. The main change from potential vegetation to current vegetation

1. Forest changes into crop land
2. Grassland changes into semidesert or desert
3. Semidesert changes into desert
4. Evergreen broadleaf tree changes into mixed forest
5. Other changes

Fig. 25.2.
Changes of four main surface parameters: albedo, surface roughness, leaf area index and total vegetation fractional coverage, from potential to current vegetation distribution

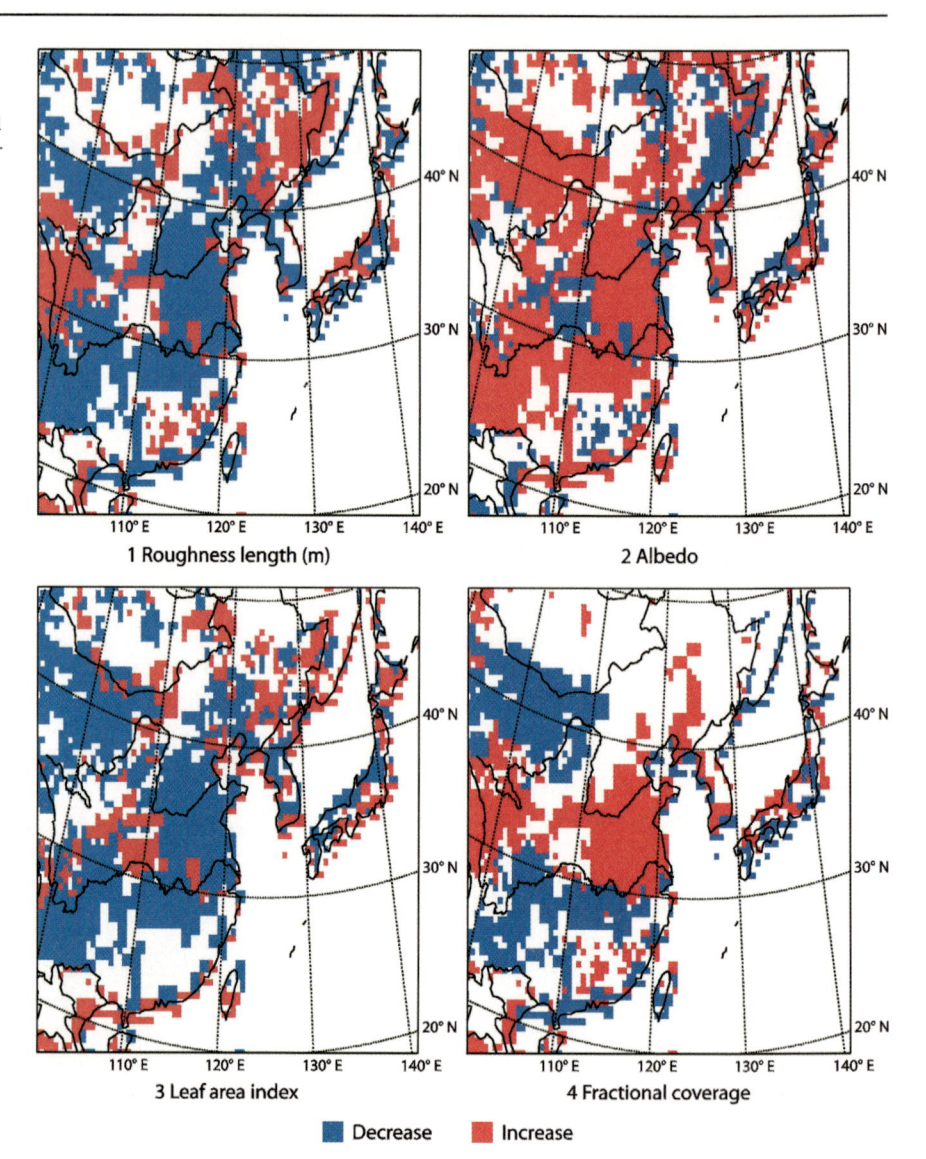

1 Roughness length (m)   **2 Albedo**

3 Leaf area index   **4 Fractional coverage**

■ Decrease   ■ Increase

expected at present, based on the prevailing climate with a biome-climate matching approach, which is widely used in the international ecological science community. This kind of vegetation cover information is named the "potential vegetation" (Vp), since all the man-made ecosystems were subtracted in the process.

Remote sensing information is now widely used to provide maps of vegetation cover with rather high temporal resolution. The current vegetation used in this paper is based on the global land cover classification data, which is part of the Global Data Sets for land-atmosphere models, ISLSCP initiative 1, developed in the International Satellite Land Surface Climatology Project (ISLSCP) (Messon et al. 1995). Although one has to realise that the actual current vegetation cover will be somewhat different to this 1987–1988 data set, these differences are relatively smaller than the differences in the potential vegetation as described in previous paragraphs. Therefore, we assume that this vegetation cover data (Vc) can be used to approximate the actual current vegetation cover. The human-induced land-cover change in this paper is defined as the difference between the potential and current vegetation as presented in Fig. 25.1. More than 80% of the region has been affected by conversion of various categories of natural vegetation into farmland, grassland into semi-desert, and by widespread land degradation. The most pronounced changes occur in Northwest China, where the grassland has been changed into semi-desert or desert and in East China where the forest has been replaced by cropland. There are also significant changes across Japan.

A pair of numerical experiments was performed under the above two land-cover conditions by using the Regional Integrated Environmental Model System (RIEMS) version 1 (Fu et al. 2000). This model consists of three main components of the climate system: a meso-

Fig. 25.3.
Mean changes of vector wind and geopotential height in the lower atmosphere (850 hPa) in summer (current-potential veg.) over East Asia (difference of wind vector (850 mb) in m s^{-1}; difference of high field (850 mb) in geopotential metre)

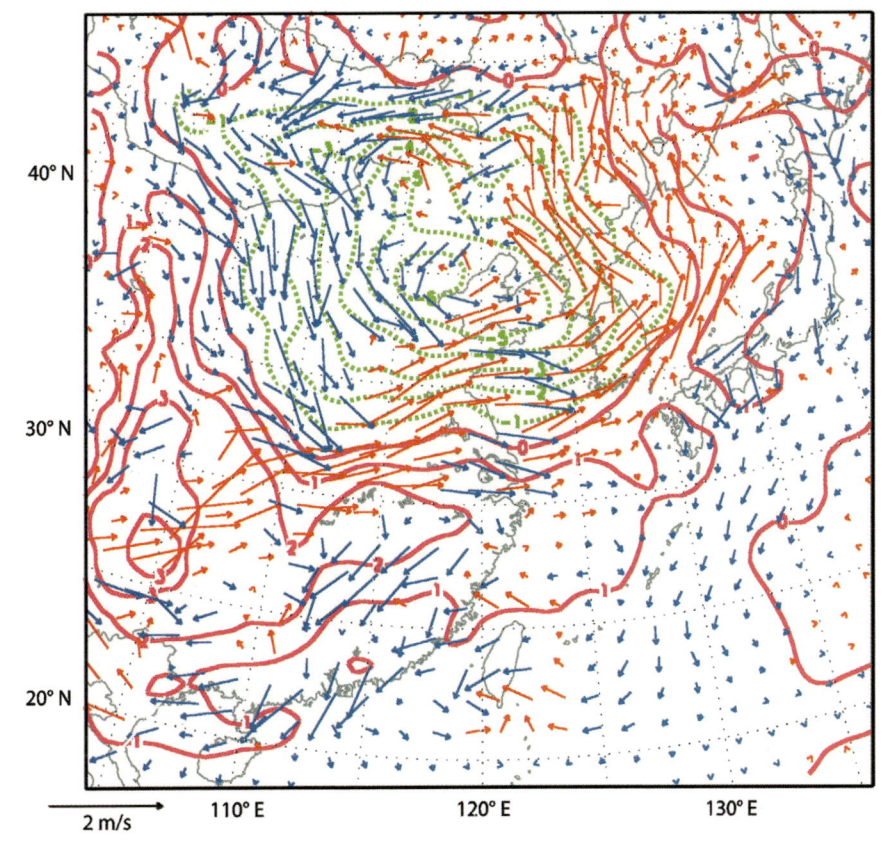

scale atmospheric dynamic model, a radiation scheme and a land surface scheme, and two components of major regional anthropogenic forcing factors: land cover changes and changes of greenhouse gases and aerosol emissions. The validation study has shown the model to perform reasonably well in its capacity to simulate the regional climate in the Asian monsoon region (Fu et al. 1998).

The differential fields of integration by the two vegetation cover data sets (current minus potential vegetation) are used to represent the impacts of changing natural vegetation, since the two simulations are identical to each other for all the conditions, including the large scale driving fields used as the initial and lateral boundary conditions and the parameters of all physical processes except for the vegetation cover. In order to maintain the linkages between the large-scale environment and the simulated region, a relaxation scheme with ten buffer zones is applied for nesting at the lateral boundary.

25.3 Changes of Surface Dynamic Parameters under Two Types of Vegetation Cover

Changes of four main surface parameters: albedo, surface roughness, leaf area index and total vegetation fractional coverage, from potential to current vegetation distribution, are shown in Fig. 25.2. In the areas where the natural vegetation (mainly forests) has been turned into farmland or where grassland has been turned into semi-desert or desert, a significant decrease of surface roughness and leaf area index but an increase in albedo are observed. The total fractional vegetation coverage is higher in farmland area than in natural forests, but it is lower in the area of semi-desert and desert in comparison with grassland, and lower in the area of mixed forests in comparison with evergreen broadleaf forests. Changes of these surface parameters have modified the exchanges of energy and water between the land surface and the atmosphere and resulted in changes in atmospheric circulation.

25.4 Changes of East Asian Monsoon by Human-Induced Land-Cover Changes

Figure 25.3 presents the mean changes of vector wind and geopotential height in the lower atmosphere (850 hPa) in summer over East Asia. The weakening of the monsoon depression is shown by the positive anomalies in the region to the south of 30° N, and the weakening of the summer monsoon is shown by the northerly anomalous flow. There is a negative departure of the height over the northern part of the domain, representing the development of a low-pressure system there, which brings about the anomalous northwest flow. There are also changes of mean meridional and zonal circulations over the region, thus enhancing descending motion flows over 35–40° N and 100–115° E respectively, which in turn would prevent the development of the summer monsoon circulation.

Both these two northerly anomalous flows and the enhanced descending motion over East Asia would prevent moisture transport northward and the development of convective activities. This would result in a drier atmosphere over most of the domain, as indicated by the differential field of the specific humidity at 850 hPa and related changes of surface climate. All components of the surface water cycle such as precipitation, runoff and soil moisture are reduced over most parts of the region. This indicates the weakening of the water cycle under conditions of deteriorated natural vegetation.

On the contrary, the winter monsoon over East Asia becomes stronger when natural vegetation cover deteriorates, as shown by the strong anomalous northerly flow in the differential fields of vector wind and geopotential height at 850 hPa. This circulation pattern would bring a dry and cold air mass from inland areas down to all regions of East Asia and result in changes of surface climate, such as the reduction of atmospheric humidity and precipitation (mostly in the southern part of the region), and cold temperatures over almost the whole region.

25.5 Conclusions

According to the above analysis, human-induced land cover changes have modified the monsoon circulation by weakening the summer monsoon and strengthening the winter monsoon over East Asia, which results in related changes of surface climate over the region. This conclusion from numerical experiments is supported by observational evidence. For example, the time evolution of the aridity index over East China since 1880 has shown a significant trend of aridification during the last 120 years, with a period of 36 years oscillation (Brucker period) superimposed on it (Fu 1994). Since the moisture condition over East China is mainly related to the intensity of the summer monsoon, this reflects the weakening of the summer monsoon in that period. A 25 000-year lake level data set for the Daihai Lake in Inner-Mongolia shows a significant reduction of its level beginning about 4 000 years ago (Wang 1990). This is also an indication of an aridification trend over Northern China and therefore the weakening of the summer monsoon since then.

It seems that the deterioration of natural vegetation due to the development of human society is perhaps one of the anthropogenic factors superimposed on the natural variability of the monsoon system.

References

An ZS, et al. (1990) A preliminary study on Paleoenvironment change of China during the last 20 000 years. In: Liu (ed) Leoss, Quaternary geology, global change. Science Press, Beijing, pp 1–26

Deng JZ, et al. (1983) General view of agriculture geography of China. Science Press

Fu CB (1994) An aridity trend in China in association with global warming. In: Zepp RG (ed) Climate-biosphere interaction: Biogenic emission and environmental effects of climate change. John Wiley & Sons, Inc

Fu CB, Wen G (1999) Variation of ecosystems over East Asia in association with seasonal, interannual and decadal monsoon climate variability. Climatic Change 43:477–494

Fu CB, Wei HL, Chen M, Su BK, Zhao M, Zheng WZ (1998) Simulation of the evolution of summer monsoon rainbelts over East China from regional climate model, Scientia Atmospherica Sinica, 22:522–534

Fu CB, Wei HL, Qian Y, et al. (2000) Documentation on a regional integrated environmental model system (RIEMS, version 1). TEACOM Science Report No. 1, START Regional Committee for Temperate East Asia, Beijing, China, pp 1–26

Messon BW, Corprew FE, McManus JMP, et al. (1995) ISLSCP initiative I – global data sets for land atmosphere models 1987–1988, vol. 1–5, published by USA NASA, GDAAC, ISLSCP, 001-USA, NASA, GDAAC, ISLSCP, 005

Wang SM (1990) Daihai Lake and climatic change. University of Science and Technology Press, China

Chapter 26

The Amazon Basin and Land-Cover Change: A Future in the Balance?

Carlos A. Nobre · Paulo Artaxo · Maria Assunção · F. Silva Dias · Reynaldo L. Victoria · Antonio D. Nobre · Thelma Krug

The Amazon Basin contains a multitude of ecosystems, biological and ethnic diversity and the largest extent of tropical forest on Earth, over 5×10^6 km², and accounts for an estimated 1/3 of the planet's animal and plant species. Currently only a small number of species are used by man. The region is abundant in water resources. Annual rainfall is 2.3 m over the Amazon Basin, and the mean outflow of the Amazon River into the Atlantic is over 200 000 m³ s⁻¹, which corresponds to 18% of the total flow of fresh water into the world's oceans. The region stores over 100 Gt of carbon in vegetation and soils. However, over the past 30 years, rapid development has led to the deforestation of over 550 000 km² in Brazil alone. Current rates of annual deforestation are in the range of 15 000 km² to 20 000 km² (INPE 2001), and the spatial pattern of deforestation in Brazilian Amazonia up to 1997 is illustrated in Fig. 26.1.

26.1 Land-Use Change

Amazonia has been inhabited by human populations from time immemorial. At the time of the arrival of European colonisers in the 16th Century, it is estimated that several million indigenous people lived in the region, and their way of life did not cause widespread changes in the ecosystems. The modern occupation of Amazonia started around 1540, but until the end of World War II, human presence in the Amazonian environment brought almost no changes to its natural vegetation cover. A large roadbuilding effort started in the late 1950s and expanded through the 1970s and 1980s, opening up large areas of the forest to agricultural development. Millions of immigrants rushed to Amazonia. The population of

Fig. 26.1.
Analysis of spatial deforestation patterns up to 1997 (*in white*) in Brazilian Amazonia superimposed on a vegetation map (courtesy of R. Alvala and E. Kalil, INPE, Brazil)

Fig. 26.2.
The colonist footprint: Average deforestation trajectories across cohorts (after Moran and Krug 2001, Global Change News)

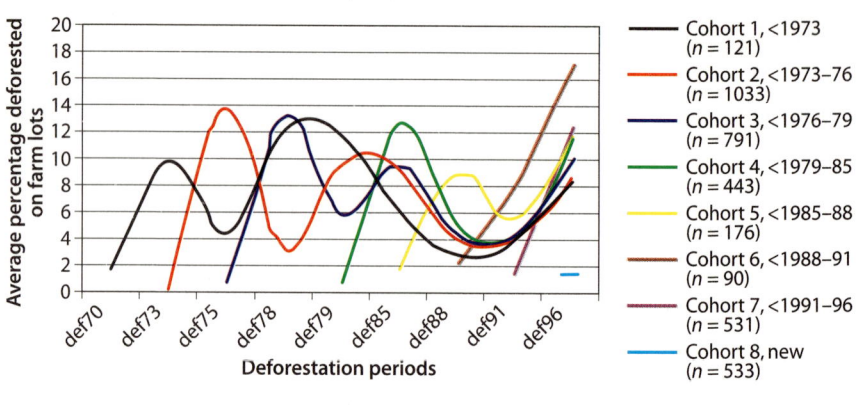

— Cohort 1, <1973 ($n = 121$)
— Cohort 2, <1973–76 ($n = 1033$)
— Cohort 3, <1976–79 ($n = 791$)
— Cohort 4, <1979–85 ($n = 443$)
— Cohort 5, <1985–88 ($n = 176$)
— Cohort 6, <1988–91 ($n = 90$)
— Cohort 7, <1991–96 ($n = 531$)
— Cohort 8, new ($n = 533$)

Brazilian Amazonia has grown from 3.5 million in 1970 to close to 20 million in 2000. That, combined with a policy of fiscal incentives for the establishment of large cattle ranches in Brazil, which was in place through the early 1990s, caused a sharp increase in deforestation rates, which present a quite complex temporal and spatial pattern as illustrated in Fig. 26.2 (Moran and Krug 2001). The underlying pressures to continue land-use change are still present: a growing population in the developing nations of Amazonia, plans for an expansion of the road network criss-crossing the region, and plans for expansion of large-scale, intensive agriculture in the region. Furthermore, the lack of sustainable agriculture in Amazonia has driven countless peasants out of their agricultural plots and into mining (gold, cassiterite, diamonds, etc.), and into a number of other activities scattered all over the basin, creating a large number of areas of spontaneous development and deforestation. Additionally, these displaced peasants and farmers contributed to the uncontrolled swelling of the Amazonian cities, where the majority of the newly urban population is poverty-stricken.

Analyses on why the various attempts on agricultural developments for the region over the last 100 years failed clearly indicate that the model of agriculture development brought to the region, imported from subtropical and drier regions of Brazil, was not appropriate. The results of the agricultural and mining development models adopted are socially, environmentally and economically adverse. They resulted in over 550 000 km^2 of loss of forests, over 200 000 km^2 of unproductive, degraded pastures, a high rate of deforestation, emissions of large amounts of greenhouse gases and aerosols due to deforestation and biomass burning, widespread mercury pollution throughout the rivers of the basin due to gold prospecting with potentially very serious long-term consequences on the heath of fish-eating populations, an exponential increase in the number of malaria infections, health problems associated with smoke and aerosols produced by biomass burning, widespread poverty for most of the rural population, low agricultural productivity, mostly illegal selective logging, etc.

26.2 Impacts on the Carbon Cycle

A number of field studies carried out over the last 20 years showed significant, but still localised changes in water, energy, carbon and nutrient cycling, and in the atmospheric composition caused by deforestation and biomass burning. For instance, the forest is important in recycling water vapour through evapotranspiration throughout the year, thus contributing to augmented rainfall and to its own maintenance. An important impact of land-use and land-cover change in Amazonia, with global consequences, is its emissions of carbon dioxide due to deforestation and biomass burning. The total annual emissions of CO_2 in Amazonia due to land use and land-cover change range between 150 to 250 Mt C (Houghton et al. 2000). In comparison, the total annual emissions of CO_2 arising from fossil fuel combustion is only about 75 Mt C for Brazil as a whole. On the other hand, carbon cycle studies of the Large-Scale Biosphere-Atmosphere Experiment (LBA) and forest inventory studies (Phillips et al. 1998) indicate that the undisturbed forest may be a sink of carbon at rates from 1 to 7 t C ha^{-1} yr^{-1} (Malhi et al. 1998; Malhi et al. 1999; Nobre et al. 2000; Araujo et al. 2001). However, it is still uncertain whether the forest functions as a sink or source of carbon to the atmosphere (Keller et al. 1996). There is a recent suggestion (Richey 2001, pers. comm.) that emission of CO_2 from rivers, streams and wetlands may be much higher than previously thought, contributing to about 1 t C ha^{-1} yr^{-1}. Figure 26.3 shows a preliminary 'synthesis' of our rather incomplete knowledge of the carbon cycle in Amazonia. Overall, if the result of the undisturbed forest behaving as a carbon sink is confirmed with further research, it sheds some new light on the role of the tropical forests for the global carbon balance, and could be considered as one more environmental service provided by the forest.

An important scientific question is whether this biotic CO_2 sink will saturate sometime this century due to global warming, that is, the undisturbed forest could become a source of carbon due to rapid decomposition

Fig. 26.3.
Preliminary synthesis of the carbon cycle for Amazonian forests. Units: t C ha^{-1} yr^{-1}. GPP = gross primary productivity; R_a = autotrophic respiration; R_h = heterotrophic respiration; VOC = volatile organic carbon compounds (source: LBA Experiment, courtesy of LBA investigators at Alterra, INPA, IH, Edinburgh University, Washington University)

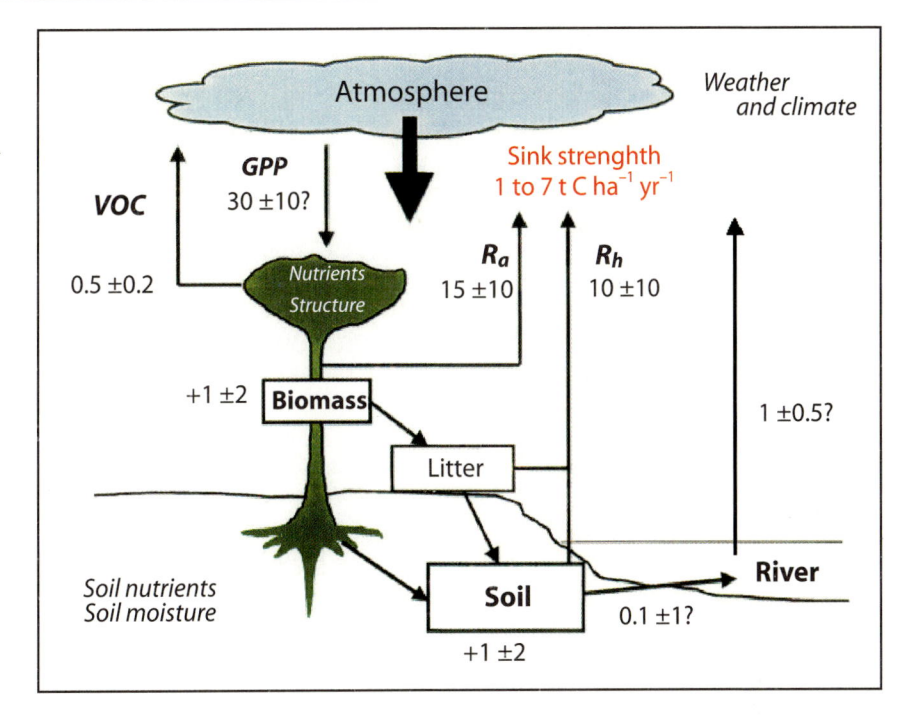

[Diagram: Atmosphere cloud connected to Biomass via VOC (0.5 ±0.2), GPP (30 ±10?), R_a (15 ±10), R_h (10 ±10); Sink strength 1 to 7 t C ha^{-1} yr^{-1}; Weather and climate arrow (1 ±0.5?); Biomass (+1 ±2) → Litter → Soil; Nutrients Structure; Soil nutrients, Soil moisture; Soil (+1 ±2) → River (0.1 ±1?)]

of soil carbon on a warmer climate. Scenarios of climate change due to global warming indicate a climate 4–6 °C warmer for Amazonia towards the end of the century (Carter and Hulme 2000). This pronounced warming can have severe impacts in terms of ecosystem maintenance. There are suggestions that large deforestation (Nobre et al. 1991) or climate change (Cox et al. 2000) may lead to drastic change of vegetation in Amazonia, primarily a tendency for its "savannisation". It is also becoming increasingly evident that forest fragmentation due to selective logging and other land-use changes are making the forest more susceptible to fires (Nepstad et al. 1999). This susceptibility would increase further with a warmer climate. The result may be more forest loss due to uncontrolled forest fires such as the largest forest fire in Brazilian history, which burned 14 000 km^2 of forest in Roraima, from January through March 1998.

26.3 Impacts on the Atmosphere

The oxidising power of the global atmosphere as measured by the hydroxyl radical (OH) has been decreasing during the 1990s. The reasons are still unknown. The tropical forest emits organic compounds that are precursors in the production of the OH radical and also may play a role in the generation of biogenic aerosols (Kesselmeir et al. 2000). It is therefore important to measure emissions of these compounds and increase understanding of their chemistry in the tropical atmosphere. There is also some observational evidence of subregional changes in the surface energy budget and boundary layer cloudiness and regional changes in the lower troposphere radiative transfer due to biomass burning aerosol loadings. There is strong reduction of solar radiation at the surface in excess of 30 W m^{-2} over several months during the dry season due to aerosol absorption and scattering (Tarasova et al. 2000). At the surface, a small amount of cooling is observed. Throughout the boundary layer, black carbon absorbs solar radiation and heats up the atmosphere. That may have implications to the stability of the boundary layer.

A larger number of cloud condensation nuclei (CCN) due to biomass burning has led to the speculation of their possible role in cloud formation and rainfall. During the rainy season, in contrast, there are very few CCN, and they are of biogenic origin (Artaxo et al. 2001). Over western Amazonia during the wet season, cloudiness, rainfall and wind systems present two rather contrasting behaviours: the so-called westerly and easterly regimes. During the westerly regimes, the low-level wind has a westerly component that is associated with the establishment of the South Atlantic Convergence Zone (SACZ), and the Amazonian clouds show characteristics of oceanic clouds; that is, relatively low cloud tops, low CCN amounts, large cloud drop size and less lightning (Silva Dias et al. 2001). During the transition season from dry to wet and during periods of prevailing easterly winds in the wet season, the clouds present characteristics of continental clouds; that is, high cloud tops, lightning, higher CCN amounts, smaller cloud drop size (Roberts et al. 2001). Figure 26.4 illustrates the differences in CCN amounts for the different seasons and wind regimes.

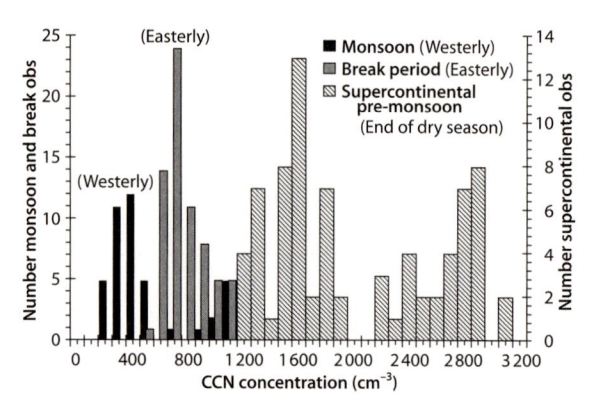

Fig. 26.4. Distribution of cloud condensation nuclei (CCN) over western Amazonia (Rondonia) for different climate regimes. Notice that CCN amounts increase substantially from the relative clean conditions during the westerly regime (surface winds with a westerly component) to the easterly regime and to the polluted atmosphere at the end of the dry season (September–October) (courtesy of E. Williams)

26.4 Impacts on Water Chemistry

In terms of alterations of water chemistry due to land-use and land-cover change, LBA studies (Victoria et al. 2000; Richey et al. 2001) have shown a higher concentration of ions in rivers cutting across pasture areas in southwestern Amazonia.

26.5 The Future

Agriculture and agro-forestry will only succeed in Amazonia when their practices incorporate and mimic the complex interdependent relationships that exist in tropical ecosystems. In reality, most basic knowledge on the biological, physical and chemical functioning of the many tropical ecosystems of Amazonia is still missing. Therefore, a necessary precondition for development of applications in sustainable agriculture and sustainable forestry in Amazonia is the establishment of research programmes to unveil the basic functioning of the tropical ecosystems, to assess the ecosystems' resilience to perturbations, primarily deforestation, fire, forest fragmentation, etc., and to understand in depth the natural cycles of water, carbon, trace-gases, nutrients and how these cycles are altered by land-use changes. The lack of this knowledge prevents the development of strategies for sustainable agriculture and forestry on a sound and robust scientific basis. As a consequence, high deforestation rates continue, because they provide small gains on the short term for the peasants, and also small gains (but ownership of large properties) to cattle ranchers. Is it strategically essential for the Amazonian countries to bring to a halt the process of occupation and development of Amazonia, which took place over the last 50 years, because it is predatory on the natural resources of the region. Development can only be successful once sustainable alternatives for the current unsustainable agricultural practices are found, tested and implemented across the Basin. Last but not least, a new focus on sustainable development in Amazonia must place a strong emphasis on education.

In summary, 70 years of failures in attempts to develop Amazonia (large-scale plantations and cattle ranching, small scale subsistence agriculture, logging, mining, and industrialisation) has led to land-use change that is causing an unprecedented imbalance in Amazonia. This imbalance is expressed mostly in terms of: biodiversity losses of unknown magnitude, and significant alterations of natural cycles of water, carbon, trace gases, aerosols, and nutrients. Is it possible to halt the "destructive path"? It must be recognised that lack of knowledge on sustainable agriculture in the tropics lies at the heart of the problem. To improve on that, in addition to applied research on tropical agroforestry, basic knowledge of ecosystem functioning is necessary (e.g., LBA), but it is not sufficient. Educating the poor and displaced for sustainable development is needed to provide the only way out of a life of continued poverty, land abandonment and deforestation.

Acknowledgements

We want to express our gratitude to all of the scientific community of the LBA Experiment. Their diligent work is enhancing significantly the knowledge basis of Amazonia, and ultimately this will prove essential to the sustainable development of Amazonia.

References

Andrea MO, Artaxo P, Brandão C, Carswell FE, Ciccioli P, Costa AL, da Culf AD, Esteves JL, Gash JHC, Grace J, Kabat P, Lelieveld J, Malhi Y, Manzi AO, Meixner FX, Nobre AD, Nobre C, Ruivo M de LP, Silva Dias MAF, Srefani P, Valentini R, Jouanne J von (2001) Towards an understanding of the biogeochemical cycling of carbon, water, energy, trace gases and aerosols in Amazonia: The LBA-EUSTACH experiments. Accepted for publication in J Geophysical Res

Artaxo P, Martins JV, Yamasoe MA, Procópio AS, Paulivequis TM, Andreae MO, Guyon P, Gatti LV, Leal AMC (2001) Physical and chemical properties of aerosols in the wet and dry seasons in Rondônia, Amazônia. Accepted for publication in J Geophysical Res

Araujo AC, Nobre AD, Kruijtz B, Culd AD, Stefani, Elber J, Dallarosa, Randow C, Manzi AO, Valentini R, Gash JHC, Kabat P (2001) Dual tower long-term study of carbon dioxide fluxes for a Central Amazonian rain forest. Submitted to J Geophysical Res

Carter T, Hulme M (2000) Interim characterizations of regional climate and related changes up to 2100 associates with the provisional SRES marker emissions scenarios. IPCC Secretariat, c/o WMO, Geneva, Switzerland

Cox PM, Betts RA, Jones CD, Spall SA, Totterdell IJ (2000) Acceleration of global warming due to carbon-cycle feedbacks in a coupled climate model. Nature 408:184–187

Houghton RA, Skole D, Nobre CA, Hackler J, Lawrence K, Chomentowski W (2000) Annual fluxes of carbon from deforestation and regrowth in the Brazilian Amazon. Nature 403:301–304

Instituto Nacional de Pesquisas Espaciais (INPE) (2001) Monitoramento da Floresta Amazônica Brasileira por Satélite: 1998–2000. São José dos Campos, SP, Brasil. Available at *http://www.inpe.br/Informacoes_Eventos/amazonia.htm*

Keller M, Clark DA, Clark DB, Weitz AM, Veldkamp E (1996) If a tree falls in the forest … Science 273:201

Kesselmeier J, Kuhn U, Wolf A, Andreae MO, Ciccioli P, Brancaleoni E, Frattoni M, Guenther A, Greenberg J, Castro, Vasconcellos P de C, Oliva T, Tavares T, Artaxo P (2000) Atmospheric volatile organic componds (VOS) at a remote tropical forest site in central Amazonia. Atmospheric Enviroment 34:4063–4072

Malhi Y, Nobre AD, Grace J, Kruijt B, Pereira MGP, Culf A, Scott S (1998): Carbon dioxide transfer over a Central Amazonian rain forest. Journal of Geophysical Research, D24:31593–31612

Malhi Y, Baldochi DD, Jarvis PG (1999) The carbon balance of tropical, temperate and boreal forests. Plant Cell Environ 22:715–740

Moran E, Krug T (2001) Predicting location and magnitude of land use and land change. IGBP Newsletter 45:4–8

Nepstad DC, Verissimo A, Alencar A, Nobre CA, Lima E, Lefebvre P, Schlesinger P, Potter C, Moutinho P, Mendonza E, Cochrane M, Brooks V (1999) Large-scale improverishment of Amazonian forests by logging and fire. Nature v. 398, n. 6727, pp 505–508

Nobre A, Malhi Y, Araujo AC, Culf AD, Dolman AD, Elbers J, Kruijt B, Randow C, Manzi AO, Grace J, Kabat P (2000) Multiyear comparative analysis of NEP and environmental factors for Manaus rainforest: "La Niña" influence on CO_2 uptake. First LBA Science Conference, 25–30 June 2000, Belém, PA, Brazil. Available at em: *http://sauva.cptec.inpe.br/posters/*

Nobre CA, Selllers P, Shukla J (1991) Regional climate change and Amazonian deforestation model. Journal of Climate v. 4, n. 10, pp 957–988

Phillips OL, Malhi Y, Higuchi N, Laurance WF, Núñez RM, Váxquez DJD, Laurance LV, Ferreira SG, Stern M, Brown S, Grace J (1998) Changes in the carbon balance of tropical forests: evidence from long-term plots. Science 282:439–442

Richey JE, Krusche AV, Deegan L, Ballester MVR, Biggs T, Victoria RL (2001) Land-use change and the biogeochemistry of river corridors in Amazon. Global Change News Letter 45:19–22

Roberts G, Andrea AO, Zhou J, Artaxo P (2001) Cloud condensation nuceli in the Amazon Basin: "Marine" conditions over a continent? Geophysical Res. Letters 28:2807–2810

Silva Dias MAF, Rutledge S, Kabat P, Silva Dias PL, Nobre C, Fisch G, Dolman AJ, Zipser E, Garstang Manzi AO, Fuentes JD, Rocha H, Marengo J, Plana-Fattori A, Sá L, Alvalá R, Andreae MO, Artaxo P, Gielow R, Gatti L (2001) Clouds and rain processes in a biosphere-atmosphere interaction context in the Amazon Region. Accept for publication in J Geophysical Res

Tarasova T, Nobre CA, Eck TF, Holben BN (2000) Modeling of gasous, aerosol, and cloudiness effects on surface solar irradiance measurede in Brazil's Amazonia 1992–1195. J Geophysical Res 105:26961–26969

Victoria RL, et al. (2000) Effects of land-use changes in the biogeochemistry of Ji-Paraná river, a meso-scale river in the state of Rondônia, southern Amazon. Proceedings of the Large-Scale Biosphere-Atmosphere Experiment in Amazonia First Scientific Conference. Belém, Pará, Brazil. June 25–28

Part IV

Looking to the Future

Part IVa

Simulating and Observing the Earth System

Chapter 27

Virtual Realities of the Past, Present and Future

John Mitchell

General circulation models of climate provide the capability of simulating the statistics of climate under a variety of conditions, real or imaginary. Comparison of simulations and observations of contemporary climate provide an indication of the model's credibility. A further check can be provided by simulating past climates, to the extent that we understand their cause. Models are increasingly being used to predict future changes in climate as a result of human activities, particularly due to increases in greenhouses gases. They can also be used to answer thought experiments (for example, what would happen to climate if we removed all the mountains in the world), allowing us further insight into the processes that control climate and climate change.

27.1 What Are General Circulation Models (GCMs)?

General circulation models are models of the climate system based on the laws of classical physics solved on a numerical grid using large computers. This is in contrast to simpler climate models, that do not explicitly represent the three-dimensional flow of the atmosphere and ocean. The atmospheric component of a GCM is similar to that in a global numerical weather prediction model. The main variables are temperature, winds, humidity, and surface pressure. Other quantities such as precipitation, radiative heating and surface fluxes of heat, momentum and moisture are then diagnosed in terms of the basic variables. Many of these processes are too small to be resolved by the model gird (typically 300 km), so they have to be approximated (parameterised) in terms of the grid scale variables. In some cases, these parameterisations can be validated against an analytical theory or higher resolution models (for example, radiative transfer), but in other cases, a more empirical approach based on observations and general physical principles is required (for example, the formation and dissipation of cloud). Those processes in the latter category lead to the largest uncertainties in a model.

A similar approach can be taken in modelling the ocean (using temperature, salinity and ocean currents as the basic variables). Other parts of the climate system (sea-ice and the land surface) have been incorporated in most climate models to date. As discussed later, attempts are now being made to include in GCMs the chemical and biological processes that influence climate.

27.2 How Do We Use Climate Models?

Before asking how well climate models perform, it is important to understand how they are used. This we illustrate using the following example, recalling that the atmospheric component in a climate model is much the same as a Numerical Weather Prediction model. If the climate model is started with the atmospheric (oceanic and surface) variables for a particular time, the atmospheric features will start to evolve in much the same way as in the real world. Thus, if in the real world there is a low pressure system west of Europe that develops and moves northeastwards, then this will be reproduced in the model. After about five to ten days, the one-to-one relationship between features in the real world and the model will begin to break down, because small errors in the starting conditions and in the model formulation grow rapidly. For this reason, deterministic weather forecasts have little accuracy beyond five to ten days.

If, however, the model simulation is continued for many years, and the climate patterns (for example, rainfall) are averaged over (say) several decades, then the averaged simulated patterns should resemble the average climatological patterns. The patterns are unlikely to match exactly, because natural (internal) variations in both the real world and the model will vary from year to year and are effectively unpredictable. It is possible to estimate the size of the uncertainty due to these weather variations, using observations or model simulations. A simulation of present-day conditions on its own is not particularly useful, as we already know what the present average climate is like. However, if the simulation above is repeated with, for example, increasing levels of greenhouse gases, then the simulated mean climate and climate variability will change. The differences between the new and the first simulation give an estimate of the climate change due to increasing greenhouse gases.

Thus, climate prediction is essentially probabilistic, whereas numerical weather prediction is deterministic. For example, a five-day weather forecast will estimate what the temperature in Amsterdam will be for each of the next five days. A climate prediction for Amsterdam will not predict temperatures for each day in 2085. It will, however, give a "best" estimate of how much warmer than present Amsterdam will be, on average, in the 2080s, and provide an estimate of the uncertainty in the forecast due to natural internal variability.

27.3 How Well Do Models Simulate Present Climate?

Climate models can now reproduce the large-scale features of the general circulation models and their seasonal variation, including the mean continental scale patterns of surface temperature and rainfall (for example, MacAvaney et al. 2001). At smaller scales, for example at the level of individual countries, the errors in temperature and particularly precipitation are larger. The models are also less successful in reproducing variations in atmospheric humidity and cloud. In the ocean, the main circulation features are represented, and the better models reproduce their strength more or less correctly (for example, the northward transport of heat in the North Atlantic). Some of the stronger current systems (for example, the Gulf Stream and Kuro Shio) are noticeably weaker than observed.

Models are also now beginning to produce credible simulations of the main aspects of climate variability. These include ENSO (the El Niño Southern Oscillation, which dominates the interannual variations in tropical rainfall) and the North Atlantic Oscillation (the year-to-year variations in the strength of westerly winds over the North Atlantic, with stronger winds giving milder winters over Western Europe). Models simulate mid-latitude storms and tropical disturbances (but not the detailed structure and strength of hurricanes). At present, models do not resolve meso-scale ocean eddies (the oceanic equivalent of atmospheric weather systems).

27.4 How Well Do Models Predict Past Climate Change?

It is possible for a model to produce a faithful representation of present climate, but give unreliable simulations of climate change. In contrast to Numerical Weather Prediction Models, which are tested several times each day, there are limited opportunities to test climate models.

The instrumental temperature record is sufficient to provide global coverage over the last 160 years. During this period, global mean temperatures have increased by 0.6 °C, which is larger than can be explained by the natural internal variations of climate. If we take into account the main anthropogenic influences on climate (increases in greenhouse gases and sulphate aerosol concentrations), then model simulations can account for

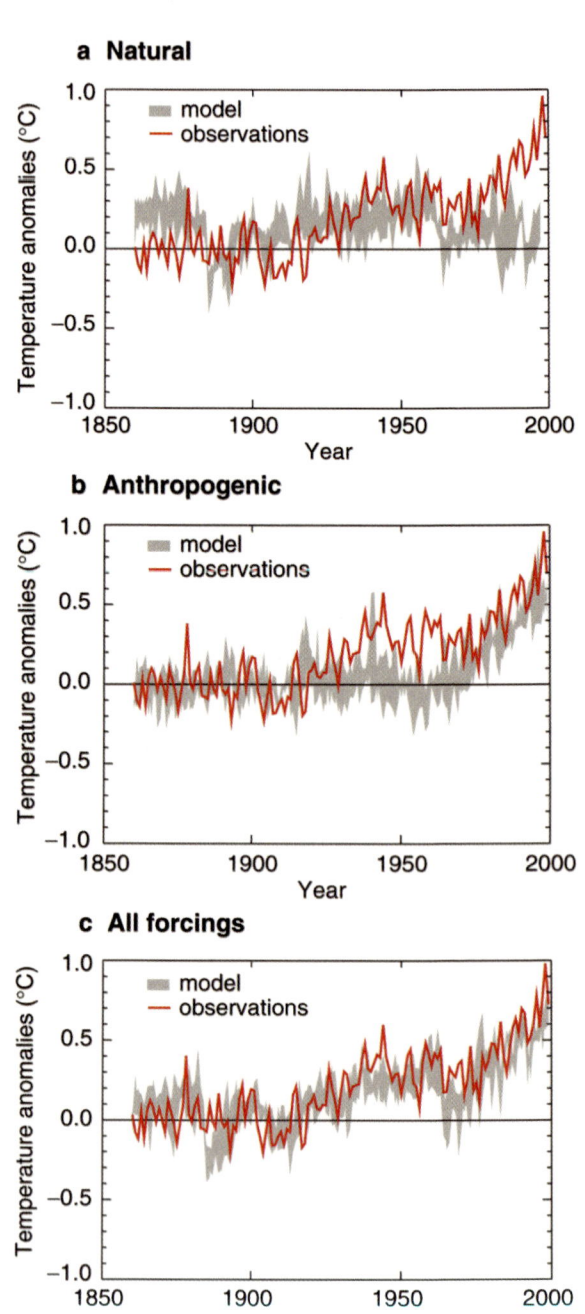

Fig. 27.1. Simulation of global mean surface temperature from 1860 to present with (**a**) only estimated natural forcing included (**b**) estimates of only anthropogenic forcing included (**c**) both natural and anthropogenic forcing. The *red curve* shows the observed temperature change, and the *grey shaded area*, the range of four model simulations (reproduced with permission from Albritton et al. 2001)

the rise in temperature over the last few decades (for example Fig. 27.1). If estimates of effects of changes in the radiative heating due to solar variations, and the radiative cooling due to major volcanic eruptions are also taken into account, then the global mean warming early in the century is reproduced more closely (Fig. 27.1).

For the runs included in Fig. 27.1, statistical tests indicate that the large-scale patterns of temperature change in the model are also consistent with the observed changes. Thus, there is some indication that the model is responding in approximately the right way. However, the climate sensitivity of current models (the global mean temperature change/unit radiative forcing) varies by about a factor of three (MacAvaney et al. 2001), and there is a similar uncertainty in the estimates of net radiative forcing over the instrumental period. Hence, it is possible that errors in the model forcing and model sensitivity have compensated; for example, the estimated net radiative forcing may be too small, but the model still gets the "right" answer because it is too sensitive to forcing.

An alternative test is to look at climate changes in the more distant past. The advantage is that there have been bigger changes in climate than over the last century. A disadvantage is that the changes have to be deduced from proxy indicators (for example, pollen, ice cores, ocean sediment cores) and hence are not accurately known. Furthermore, there is some uncertainty in the factors leading to the past changes in climate. Nevertheless, models have been used to simulate the mid Holocene (6 000 years before present when there were changes in monsoon strength due to perturbations in the Earth's orbit) and 21 000 years before present during the last glacial maximum (for example, Hewitt et al. 2001). In general, models reproduce the gross nature of the reconstructed changes (for example Joussaume et al. 1999), although the regional details vary greatly from model to model.

27.5 What Are the Predictions for the Future?

The predictions for the future depend on the assumed emission scenarios and the climate model used. All the recent IPCC (Intergovernmental Panel on Climate Change) emissions scenarios (Nakocenovic et al. 2000) ensure that greenhouse gas concentrations will continue to increase, at least for the next few decades. Thus, warming to a greater or lesser degree is predicted for some decades to come.

The warming is predicted to be greater over land than over the sea, much slower in the regions of deep ocean mixing in the Southern and North Atlantic oceans, and more pronounced in high northern latitudes in winter (see Cubasch et al. 2001). With warming, snow cover and sea-ice extents generally decrease, and global mean humidity and precipitation increase. There is a wide variation from model to model in the predictions of regional precipitation changes. Nearly all predictions indicate that global warming leads to increases in precipitation in high northern latitudes, extending into mid-latitudes in winter, and small decreases in the subtropics. There is an indication of an overall increase in rainfall in the tropics, though locally there are regions of increase and decrease that differ from model to model. It also seems likely that the intensity of rainfall will increase.

Predicted changes in variability with warming are less certain – there is no clear indication of whether or not the frequency and intensity of ENSO or the North Atlantic Oscillation will change. It is also not clear whether or not the North Atlantic thermohaline circulation will weaken (Fig. 27.2) – all predictions show a warming over Western Europe.

The recent IPCC Third Assessment Report considered the response to six illustrative emission scenarios

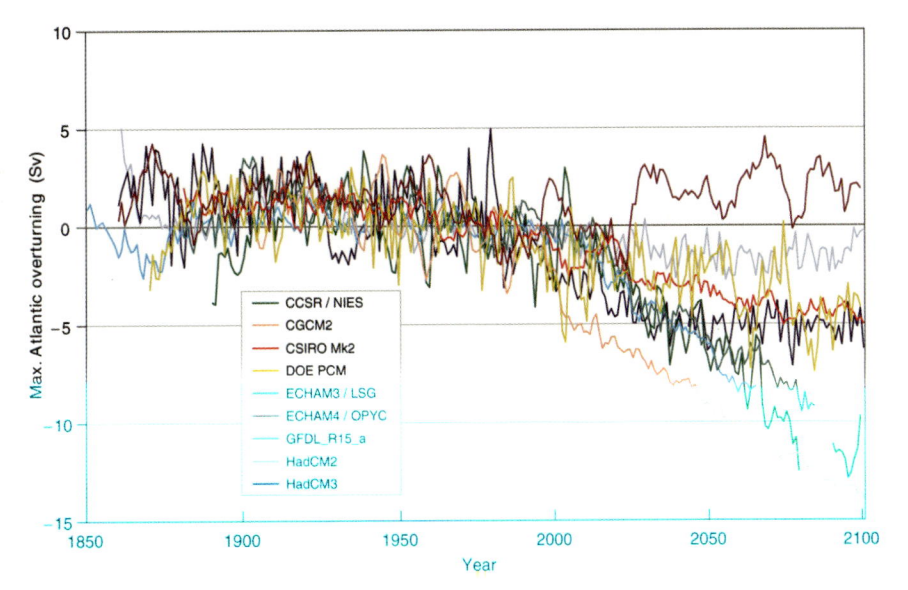

Fig. 27.2.
Change in the strength of the North Atlantic meridional overturning circulation (svds) in a number of simulations with increases in greenhouse gases (reproduced with permission from Cubasch et al. 2001)

Fig. 27.3.
Global mean temperature changes (°C) for the six illustrative SRES scenarios derived using a simple model tuned to a number of complex models with a range of climate sensitivities (also shown are changes for the earlier IS92 scenarios using the same method). The *darker shading* represents the envelope of the full set of 35 SRES scenarios using the average of the model results. The *lighter shading* is the envelope based on all 7 models (with a climate sensitivity range of 1.7 to 4.2 °C) (reproduced with permission from Cubasch et al 2001)

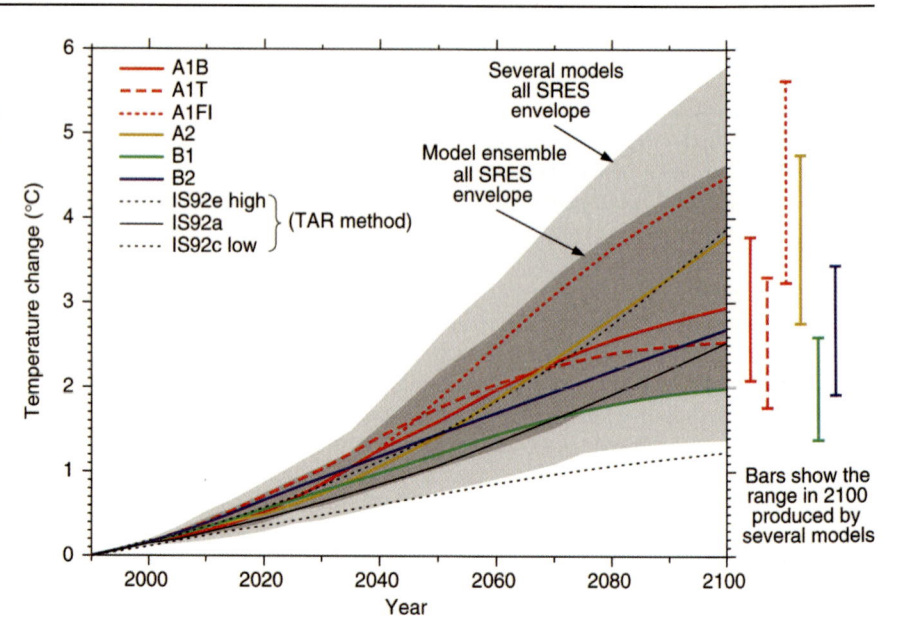

covering a range of noninterventionist social and economic assumptions. These give a range of predicted global mean warming of 1.4 to 5.8 °C from 1990 to 2100 (Fig. 27.3). Half of the uncertainty range arises from uncertainties in climate model predictions with the remainder arising from the difference in emission scenarios.

27.6 What Developments Are Likely in the Future?

Most GCMs to date do not represent chemical or biological processes. Thus, changes in the chemical composition of the atmosphere have been calculated "offline", using separate chemical transport models, and vegetation is assumed to remain constant. In practice, both atmospheric chemistry and vegetation are likely to change as climate changes. For example, as noted above, all models produce an increase in atmospheric humidity with warming, and hence an increase in the atmospheric concentration of the hydroxyl radical, which plays a key role in atmospheric chemistry. For example, recent climate simulations using a climate model that included atmospheric chemistry gave a smaller increase than expected in methane, and hence a slightly smaller climate change (Johnson et al. 2001).

Similarly, changes in temperature and precipitation can affect biological processes, including those that affecting the atmospheric concentrations of carbon dioxide. Again, initial climate experiments incorporating the carbon cycle, albeit in a simplified manner, indicate that there may be appreciable climate-biosphere feedbacks (Cox et al. 2000; Friedlingstein et al. 2001). For example, although simulations indicate that the terrestrial biosphere initially stores more carbon dioxide, this may not always be the case in the future. The loss of carbon dioxide by soil respiration is likely to increase with temperature, and changes in climate may lead to dieback of vegetation in key areas such as tropical forests. One study (Cox et al. 2000) suggests that these two effects could lead to a substantial increase in atmospheric carbon dioxide concentrations in addition to that resulting from direct anthropogenic emissions.

There is much more to be done to confirm the results of these preliminary experiments – they need to be repeated with different models using different sets of assumptions. The underlying processes need to be validated against observations, where possible. The inclusion of chemical and biological processes in climate models will become much more prevalent in the next few years.

On longer time scales, changes in the major ice sheets are likely to become important, as at the last glacial maximum. Changes in the volume of the ice sheets also affect sea level, so they are of interest for future predictions as well as for reconstructions of the past. This represents a major challenge to climate modellers, as the critical ablation zones are typically much smaller than can be resolved by current models, yet the changes occur over thousands of years. The first attempts to include ice sheets are likely to be highly parameterised.

27.7 How Can We Deal with Uncertainty in Predictions?

In the last section, I described some ways in which climate models are likely to become more complex over the next decade. However, increased complexity does not necessarily mean increased accuracy. Indeed, the addition of interactive chemistry and biology is more likely to increase the current large range of uncertainty rather than reduce it. Although one of the aims of climate research is to reduce uncertainty, it is likely that we will have to learn to live with large uncertainties in climate predictions for the time being.

There are several sources of uncertainty. The first arises from the natural internal variability of the climate system – particularly the atmosphere – which limits the accuracy of any single prediction. In principle, one can obtain a probability distribution of the predicted change at a time in the future by running an ensemble of prediction experiments. Each ensemble member is started from slightly different initial conditions. The spread of the predicted values at any time gives an indication of the uncertainty, and the mode of the distribution gives the most likely outcome. This approach is already used for seasonal weather prediction.

The second source of uncertainty arises from the difference in emission scenarios. The IPCC SRES (Special Report on Emission Scenarios) scenarios cover a wide range of emissions, but are all regarded as equally probable noninterventionist scenarios. In other words, they are all "business-as-usual" scenarios against which any mitigation policy needs to be evaluated. In principle, each SRES scenario could be weighted by its relative probability, and those weights used in producing a probability forecast of the climate outcomes and their impacts, but this has not been done.

The third source of uncertainty arises from the uncertainty in climate model predictions. For example, current models give a range of 2 to 5 °C for the long-term global mean surface warming due to doubling of the atmospheric carbon dioxide. Most of this uncertainty can be traced to differences in cloud feedbacks, through differences in the representation of clouds and related process such as boundary mixing and atmospheric convection. The long-term hope is that these uncertainties will be reduced through gradual increases in understanding and through model improvement. In the short term, at least two approaches have been advocated.

The first is to use recent observations to constrain future predictions (Allen et al. 2001). In its simplest form, this would involve scaling model simulations so that the simulation of the recent past global mean temperature agrees with observations. In practice, a more complex statistical approach can be used to give a best estimate with uncertainty limits, and to allow for errors in the different factors producing the changes in climate. However, this approach assumes that the real world response is approximately linear in the recent past and into the future, and that the model patterns of response are essentially correct. Whereas the first condition might conceivably be true, the second condition almost certainly is not.

The second approach is to perturb the parameters used in the models and run a very large ensemble of prediction simulations to produce a probability distribution of future changes. Combinations of parameters that produce demonstrably unrealistic simulations of current climate and perhaps recent climate change could be omitted to reduce the range of uncertainty. Ideally, this approach should be applied to as many different models as possible, but this may be precluded by lack of computing resources. Future work could be concentrated on reducing the uncertainty in those processes that give rise to the greatest uncertainty in the predictions.

27.8 Concluding Remarks

Two factors emerge as being priorities for the future. First, there is an ever-growing need for increased computing power to run models at higher resolution, to include new processes and to run large numbers of ensembles. Secondly, there remains the need to continuously improve the processes already included in models. The need is both to improve the simulation of present-day climate in order to provide credible simulations of the biosphere and atmospheric chemistry, as well to reduce the uncertainty in predictions, particularly the smaller scales that are relevant to evaluating mitigation of and adaptation to anthropogenic climate change. This second priority is highlighted by the magnitude of uncertainty in cloud-climate feedbacks, which has remained unchanged over the last decade.

References

Albritton DL et al. (2001) Summary for policy makers. In: Houghton JT, Ding Y, Griggs DJ, Noguer M, van der Linden PJ, Dai X, Maskell K, Johnson CA (eds) Climate change 2001: The scientific basis. Contribution of Working Group I to the Third Assessment Report of the Intergovernmental Panel on Climate Change. Cambridge University Press, Cambridge New York

Allen MR, Stott PA, Schnur R, Delworth T, Mitchell JFB (2000) Uncertainty in forecasts of anthropogenic climate change. Nature 407:617–620

Cox PM, Betts RA, Jones CD, Spall SA, Totterdale IJ (2000) Acceleration of global warming due to carbon cycle climate feedbacks in a coupled model. Nature 408:184–187

Cubasch U, Meehl GA, Boer GJ, Stouffer RJ, Dix M, Noda A, Raper S, Senior CA, Yap KS (2001) Projections of future climate change. In: Houghton JT, Ding Y, Griggs DJ, Noguer M, van der Linden PJ, Dai X, Maskell K, Johnson CA (eds) Climate change 2001: The scientific basis. Contribution of Working Group I to the Third Assessment Report of the Intergovernmental Panel on Climate Change. Cambridge University Press, Cambridge New York

Friedlingstein PL, Bopp P, Ciais P, Dufresnem J-L, Fairhead L, Le Treut H, Monfray P, Orr J (2001) Positive feedback between future climate change and the carbon cycle. Geophysical Res Lett 28:1543–1546

Hewitt CD, Broccoli AJ, Mitchell JFB, Stouffer RJ (2001) A coupled model study of the Last Glacial Maximum: Was the North Atlantic relatively warm? Geophysical Res Lett 28:1571–1574

Johnson CE, Stevenson DS, Collins WJ, Derwent RG (2001) Role of climate feedback on methane and ozone studied with a coupled Ocean-Atmosphere-Chemistry model. Geophysical Res Lett 28:1723–1726

Joussaume S, Taylor KE, Braconnot P, Mitchell JFB, Kutzbach JE, Harrison SP, Prentice IC, Broccoli AJ, Abe-Ouchi A, Bartlein PJ, Bonfils C, Dong B, Guiot J, Herterich K, Hewitt CD, Jolly D, Kim JW, Kislov A, Kitoh A, Loutre MF, Masson V, McAvaney B, McFarlane N, de Noblet N, Peltier WR, Peterschmitt JY, Pollard D, Rind D, Royer JF, Schlesinger ME, Syktus J, Thompson S, Valdes P, Vettoretti G, Webb RS, Wyputta U (1999) Regional climates for 6000 years ago: Results of 18 simulations from the Palaeoclimatic Modelling Intercomparison Project. Geophys Res Lett 26:859–862

MacAvaney B, Covey C, Joussaume S, Kattsov V, Kitoh A, Ogana W, Pitman AP, Weaver AJ, Wood RA, Zhao Z-C (2001) Model evaluation. In: Houghton JT, Ding Y, Griggs DJ, Noguer M, van der Linden PJ, Dai X, Maskell K, Johnson CA (eds) Climate change 2001: The scientific basis. Contribution of Working Group I to the Third Assessment Report of the Intergovernmental Panel on Climate Change. Cambridge University Press, Cambridge New York

Nakicenovic N, Alcamo J, Davis G, de Vries B, Fenhnn J, Gaffin S, Gregory K, Gruber A, Jung TY, Kram T, La Rovere EL, Michaelis L, Mori S, Morita T, Pepper W, Pitcher H, Price L, Raihi K, Roehrl A, Rogner H-H, Sankovski A, Schlesinger M, Shukla P, Smith S, Swart R, van Rooijen S, Victor N, Dadi Z (2000) IPCC special report on emissions scenarios. Cambridge University Press, Cambridge New York, 599pp

For a web site on quantifying uncertainty in climate predictions, see *www.climateprediction.com*

Chapter 28

Coping with Earth System Complexity and Irregularity

Hans Joachim (John) Schellnhuber

28.1 The Challenge

Some 500 years after the Great Copernican Revolution, humanity sets out to explore the far ends of the universe and to unravel the last secrets of cosmic dynamics: sophisticated sensors are being installed to detect gravitational waves triggered by black holes, super-telescopes watch the birth and collision of entire galaxies, and powerful radiation signals are emitted to approach extraterrestrial fellow civilisations in the Milky Way. While most of these activities will be splendid successes, no sign of primitive, let alone intelligent, life outside planet Earth has been tracked down so far. This is certainly irritating, as the universal laws of nature should warrant the emergence of fully-fledged ecospheres in many instances of space-time.

The (preliminary) failure of ET hunting may have to do with the possibly unique entanglement of astrogeophysical conditions that our planet provides for the self-organisation of matter far from thermodynamical equilibrium. Many peculiar factors like the continuous residence of Earth in the so-called habitable zone (Franck et al. 2000), the stimulating presence of a perfectly designed moon, or the magnetic shielding against cosmic particles bombardment generated by interior convective motion seem to combine in a most singular way. The planetary machinery that has developed under those conditions defies the power of standard scientific analysis: it is composed of innumerable heterogeneous parts and pieces, it is dominated by strongly varying forces, and the involvement of crucial processes at all scales introduces nasty elements of stochasticity. Researchers love to study just the opposite type of specimens, however, namely simple, linear and deterministic ones.

The cognitive situation is exacerbated by the fact that the ecosphere, as the natural part of the *Earth System*, is complemented by the "human factor", which has begun to hold sway over the planetary dynamics (Schellnhuber 1999; Crutzen and Stoermer 2000). This factor does not only massively increase the overall complexity of the system, it is also the source of fundamental indeterminacy through individual and collective volition. In fact, science itself creates novel degrees of freedom by constructing images of plausible global futures for inspection and selection by decision makers. Thus the Earth System development is heavily underdetermined and cannot be foreseen, in principle.

And yet Earth System science is emerging now that strives to tame the intricacies and irregularities encountered through ambitious international monitoring and modelling programmes (Sahagian and Schellnhuber 2001). Is this a megalomaniac and ultimately futile effort? Will comprehensive simulation machines coupling the best sub-models representing the atmosphere, hydrosphere, biosphere, lithosphere, and anthroposphere, respectively, pave the master road towards genuine systems understanding? Or are there smart alternatives to brute-force mimicking that can provide some of the insights urgently needed for avoiding nonsustainable development? In Sect 28.2, the specific intellectual challenges to be tackled will be sketched and illustrated, while Sect. 28.3 will try to demonstrate how the insidiousness of reality may be outwitted. Conclusions regarding the feasibility of good Earth System stewardship are then drawn in Sect. 28.4.

28.2 Great Cognitive Barriers

Nonlinearity, which pervades all parts of the Earth System, can be simply defined as the disproportional relationship between cause and effect. Even mild and regular nonlinearities involved in mathematical or physical processes – like quadratic functions or forcings – can already generate quasi-unpredictable outcomes of spectacular intricacy. The modern theory of dynamical systems has demonstrated, for instance, how the iteration of simple stretching and folding operations produces so-called "strange attractors" with fractal texture, i.e., structure at all scales (Wiggins 1990). But also rather malignant nonlinearities have to be considered like singular relationships of discontinuous or wildly oscillating character.

Important examples of that class are critical threshold functions that give rise to hysteretic, or "no-return", systems behaviour. Criticality is actually quite characteristic of the dynamics of the total Earth System and most of its parts: the melting of inland ice-sheets towards the end of quaternary glaciation periods, for instance, is a phenomenon that can be reversed only over many millennia, and major ocean currents can be flipped between distinct modes of operation by perturbations of sufficient strength only (Ganopolski and Rahmstorf 2001).

Complexity, which is probably the dominant feature of the Earth System, evades a general definition. The notion is often merely used to qualify very complicated items, yet there are rigorous algorithmic concepts to quantify the informational complexity encoded in given mathematical prescriptions. The so-called Ω numbers, representing the probability that arbitrary computer programmes will eventually halt, play a crucial role in this context (Calude and Chaitin 1999). In the real world, complex systems may be characterised by rich combinatorial constructions from basic building blocks (as in the case of the human genome) or by fat "causal mattresses" instead of slim causal chains (as in the case of ecological communities).

Multi-cause-multi-effect processes certainly contribute to the intricacy of planetary dynamics, which is largely organised in teleconnection patterns (like the ENSO phenomenon) and feedback loops (like the tundra-taiga-snow-albedo scheme (Foley et al. 1994)). One of the hottest research topics in present global change science is the spatiotemporal flux of carbon across the various Earth System compartments: many factors and forces seem to cooperate to achieve the natural quasi-stability of this flux system as reflected in the famous Vostok ice-core records (Petit et al. 1999). The modern involvement of the human factor (through fossil fuel burning, etc.) very much complicates matters, however, and may actually push global carbon dynamics outside the self-control range (Smith et al. 2001).

Indeterminacy, which will be the unpleasant companion to false and true Earth System managers for all eternity, generally means that a complete formal description of a given system is impossible, with the important consequence that its behaviour cannot be precisely predicted. The scientific enterprise is confronted with many types and degrees of indeterminacy (or "uncertainty", as popular jargon puts it). The mild end of the spectrum is represented by inaccurate or fragmentary knowledge of well-defined systems parameters due to insufficient measurement or analysis. This form of indeterminacy is reducible, in principle, by diligence and skill. Things become worse, if the specimen in question is governed by innumerable intrinsic and extrinsic variables or is prone to irregular dynamics. Under these circumstances, quasi infinite empirical and theoretical research capacities would be needed to beat stochasticity and chaos, respectively. So, this form of uncertainty may be qualified as irremediable for all practical purposes (Schellnhuber and Wenzel 1998).

The rough end of the indeterminacy spectrum is occupied by quantum behaviour and human volition, which exclude the prediction of individual events *a priori*. Earth System science is, of course, more concerned with the socioeconomic decision-making process that can make all the difference for the development of regions, continents and the entire planet. This process is self-referential in the sense of playing strategic games with scenarios about the impacts of its own outcome. The so-called integrated assessment approach to optimal climate protection through international agreements (Schneider 1997) is confronted with this very indeterminacy problem.

So, what can be done about all those intimidating, if not discouraging, features of the Earth System? The next section tries to provide a couple of answers.

28.3 Breaches and Bypasses to Understanding

First of all, it seems appropriate to comment on the potential represented by the straightforward cutting-edge activities pursued by the various specialist communities contributing to global change research. Reports are coming in almost daily about breath-taking new developments regarding palaeo data mining (Alverson et al. 2000), Earth System monitoring via the most advanced remote-sensing equipment (Justice et al. 1998), artificial staging of environmental dynamics in miniaturised reality (Norby et al. 2001), or coupling of state-of-the-art modules for the relevant Earth compartments to patch up planetary simulators in virtual reality (Cox et al. 2000). Unfortunately, there is no guarantee that the ever finer chopping of the planetary machinery into bits of evidence and the formal recombination of these chunks will produce a genuine systems-level understanding – not to mention the exploding costs of such a programme. When it comes to tackle, in particular, the challenges sketched above, more simple, eclectic and opportunistic cognitive strategies – based on protomorphic wisdom, intuition and careful simplification – may do a better job.

For example, the analysis of the unfolding of strongly nonlinear events like climate disruptions, heart attacks, or even stock market crashes can be considerably advanced by hunting up and using the phenomenological scaling laws involved (Bunde et al. 2002). Scaling behaviour reflects a fundamental, yet "noisy" self-similarity of certain underlying natural or socioeconomic processes (Mandelbrot 1988). As a consequence, weak forms of long-term weather or market forecasting may be derived. Whether novel techniques of calculus, using the fractional differintegration of ubiquitous "monster" functions that vary at all scales will facilitate this task, needs to be further explored (Schellnhuber and Seyler 1992).

When it comes to investigating global ecosphere variability and resilience, the very existence of critical thresholds or complicated invariant sets in parameter space can be turned into a heuristic blessing, as these structures largely organise the qualitative systems dynamics. It often suffices to reveal the multiple equilibria or attractors of the unperturbed system for understanding the seemingly erratic motion triggered by external perturbations like Milankovitch forcing or anthropogenic interference (Hasselmann 1999; Ganopolski and Rahmstorf 2002). Therefore GAIM, as the IGBP Task Force for Earth System analysis, has embarked on the search for the crucial switch and choke elements in the planetary machinery that might be activated either by internal fluctuations or by human disturbance. Figure 28.1 gives a very preliminary account of those areas on the globe, which seem most relevant in this context.

The bistability of the West-Saharan vegetation cover serves as a nice illustration here: recent studies indicate that this region could be "re-greened" in the course of global warming (Claussen et al. 2002). This type of knowledge is actually imperative for global environmental management, particularly for *anti-geoengineering* in the sense of avoiding pushing the critical control buttons in the Earth System.

In order to reveal the sheer presence of such buttons, simulation models of intermediate complexity become increasingly important. These models do not strive for straightforward high-resolution mimicking of the climate system or the total Earth machinery, but try to capture the salient features of the original, much like a skilful caricature. This is often achieved by constructing reduced-form representations of the full set of dynamical equations involved through separation of "slow" and "fast" variables (Saltzman 1988), filtering techniques for obtaining effective entities (Petoukhov et al. 2000), or careful averaging over irrelevant details. Intermediate-complexity models can be employed as fast time machines that redraw the quaternary biography of global changes with a broad brush, and search the scenario space of future anthropogenic forcing for high-risk items.

Another promising – yet somewhat "softer" – way of coming to grips with genuine complexity is to describe system dynamics in terms of spatiotemporal and functional patterns. In many cases, a short list of archetypal patterns can be constructed that encapsulates the major part of the relevant information. A modern technique to extract such typologies from an ocean of data is the neural-network approach (Kohonen 2001), which was inspired by the architecture of the human encephalon.

In the context of global change research, the so-called syndromes analysis (Petschel-Held et al. 1999a) provides a comprehensive semi-quantitative picture of the dynamical degradation patterns that characterise contemporary nature-society interactions across the planet. These syndromes are identified by inspecting innumerable place-based case studies for typical causal webs of cofactors that may drive unfavourable environmental developments. As an illustration, Fig. 28.2a shows such a web for the "Sahel Syndrome" that describes the nonsustainable, self-enforcing marginalisation of subsistence land-users.

From the network of crucial functional interrelationships, a number of semi-quantitative measures can be derived via fuzzy-logic methods (Cassel-Gintz et al. 1997). Figure 28.2b depicts, for instance, the relative susceptibility ("disposition") of the Earth's regions to Sahel-Syndrome infection. The map is constructed from a

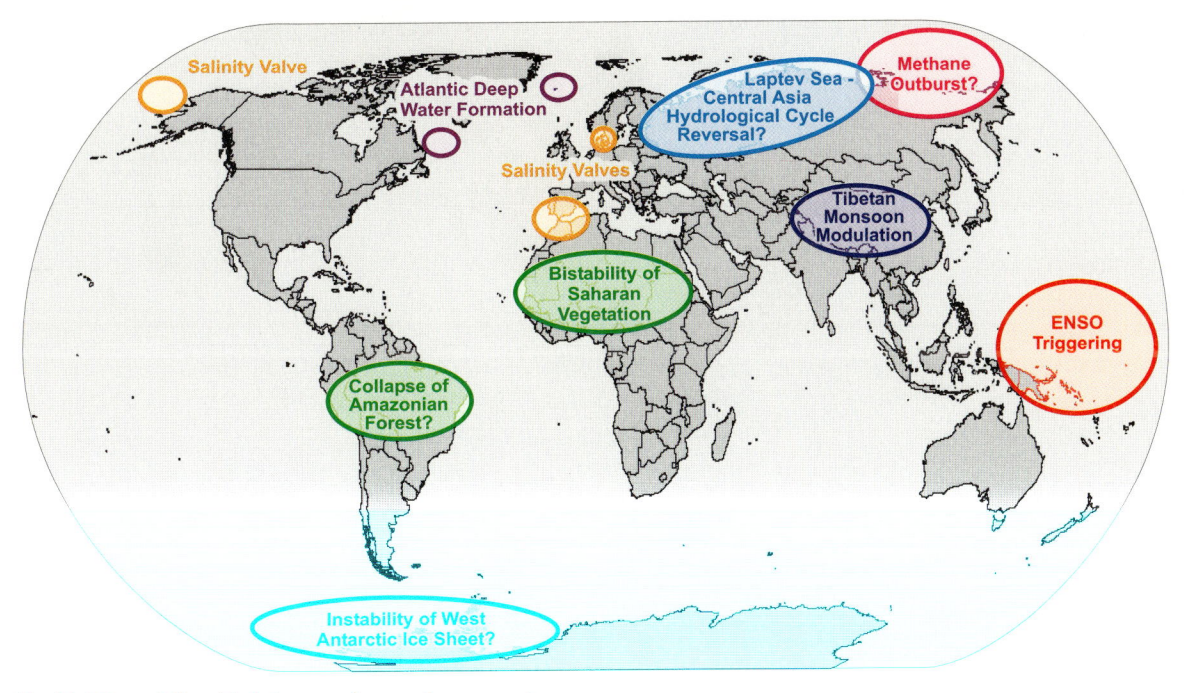

Fig. 28.1. Potentially critical elements for ecosphere operation

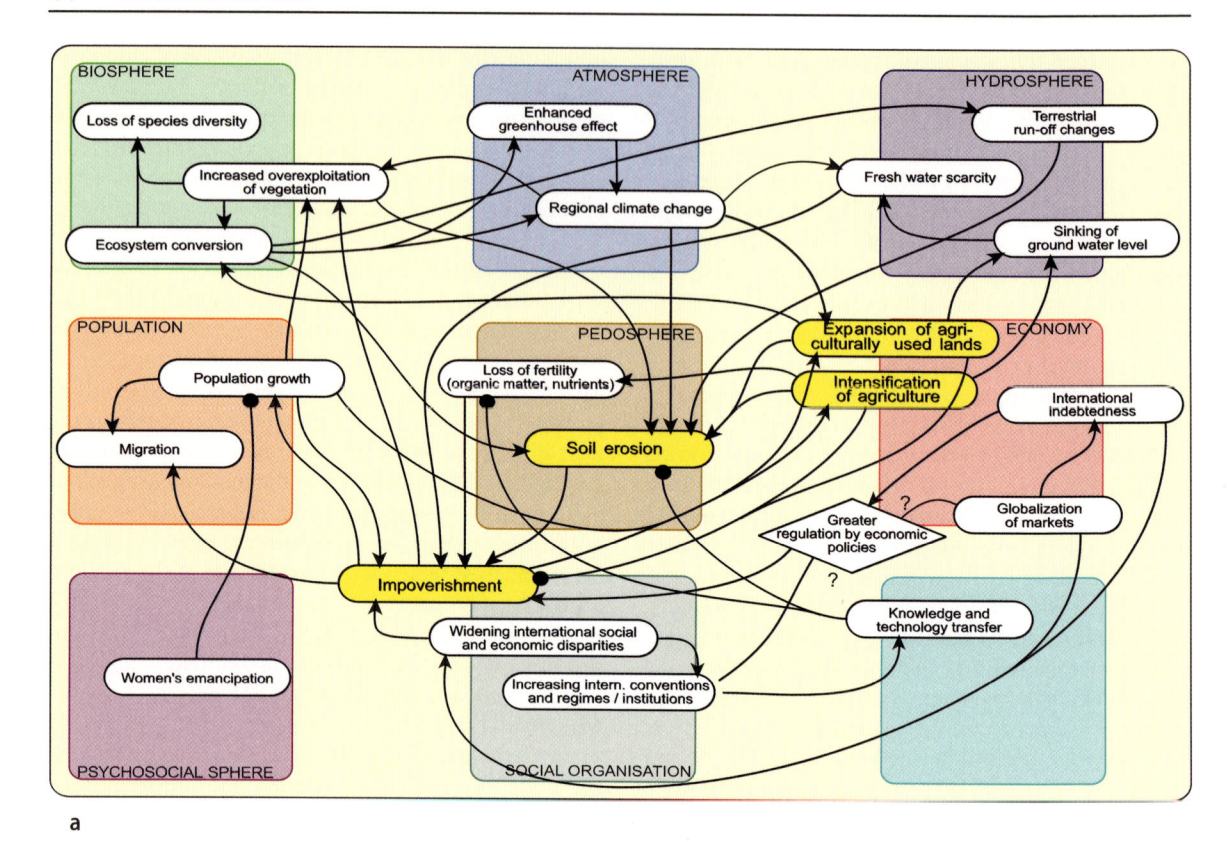

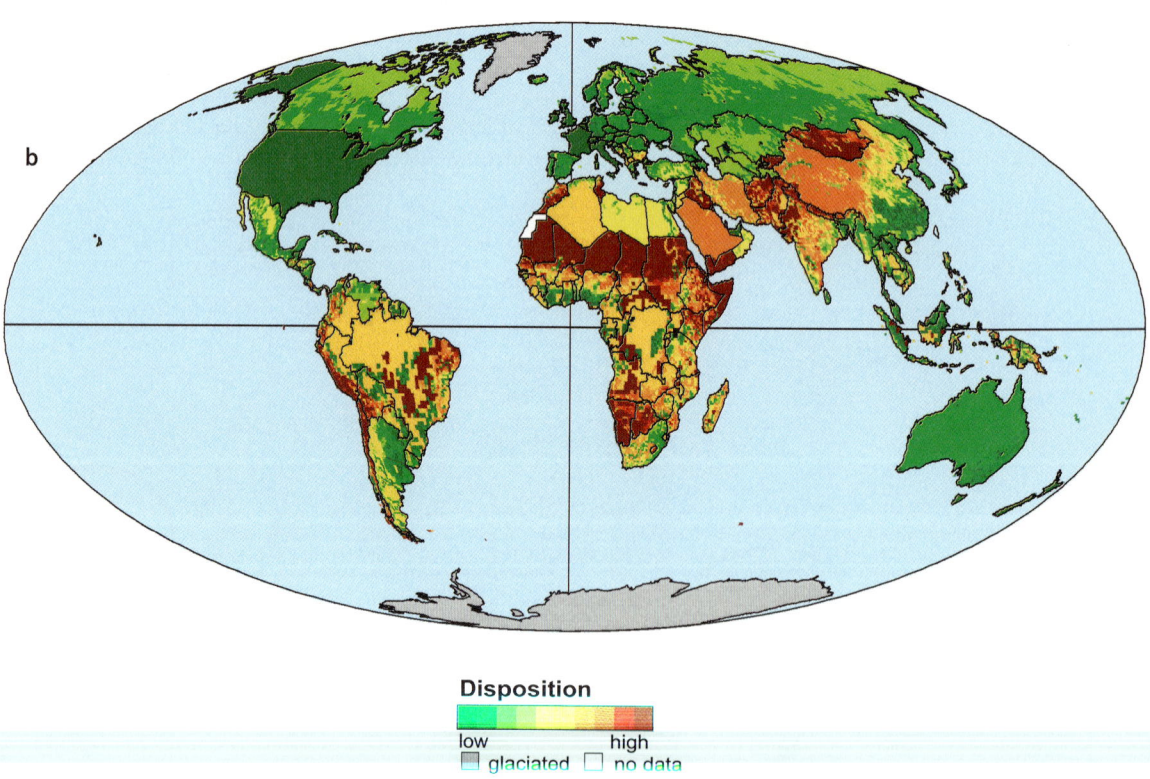

Fig. 28.2. The Sahel Syndrome. **a** Archetypical network of interrelations, where the core symptoms (*yellow*) combine to form a vicious circle; **b** geographically explicit representation of the world-wide differential disposition for this syndrome

large number of pertinent global data sets quantifying, i.e., climatic conditions, soil quality, ecosystem structure, degree of economic development, and social organisation (Lüdeke et al. 1999).

The information about the elements of an inference diagram like the one shown in Fig. 28.2a is generally quite fragmentary, unbalanced and vague, so the fuzzy-logic approach seems appropriate for constructing static objects like the disposition measure of Fig. 28.2b. When it comes to transforming the limited systems knowledge into something like a dynamical model for anticipating the temporal evolution of syndromes or other functional patterns, however, then a crisp formalism employing quantitative differential equations would be pure make-believe. There is, however, an honest alternative for "weak" prediction, at least: Symbolic dynamics as implemented through qualitative differential equations (Kuipers 1994), for instance, utilises only elementary information about the monotonicity properties of the functions involved, about possible turning points or discontinuities, or about topological characteristics of the system components. This strategy generates an "evolution tree", embracing all the distinct qualitative developments consistent with the rudimentary knowledge available. Figure 28.3a gives an example of such a tree, resulting from a case study based on a qualitative model of the Sahel Syndrome (Petschel-Held et al. 1999b). The locations of the relevant studies are indicated in Fig. 28.3b.

Over the last decades, many other ways of dealing effectively with irremediable uncertainty have been devised, particularly by operations research and artificial intelligence activities. But what can one do about *a priori* indeterminacy as created by human volition acts, for instance? Considerable efforts are made, in fact, to simulate socioeconomic behaviour in everyday situations (like stop-and-go traffic) through agent-based models, using either very few perfectly designed digital Frankensteinian creatures or a huge mass of utterly primitive electronic entities. This is a promising road, yet a radical shortcut to quasi-prediction may exist: The formal components of integrated assessment models (for climate protection strategies, etc.) could be complemented by real human elements, i.e., a sufficiently large crowd of individuals selected representatively for "playacting" various classes of decision makers relevant in the given context. The internet provides the opportunity to engage actually tens

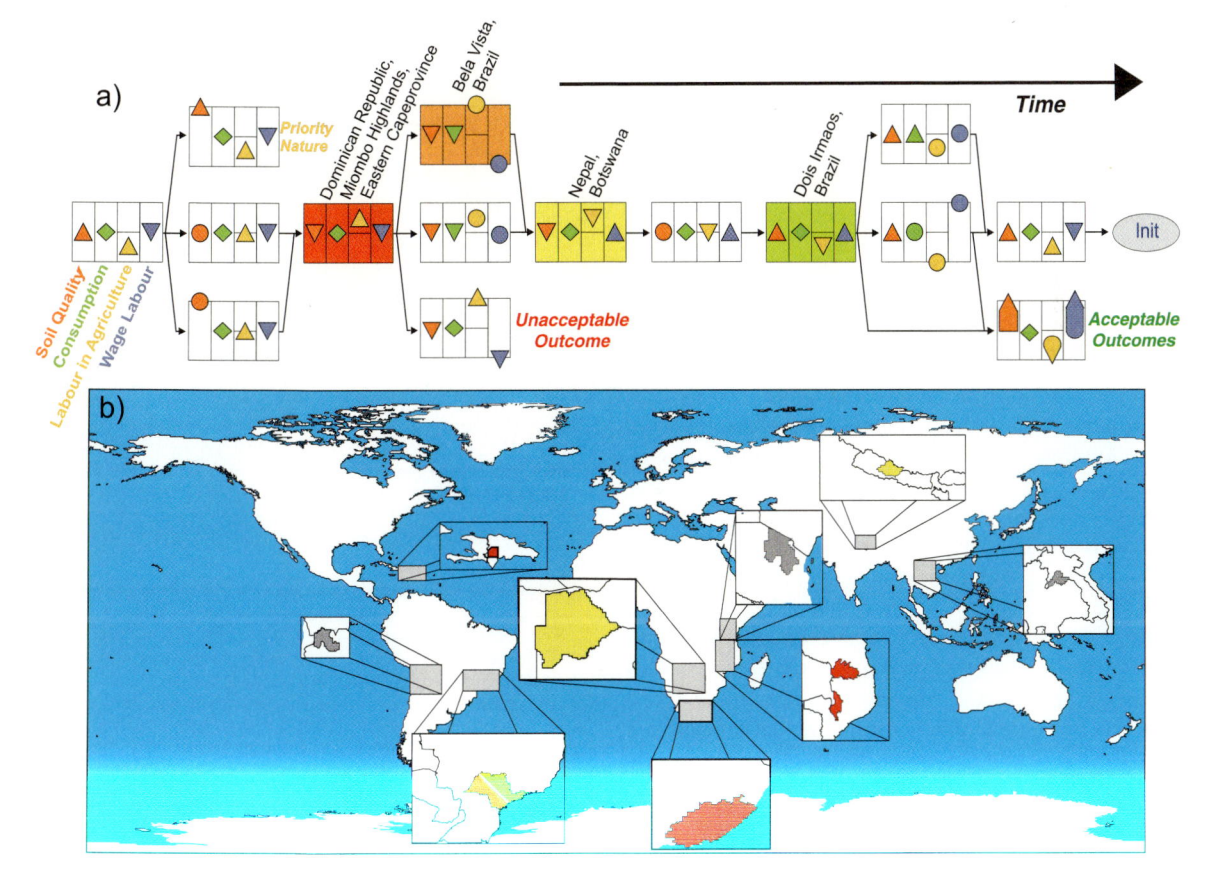

Fig. 28.3. Symbolic dynamics of the Sahel Syndrome, reproducing various development patterns as observed in numerous country case studies

of thousands of volunteers or remunerated participants to generate robust projections into the future of human-environment interaction. In fact, several scientific initiatives to explore the potential of this approach have been launched already.

By way of contrast, it should be emphasised that global change science has only recently begun to deal seriously and systematically with reducible uncertainties as arising from mediocre data quality control, poor harmonisation of disciplinary input, or sub-optimal algorithmic techniques. The GAIM model intercomparison projects performed during the last few years (Sahagian 2000) indicate how this "standard job" could be done properly.

28.4 Adaptive Planetary Stewardship

The message from the preceding reflections and illustrations is that science can cope with Earth System complexity and irregularity in many routines as well as in novative ways. Pursuing and exploring the potential of the latter will actually advance the scientific enterprise as a whole – who can tell today what "cyberspace simulations", combining electronic processors and human switchboards, may ultimately achieve? And yet there is not the slightest hope that the pertinent research community will provide the responsible politicians with the full body of evidence necessary for sustainable environmental management *in time*:

For the Earth System is currently operating in a non-analogous state with unprecedented magnitudes and rates of change provoked by human interference. So deliberate strategies of good global stewardship – replacing the present mismanagement strategy of planetary business-as-usual – are urgently needed. In fact, the international research programmes should lay the cognitive foundations for *deterministic* stewardship within this decade, where irreversible decisions (about greenhouse gas emissions, biodiversity protection measures, soil conservation policies, etc.) have to be made already. This is clearly beyond the capacities of Earth System Science ...

Adaptive management for sustainable development is the only reasonable remaining option. Described in non-technical terms, this means that environment and development decisions are tentatively made on the basis of imperfect information but perpetually revised according to the influx of relevant new data and insights. One crucial political precondition for this concept is a most flexible design of international institutions like the pertinent protocols. Science must help to make adaptive global stewardship a success, not only by advancing the knowledge about the utterly intricate Earth System, but also by developing a "science of dealing with scientific incapacity." Strengthening the emergent fields of nonlinear and fuzzy control is an important step in this direction.

References

Alverson K, Oldfield F, Bradley R (eds) (2000) Past global changes and their significance for the future. Quaternary Science Reviews 19:1–5
Bunde A, Kropp J, Schellnhuber HJ (eds) (2002) The science of disasters. Springer, Berlin
Calude CS, Chaitin GJ (1999) Nature 400:319
Cassel-Gintz M, Lüdeke MKB, Petschel-Held G, Reusswig F, Plöchel M, Lammel G, Schellnhuber HJ (1997) Clim Res 8:135
Claussen M, Brovkin V, Ganopolski A, Kubatzki C, Petoukhov V (2002) Climate change in Northern Africa: The past is not the future. Climatic Change, in press
Cox P, Betts R, Jones C, Spall S, Totterdell I (2000) Nature 408:184
Crutzen PJ, Stoermer EF (2000) Global Change Newsletter 41:17
Foley JA, Kutzbach JE, Coe MT, Levis S (1994) Nature 371:52
Franck S, von Bloh W, Bounama C, Steffen M, Schönberner D, Schellnhuber HJ (2000) J Geophys Res – Planets 105:1651
Ganopolski A, Rahmstorf S (2001) Nature 409:153
Ganopolski A, Rahmstorf S (2002) Abrupt glacial climate change due to stochastic resonance. Phys Rev Lett 88(3):038501
Hasselmann K (1999) Nature 398:755
Justice CO, Vermote E, Townshend JRG, DeVries R, Roy DP, Hall DK, Salomonson VV, Privette JL, Riggs G, Strahler A, Lucht W, Myneni RB, Knyazikhin Y, Running SW, Nemani RR, Wan Z, Huete AR, van Leeuwen W, Wolfe RE, Giglio L, Muller J-P, Lewis P, Barnsley MJ (1998) IEEE Trans Geosci Rem Sens 36:1228
Kohonen T (2001) Self-organizing maps. Springer, New York
Kuipers B (1994) Qualitative reasoning: Modeling and simulation with incomplete knowledge. MIT Press, Cambridge, Massachusetts
Lüdeke MKB, Moldenhauer O, Petschel-Held G (1999) Environ Mod Assess 4:315
Mandelbrot BB (1988) The fractal geometry of nature. Freeman & Co., New York
Norby RJ, Kobayashi K, Kimball BA (2001) New Phytologist 150:215
Petit JR, Jouzel J, Raynaud D, Barkov NI, Barnola J-M, Basile I, Benders M, Chappellaz J, Davis M, Delayque G, Delmotte M, Kotlyakov VM, Legrand M, Lipenkov VY, Lorius C, Pépin L, Ritz C, Saltzman E, Stievenard M (1999) Nature 399:429
Petoukhov V, Ganopolski A, Bovkin V, Claussen M, Eliseev A, Kubatzki C, Rahmstorf S (2000) Clim Dyn 16:1
Petschel-Held G, Lüdeke MKB, Reusswig F (1999a) In: Lohnert B, Geist H (eds) Coping with changing environments. Ashgate et al., Singapore, p. 255
Petschel-Held G, Block A, Cassel-Gintz M, Kropp J, Lüdeke MKB, Moldenhauer O, Reusswig F, Schellnhuber HJ (1999b) Environ Mod Assess 4:295
Sahagian D (2000) Global Change Newsletter 41:11
Sahagian D, Schellnhuber HJ (2001) Research GAIM 4, (2, Winter 2001), 2
Saltzman B (1988) In: Schlesinger ME (ed) Physically based modelling and simulation of climate and climatic change – Part II. Kluwer, Dordrecht, p. 737
Schellnhuber HJ (1999) Nature 402, Supp. 2 Dec 1999, C19
Schellnhuber HJ, Seyler A (1992) Physica A 191:491
Schellnhuber HJ, Wenzel V (eds) (1998) Earth system analysis. Springer, Berlin
Schneider SH (1997) Environ Mod Assess 2:229
Smith JB, Schellnhuber HJ, Mirza MMQ (2001) Vulnerability to climate change and reasons for concern: A synthesis. In: McCarthy JJ, Canziani OF, Leary NA, Dokken DJ, White KS (eds) Climate change 2001. Impacts, adaption and vulnerability. Cambridge University Press, Cambridge, p. 913–967
Wiggings S (1990) Introduction to applied nonlinear dynamical systems and chaos. Springer, New York

Chapter 29
Simulating and Observing the Earth System: Summary

Chris Rapley

29.1 Earth System

As far as we are aware, the Earth is the most complicated object in the Universe. Describing, understanding and predicting changes within the Earth System present daunting research challenges. One of the great scientific successes of the last two decades has been the development of the International Global Change Research Programmes (IGCRPs), which collectively address the study of the Earth as a system. The contribution of these programmes has been to:

- Create and sustain active networks of the world's leading scientists in relevant fields;
- Identify and publicise priority issues and questions;
- Agree action plans linking and adding value to national research activities;
- Synthesise valuable scientific and policy-relevant insights through the integration and analysis of a wide range of individual results;
- Develop new areas of science, global change palaeoscience being an important example.

A key objective of the IGCRPs has been the breakdown of traditional disciplinary barriers. Reductionism, although a necessary feature of any scientific endeavour, has a tendency to isolate the various specialisms, undermining a "systems" view. The very wide range of disciplines represented by the participants in the Global Change Open Science Conference is a tribute to the progress made by the IGCRPs in overcoming this deeply rooted problem. The fact that they are working closely together to agree on an integrated approach for their future activities is especially significant.

However, the road embarked on by those who established the IGCRPs is a long one, as yet incomplete. Figure 29.1 provides a diagrammatic overview of the Earth System and illustrates that in addition to the planetary components (atmosphere, hydrosphere, cryosphere, lithosphere/solid Earth, biosphere, and anthroposphere), the system includes the Sun, the solar wind and a variety of other astronomical phenomena. A true "systems" approach must ensure that the relevant experts are involved fully, not only because these influences are important, but because their study (through comparative planetology, for example) has the potential to provide important new insights into the functioning of the Earth.

Other disciplines not represented in Amsterdam, such as demography, health science and geology must be encouraged to find their roles also. We need to ask our-

Fig. 29.1.
A diagrammatic overview of the Earth System: as well as planetary components, the system includes a variety of astronomical phenomena

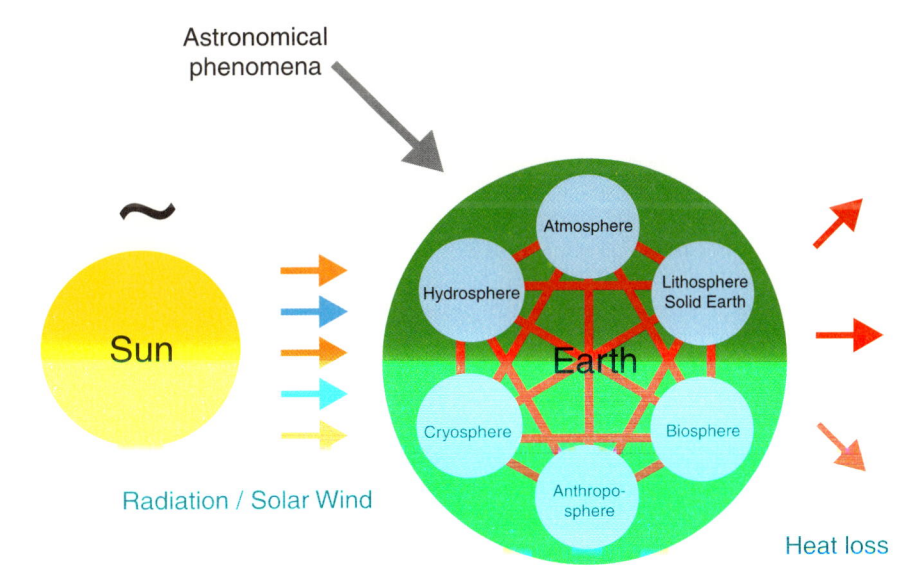

selves "Who should be here and is not?" It is crucial that the emphasis given to the study of specific aspects of the Earth System should result from rational appraisal, and not be the consequence of historical or organisational happenstance.

The presentations in the session "Simulating and Observing the Earth System" included issues discussed in several of the parallel sessions (in particular A4 "Ground Truthing the Earth System", B4 "Earth Systems Analysis", and C4 "Nonlinear Responses to Global Change"). Key points are that the Earth System is composed of a host of component subsystems that interact in ways which are often:

- Nonlinear (with trigger points and thresholds);
- Nondeterministic;
- Not amenable to standard scientific analysis.

In addition, the outcomes of Earth System processes often result from small differences between large numbers. Examples include the heat balance of the planet, annual uptake of carbon dioxide in the atmosphere, and the mass balance of the Antarctic and Greenland ice sheets. This presents a real challenge to the precision, accuracy and completeness of observations and models.

Furthermore, Earth System interactions occur over a range of spatial and temporal scales, covering at least thirteen orders of magnitude, with the consequence that however powerful digital computers become in the future, and however detailed our knowledge at the microscale, it will always be necessary to parameterise processes in Earth System models, with the attendant limitations on realism and reliability.

However, the Earth System includes dominant "features", such as specific switch/choke points and functional "modes". This provides hope that a "crafted" approach to planetary simulation, focussing on the key processes and interactions, will be tractable.

29.2 Simulators

There is general agreement that a hierarchy or spectrum, of Earth System models is required, each type addressing a different purpose ("horses for courses"). Categories range from the simple toy and conceptual variety, through process models, Earth Models of Intermediate Complexity (EMICS), and comprehensive Earth System simulators. We have seen evidence of impressive progress in all of these. Exotic approaches, such as the syndrome and eigenfunction analyses are also being developed.

Issues limiting progress are:

- Computer power: even assuming this continues to increase at historical rates;
- Architecture: numerical models, especially of the comprehensive variety, are relatively inaccessible to the research community at large. There is a need for new architectures providing the opportunity for other researchers to link, test and run modular elements, taking new approaches, or simulating aspects of the system not currently addressed;
- Knowledge and understanding of the Earth System: computer models "only contain what they contain." They cannot include what we are ignorant of, what we do not understand or behaviours latent in the system but currently not active.

29.3 Observations

David Karl reminded us that we should "study nature, not books". John Schellnhuber pointed out the power of a "macroscopic" view of the Earth. In spite of the best efforts of the World Meteorological Organisation, the research and operational space agencies, a host of organisations responsible for observing one aspect or another of the global system, and the science community at large, observations, or the lack of them, remain the prime limitation in predicting the future of the planet.

Ray Bradley made a powerful plea for the development of a Global Palaeoclimate Observing System. However, the long, difficult and as of yet unsuccessful gestation of the Global Climate Observing System (GCOS) suggests that this will not be easy or quick to create.

Key objectives for a global observing system include the need to integrate observations from space and *in situ* (the Integrated Global Observing Strategy (IGOS) and the Global Monitoring for Environment and Security (GMES) initiatives seek to achieve this). Also, it is essential to ensure that operational systems are sustained over the long-term, and to take full advantage of initiatives that are already under way. Examples of the latter include the European Space Agency's Earth Watch and Earth Explorer missions, and the ARGO ocean buoy network.

The primary impediments to progress are institutional. In the case of a palaeo-observing system, the absence of an institution to foster and finance such an initiative is the problem. In the case of an operational Global Observing System, it is the opposite; institutions such as the space agencies and operational observing agencies exist, but they are arguably the wrong institutions. They generally have a history of limited cooperation or rivalry. Indeed, the charters of some inhibit the type of developments required.

29.4 The Way Forward

In setting out on a purposeful journey, it is important to have a clear understanding of the destination. Success will depend on the answers to questions such as:

- What would it be like?
- What would convince you?
- What would you settle for?

The overarching aim of simulating and observing the Earth System could be articulated as:

"*The production of an integrated operational observing and simulation system, which routinely affects human individual and collective behaviour in ways beneficial both to society and to the Earth System.*"

Those inclined to think this too ambitious should note that it is satisfied to a degree by the existing meteorological forecasting services.

On what time scale do we need such an outcome? Figure 29.2 shows the predictions for global mean surface temperatures over the next century taken from the latest report of the Intergovernmental Panel on Climate Change (IPCC). Vertical bars indicating 15 year time intervals have been added. Even the most optimistic scenarios indicate significant change on a 15 year time scale. The IGCRPs have demonstrated that our knowledge and understanding of the Earth System can be transformed in 15 years. Less easily changed are the "rigid" aspects of human society such as major infrastructure and attitudes, where a time scale of 15 years may be hard to meet. An aim for an effective operational system in 15 years time would seem appropriate if the consequent actions are to be of value.

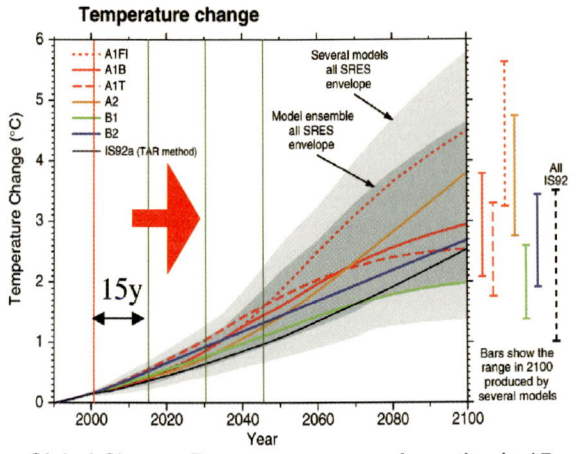

- Global Change Programmes - transformation in 15y
- Societal rigidities - major infrastructure change ~15y

Fig. 29.2. Even the most optimistic scenarios predict that global mean surface temperatures over the next century will change significantly on a 15 year time scale. *Vertical bars* indicate 15 year time intervals (adapted with permission from Albritton et al. 2001, © IPCC 2001)

The main opportunities for such an initiative are:

- Ongoing growth of computing power;
- Ongoing expansion of digital communications and the Internet;
- The existing Global Change Institutions and their programmes;
- Existing national investments in Earth System science;
- Political recognition of the importance of the enterprise (e.g., UNCED, Kyoto).

The impediments to progress include:

- Institutional barriers;
- The often equivocal commitment of national and international funding agencies;
- Limits of the researcher pool.

The latter include an insufficient quantity of researchers and the need to strengthen disciplinary skills as well as to overcome ongoing reluctance/resistance to engage in interdisciplinary research projects.

The way forward can be summarised as follows:

- Continue to integrate/refocus international programmes – but faster!
- Increase commitment from national/international programmes;
- Expand, integrate and sustain observing systems;
- Create internationally coordinated initiatives in researcher training and development with continuing elimination of the barriers between disciplines as a specific aim;
- Engage in a major thrust to influence human behaviour in ways beneficial to the Earth System and society, using information derived from an operational system of Earth simulators and observations.

In his opening address, Berrien Moore asked "Are we listening to the planet?" Unfortunately, the signal that the vast majority of humans receive is currently muffled and intermittent. Our objective must be the provision of a planetary "Hi-Fi" system, which nobody can ignore. I am confident that when connected to such a system, people will respond.

Reference

Albritton DL, et al. (2001) Climate Change 2001; The Scientific Basis; Summary for Policymakers and Technical Summary of the Working Group I Report, Cambridge University Press

Part IVb

Does the Earth System Need Biodiversity?

Chapter 30

Marine Biodiversity: Why We Need It in Earth System Science

Katherine Richardson

The term "biodiversity" is used to described the variety that is found in living organisms and their habitats. It is not a term that is easy to define or measure in scientific terms, and many scientists were less than delighted when politicians in the wake of the 1992 Rio Conference popularised the term with the launching of the Biodiversity Convention. Indeed, many scientists are still frustrated by their governments' attempts to focus research programmes on popular topics such as biodiversity. Nevertheless, this recent research focus has lead to a quantum leap in knowledge concerning how *nonhuman life on Earth is absolutely essential to the functioning of the Earth System*.

We now realise that life has evolved entirely differently in the oceans than on land. On land, the driving force of evolution has been competition for resources and space. This is exemplified by a rainforest, where every conceivable light and nutrient niche is utilised by one or another plant species. Interestingly, while these plants in the rainforest look very different to us, they are, in fact, much more closely related to each other taxonomically than the plants in the ocean. However, there is an even more important difference between life on land and in the oceans. Evolution in the oceans has, apparently, not been driven by competition for resources in the same way as evolution on land. If this were the case – and evolution had been as successful in the ocean as on land so that all possible light niches were filled – then the "ocean surface would be covered by an oily scum that would, as well as changing the planetary heat budget, severely reduce evaporation and hence rainfall on the continents, where life as we know it could not then have evolved" (Smetacek 2001). This means that *if life had evolved differently in the oceans, the Earth System as we know it today would not exist. Life in the oceans has contributed to making the Earth inhabitable for organisms such as human beings.*

Perhaps the best example of the importance of marine organisms in creating the contemporary Earth System are the tiny cyanobacteria. These primitive plant-like organisms evolved, about 3.5 billion years ago, the capacity to carry out a form of photosynthesis where oxygen is a by-product. Together with their photosynthesising descendants, these tiny organisms have completely changed the composition of the atmosphere on this planet, thus allowing the evolution of oxygen requiring species such as our own. *We often think of our own species as that which has had the most dramatic impact on the Earth's environment, but compared to the impact of the microscopic cyanobacteria, the global environmental changes brought about by humans are still tiny.*

For most people, the term "biodiversity of the oceans" conjures up images of leviathans or fish – the largest and most visible inhabitants of the seas. Indeed, conservation groups have for years been focusing public attention on large species whose numbers may be threatened. As demonstrated with the example of the cyanobacteria above, however, it is the very smallest of the oceans' inhabitants that have the most important role in the establishment and functioning of the Earth System. This can be easily explained using the carbon cycle as an example. The oceans contain about 50 times as much carbon as the atmosphere. The oceans are also in contact with the atmosphere, and exchange of carbon between the oceans and atmosphere is constantly occurring. This means that if we want to understand the carbon cycle in the atmosphere, we have to understand what is happening in the ocean as well. Of the total carbon found in the oceans, less than 5% is bound up in the biota, and there is approximately the same amount of carbon bound up in all of the different size groups of organisms found in the sea. In other words, there is about the same amount of carbon bound up in bacteria as in whales. Each individual bacterium contains much less carbon than a whale. However, there are so very many bacteria compared to whales in the oceans that the sum total of bacteria carbon is similar to the sum total of whale carbon. Through their metabolism, small organisms turn over elements much more quickly than large. *In an Earth System perspective, the role of the oceans' biota is to catalyse the movement of elements between compartments within the System, and the microscopic plants and animals of the oceans do this much more effectively than the larger organisms that attract so much attention.*

This paper focuses on the role of microscopic plants (phytoplankton) that float freely in the world's oceans in the functioning of the Earth System. As mentioned above, these plants are much less closely related to one another than the plants on land. Nevertheless, they have many similarities. All of the taxonomic groups to which they belong carry out photosynthesis (and together fix through photosynthesis on the order of 50 Gt carbon per year!). All of these organisms provide food and energy for other marine organisms, and all contribute to the cycling of nitrogen, carbon, phosphorus, and other elements in the Earth System. Some of these taxonomic groups of phytoplankton, however, play unique roles in elemental cycling. Some cyanobacteria, for example, are capable of fixing atmospheric nitrogen. In fact, *most of the nitrogen found in the world's oceans got there originally through the biological fixation of nitrogen by cyanobacteria.*

Another group of phytoplankton, the coccolithophorids, makes calcium carbonate scales. These sink and can ultimately become embedded in the seafloor. Through geological time, these deposits can become chalk or limestone. No one knows exactly how much calcium carbonate is produced by these organisms. However, *it is estimated that coccolithophorids, together with the other major calcium carbonate producers in the oceans, the corals and the foraminifera, fix on the order of a Gt carbon in calcium carbonate every year.*

Several groups of phytoplankton produce (in varying amounts) greenhouse gases such as dimethylsulphide (DMS). DMS is a volatile sulphur compound and is released to the atmosphere from the surface waters of the oceans. There it reacts with particles to form cloud condensation nuclei around which clouds form. Clouds then affect the amount and kind of light reaching the Earth's surface. They also affect temperature at the Earth's surface. In some cases, they may have a cooling effect as they may reflect heat entering the atmosphere. In other cases, clouds may have a net warming effect, as they may help to retain heat leaving the Earth's surface in the near-Earth atmosphere. Some scientists have argued that the net effect of DMS production by the ocean's plant life is a cooling of the atmosphere and that the ocean conditions predicted as a result of global warming (i.e., a warmer, more stable ocean) will stimulate the growth of DMS-producing organisms. *Thus, it is argued by some that the DMS-producing phytoplankton represent a kind of "feedback system" within the Earth System where, as the Earth warms, the stimulation of DMS producers causes greater cloud cover and cooling of the Earth.*

Another group of phytoplankton, the diatoms, has silica-containing cell walls. The incorporation of silica into diatom frustules and the subsequent decomposition or sinking and burial of these in the sea floor controls the Si cycle in the oceans.

These examples demonstrate the important functions relating to elemental cycling that the microscopic phytoplankton of the oceans carry out in the Earth System. There are about 4000–5000 known species of phytoplankton, and they are assigned to 16 different taxonomic classes. They vary in size from less than 1 micron in length to over a 1 mm. *Thus, while phytoplankton all seem very small to the human eye, the relative volume difference between the smallest and the largest is similar to the relative volume difference between a mouse and an elephant! This is marine biodiversity at its most magnificent. These and other microscopic marine organisms have helped to create and continue to maintain the Earth System as we know it. Thus, the Earth System NEEDS marine biodiversity, but in addition to being a threat to the leviathans of the ocean, global change is also a threat to these tiny phytoplankton.*

It is easy to identify general threats to entire groups of phytoplankton. Examples are: increasing atmospheric CO_2 concentrations that will increase CO_2 concentrations in surface waters of the oceans and decrease the pH. The lower pH will make the job of producing calcium carbonate chemically more difficult and provide less ideal conditions for calcite-producing organisms. Increasing temperatures will create a semipermanent layer of warm water at the surface of the oceans, thus making the water column more stable. Diatoms cannot thrive in a stable water column, as they require turbulence to keep them suspended in the well-lit surface waters and so on.

Identifying threats to individual species is more difficult, as our understanding of what controls species dominance and succession is still very primitive. We do know, however, that some individual species can be disproportionately important in elemental cycling. For example, about 75% of the Si being deposited in the world's

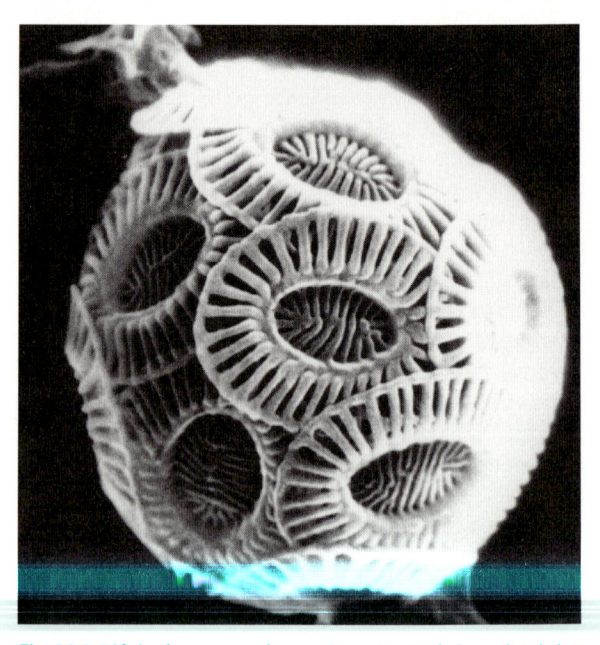

Fig. 30.1. Life in the oceans plays an important role in maintaining geochemical balance in the Earth System. Annual blooms of this coccolithophorid species, *Emiliania huxleyi*, are important in fixing carbon in the ocean, as individuals produce calcium carbonate platelets (liths) (photo: courtesy of Helge Thomsen, Danish Fisheries Research Institute)

oceans is deposited in the Southern Ocean – along the Antarctic Polar Frontal Zone, and approximately 1/3 of the Si being deposited here is associated with one single species, *Fragilariopsis kerguelensis*. The coccolithophorid, *Emiliania huxleyi* (Fig. 30.1), provides another example of a species with particular importance in global elemental cycling. This species forms huge blooms over much of the world's oceans, and its sinking coccoliths are an important mechanism of carbon transport from the surface to the deeper layers of the oceans. The geological record shows us, however, that prior to the Holocene, it was another species, *Gephyrocapsa* that was the major bloom forming coccolithophorid. *Gephyrocapsa* is still found but now plays a very minor role in carbon transport compared to *Emiliania*.

Thus, individual species are different and contribute differently to the functioning of the Earth System at any given time. At present, we know that different taxonomic groups of microscopic marine organisms contribute vital services to the functioning of the Earth System, but we do not know whether all species are equally important in terms of the Earth System and its function or which species might be more important than others. There is not question, then, that marine biodiversity – expecially the diversity in the organisms we cannot even see with the naked eye – is an integral component of the Earth System. This biodiversity deserves both protection and further investigation in our efforts to understand and maintain the functioning of the System.

Reference

Smetacek V (2001) A watery arms race. Nature 411:745

Chapter 31
Does Biodiversity Matter to Terrestrial Ecosystem Processes and Services?

Sandra Díaz

31.1 Is Biodiversity Important to the Functioning of the Earth System?

Terrestrial ecosystems, such as forests, savannahs, or grasslands produce living plant and animal mass, transform it back into inorganic components, and circulate water between the soil and the atmosphere. Because of these processes, ecosystems provide essential services to humankind. They provide food, medicines, and fibres for clothing and construction. They provide clean drinking water by capturing rainfall and filtering contaminated water. They hold the fertile soil in place and decrease the risk of floods. They regulate local air temperature and provide opportunities for recreation and aesthetic enjoyment. Unfortunately, these ecosystem services are often taken for granted, and most of us realise their importance when they cease to exist.

The present and future trends in natural resource use represent threats of unprecedented importance to natural ecosystems. On the one hand, projected changes in climate have provoked serious concern about what the direct effects of changes in temperature and precipitation would be on the geographical distribution and functioning of ecosystems such as boreal forests, semideserts, and alpine meadows. On the other hand, in view of the accelerated extinction of species because of land-use practices, some have started to wonder whether losing species from a given ecosystem could alter its functioning. The key question here is whether the bio-diversity of an ecosystem is important to its functioning and to the quantity and quality of the services it provides.

31.2 What is Biodiversity?

Biodiversity refers to the number and relative abundance of species, as well as the range of traits, present in a given ecosystem. For example, biodiversity includes the number and relative abundance of different species of trees, birds, mammals, mites, or insects. But it also consists of the range of sizes of plants (whether they are all small and herbaceous or tall and woody), whether they die, go dormant, or stay put during the unfavourable season, and whether birds eat fruits or insects, migrate or live in the same site during the whole year. Very often, natural and seminatural ecosystems consist of a high number of organisms, but typically some of them are very abundant, and some are rare.

31.3 The Most Abundant Plants Are Important for Ecosystem Functioning at Any Given Moment

We have good evidence that *the traits of the locally most abundant species* strongly influence the way in which an ecosystem works at a certain moment in time (Fig. 31.1a,b). If we have a single list of species with some of them being grasses and some being trees, we may end up with very different ecosystems depending on their relative abundances. If the grasses are much more abundant, we will get a very open savannah; if the trees predominate, we will get a closed forest. A forest and an open savannah function very differently and have very different services to offer. Under the same climate and on the same soil, a clover plot will have different biomass, will feed herbivores differently and will provide different roosting opportunities for animals, than a eucalypt plot.

The fact that ecosystems can be strongly influenced by just one very abundant species is well illustrated by biological invasions. Often the invading species is only one, but it can have huge implications for ecosystem processes. Examples have been the invasion of the prickly pear *Opuntia* in Australia some decades ago, which drastically changed the pastoral value of an extensive land surface, or the invasion of Mediterranean grasslands by tall tussock grasses, which have substantially increased the frequency of fires.

Ecosystem functioning is also likely to be influenced by the *range of different characteristics* exhibited by the component species (Fig. 31.1c,d). For example, as compared with a purely evergreen forest, a forest containing a mixture of deciduous and evergreen trees represents more opportunities for understorey herbs to grow and flower before the canopy closure in summer. It provides a patchwork of sunnier, hotter, drier sites and shadier, cooler and wetter sites, which can be differentially used by different plant and animal species. The falling leaves from different trees can form thick or thin layers and can decompose fast or slowly, thus representing different resources and conditions for soil organisms.

31.4 The Number of Functionally Similar Species Is Important in Facing Environmental Change

The strong influence of the most abundant plants on ecosystem functioning does not mean we can afford to underestimate the importance of species richness. Although there are very many different species of plants, they are not equally different from each other. Some scientists believe that several species can perform more or less the same ecosystemic 'job'. Examples are legumes that grow fast and incorporate nitrogen to the soil, tussock grasses that monopolise space and increase the probability of fire, trees that provide food and nesting places for certain kinds of birds, or slow-growing mosses that isolate the soil from air temperature and make it very acidic, thus slowing down decomposition. When there is more than one species doing the same job (i.e. belonging to the same "functional type") in a given system, some scientists have argued that there is *redundancy* with respect to that particular ecosystem function (Fig. 31.1e,f).

The concept of functional redundancy is to be taken with great caution, though. It is not related to dispensability, and certainly has nothing to do with "job redundancy" (i.e. allegedly too many participants doing a job that could be done just as well by only one of them). Functional redundancy is much closer to the reliability engineering concept: different components doing the same job, far from being unnecessary repetition, provide insurance to the system. In case one component fails, a few others can replace it, so that the functioning of the system is not seriously disturbed. Some plant species can be functionally equivalent at a certain time, but they may show different responses to a rapidly changing environment. For example, some of them may tolerate untimely frosts, higher grazing pressure, the competition of an invasive species, or the attack of a pest better than others. So the overall "job" these species perform in an ecosystem may not be dramatically changed if one or a few of them disappear.

31.5 Implications for Conservation and Sustainable Management

All these ideas suggest that species number *per se* (that is, without considering who is there and how many different characteristics are represented) is not always necessarily and directly linked with "good ecosystem performance." The continued presence of some key species, and – most importantly – the preservation of their interactions, can be much more crucial than the maintenance of high numbers of species for its own sake. However, this is not suggesting we should stop worrying about accelerated biodiversity loss. Usually, sophisticated pieces of equipment performing important functions, such as planes, are constructed with a high level of built-in redundancy. Just as very few people would recommend getting rid of three engines in a four-engine plane, the argument of letting species go extinct because other members of the same functional type seem able to perform more or less the same ecosystemic role *at present*, does not seem wise.

From the perspective of local to regional conservation and sustainable management, the main consequence of the argument developed here, and that of the more traditional approach focused on the importance of species numbers, are very similar: try to preserve the highest number of species, and try to keep their relative abundances and interactions as intact as possible. What is new, then? I believe we have made a modest step forward in our understanding of why and how diversity

Fig. 31.1.
Three important components of biodiversity

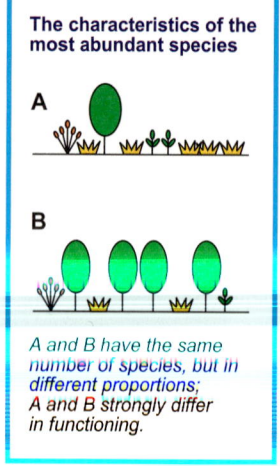

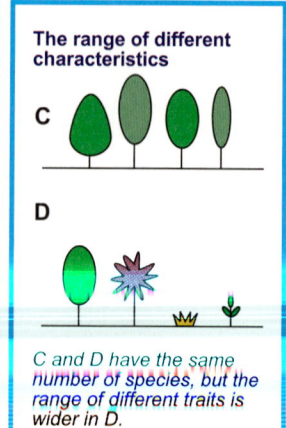

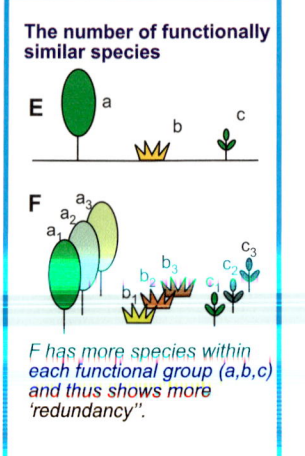

A and B have the same number of species, but in different proportions; A and B strongly differ in functioning.

C and D have the same number of species, but the range of different traits is wider in D.

F has more species within each functional group (a,b,c) and thus shows more 'redundancy''.

matters for ecosystem functioning. This should hopefully lead us to some practical progress, in the same way having some fragmentary pages of the manual of instructions is a step forward with respect to having just a list of pieces when trying to assemble or repair a complex piece of equipment.

Because biodiversity effects on ecosystem processes are most obvious at the local scale, the answer to the question, 'does the planet – the Earth System – need biodiversity?' may not be obvious at first inspection. However, the two aspects of biodiversity explained above may be of crucial importance for the maintenance of the Earth System's functioning. On the one hand, the results of the combined impacts of climate and land-use changes on ecosystems are very difficult to predict. Functional redundancy may help buffer the biosphere against 'surprises' and 'catastrophes' (i.e., unexpected and accelerated changes). On the other hand, because the characteristics of dominant species matter to ecosystem functioning, we need more than just "some green stuff" to preserve the cycling of energy and materials at the planetary scale. We cannot rely on a uniform set of very common, 'weedy' species to maintain ecosystem functioning in different regions. The current evidence indicates that the preservation of the local arrangements of abundant and rare species and their interactions could be important not only for local ecosystem services, but also for the large-scale functioning of the Earth System.

Chapter 32

Biodiversity Loss and the Maintenance of Our Life-Support System

Michel Loreau

Early global change research focused on the interactions between environmental changes (mainly climate change), ecosystem functioning and human societies. When biodiversity was added to the picture, it was to the extent that it is affected by these other components of global change (Fig. 32.1). But why does biodiversity matter to us? There are at least three classes of reasons why it does. First, it provides us with a number of goods that have direct economic value, such as food, new pharmaceuticals, genes that improve crops, and organisms that perform biological control. Second, it is intricately linked to human well-being for aesthetic, ethical, cultural and scientific reasons. And third, it may contribute to the provision of ecological services that are generally not accounted for in economic terms, such as primary and secondary production, the regulation of climate, the maintenance of atmosphere quality, the regulation of the hydrological cycle, the maintenance of water quality, and the maintenance of soil fertility. During the last decade, the effects of biodiversity on the other components of global change have received increasing attention (Fig. 32.1). In particular, there has been an explosive growth of research into the potential effects of biodiversity loss on ecosystem functioning and thereby on the provision of ecological goods and services to human societies (see review in Loreau et al. 2001).

32.1 How Does Biodiversity Affect Ecosystem Functioning at Small Scales?

To investigate the effects of biodiversity on ecosystem processes, a new wave of experimental studies has manipulated species diversity using synthesised model ecosystems in both terrestrial and aquatic environments. While the first study that experimentally manipulated diversity did so across several trophic levels (Naeem et al. 1994), later studies focused mainly on effects of plant taxonomic diversity and plant functional-group diversity on primary production and nutrient retention in grassland ecosystems (e.g., Tilman et al. 1996, 1997a; Hooper and Vitousek 1997; Hector et al. 1999). Because plants, as primary producers, represent the basal component of most ecosystems, they represent the logical place to begin detailed studies. Several, though not all, experiments using randomly assembled communities found that plant species and functional-group richness both have a positive effect on primary production and nutrient retention. The largest of these experiments to date, the BIODEPTH project, showed a consistent positive effect of diversity on plant aboveground biomass production across eight sites with widely different soils and climates in Europe (Hector et al. 1999; Fig. 32.2). Each halving of the number of plant species reduced productivity by approximately 80 g m^{-2} on average, a figure that might not look spectacular on a m^2 basis, but that does when extrapolated to the total surface area covered by European grasslands (about 48 million t). Similarly, the omission of a single functional group reduced productivity by approximately 100 g m^{-2} on average.

It is fair to mention, however, that the interpretation of these experiments has been controversial (e.g., Huston et al. 2000; Hector et al. 2000), because their results can be generated by different mechanisms. These mechanisms may be grouped into two main classes (Loreau 1998, 2000). First are local deterministic processes, such as niche differentiation and facilitation, which increase

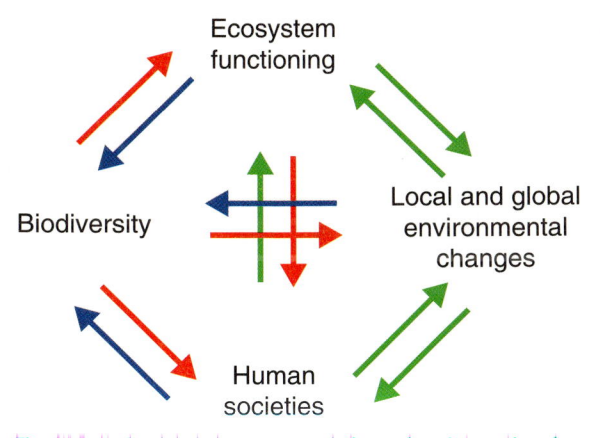

Fig. 32.1. Early global change research focused on interactions between environmental changes, ecosystem functioning and human societies (*green arrows*). When biodiversity first entered the picture, it was to the extent that it is affected by the other components of the Earth System (*blue arrows*). During the last decade, new research has focused on how biodiversity itself affects the other components, in particular how it influences ecosystem functioning and the provision of ecological goods and services to human societies (*red arrows*)

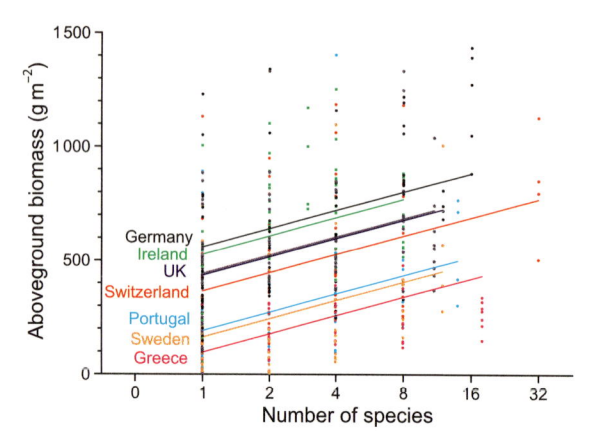

Fig. 32.2. Aboveground biomass production on average increases with plant species richness in grassland ecosystems. Results from the second year of the BIODEPTH experiment across eight sites in Europe (modified from Hector et al. 1999)

the performance of communities above that expected from the performance of individual species grown alone. I shall subsume them under the term "complementarity". Second are local and regional stochastic processes involved in community assembly, which are mimicked in experiments by random sampling from a species pool. Random sampling coupled with local dominance of highly productive species can also lead to increased average primary production with diversity, because plots that include many species have a higher probability of containing highly productive species. As sampling processes were not an explicit part of the initial hypotheses, they have been viewed by some as "hidden treatments" (e.g., Huston 1997), whereas others have viewed them as the simplest possible mechanism linking diversity and ecosystem functioning (e.g., Tilman et al. 1997b).

New theoretical advances are making the resolution of this controversy possible. First, it is becoming clear that complementarity and sampling are not mutually exclusive mechanisms as previously thought. Communities with more species have a greater probability of containing a higher phenotypic trait diversity. Ecological "selection" of species with particular traits that leads to dominance and complementarity among species with different traits are two ways by which this phenotypic diversity maps onto ecosystem processes (Loreau 2000). These two mechanisms, however, may be viewed as two poles on a continuum from pure dominance to pure complementarity. Intermediate scenarios involve complementarity among particular sets of species or functional groups, or dominance of particular subsets of complementary species. Any bias in community assembly that leads to correlations between diversity and community composition may involve both dominance and complementarity.

Second, a new methodology now exists to partition the net effect of biodiversity into a selection effect and a complementarity effect, using comparisons between the

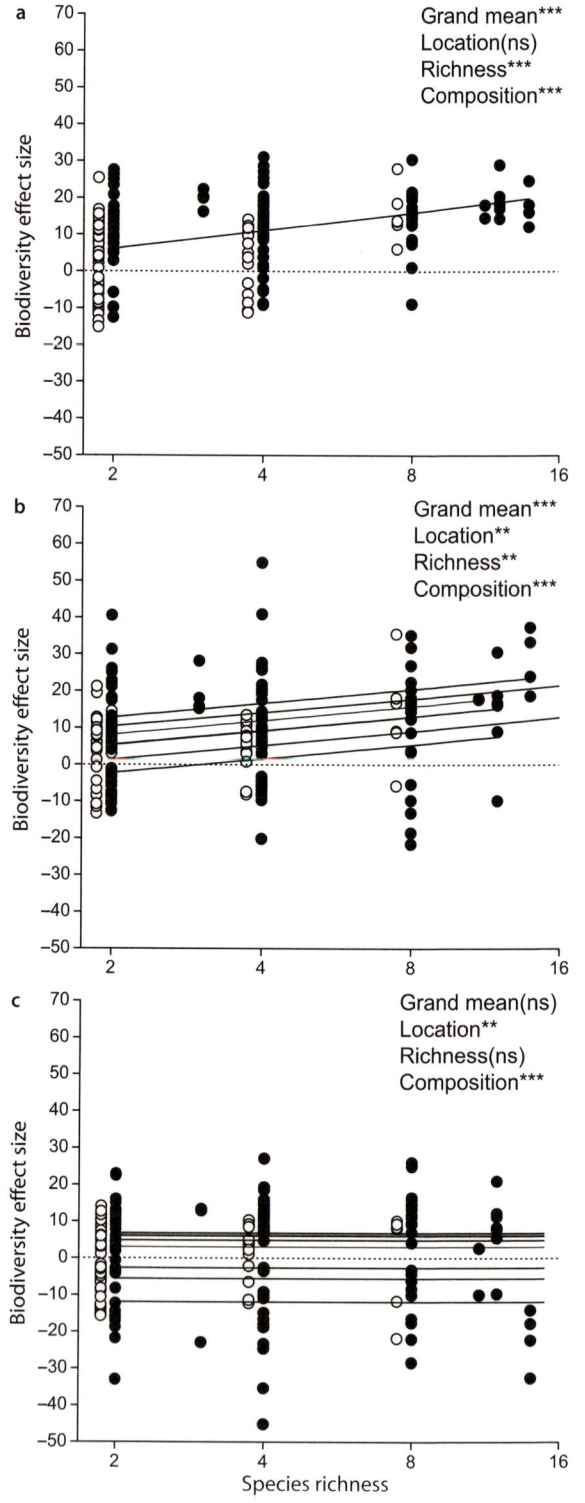

Fig. 32.3. Partitioning of biodiversity effects in the results from the BIODEPTH experiment (Fig. 32.1). **a** Net, **b** complementarity and **c** selection effects are based on comparisons of mixture yields and monoculture yields. Their values (in $g\ m^{-2}$) are square root transformed but preserve the original positive and negative signs. *Open circles* are plots that do not contain any legume species; *filled circles* are plots that contain legumes. *: $P < 0.05$, **: $P < 0.01$, ***: $P < 0.001$, ns: not significant (modified from Loreau and Hector 2001)

yield of a mixture and its expected value based on monoculture yields (Loreau and Hector 2001). Application of this methodology to the data from the BIODEPTH experiment showed that the selection effect was variable, ranging from negative to positive values in different localities, but was zero on average (Fig. 32.3). Therefore it can be rejected as the sole mechanism explaining results from this experiment. In contrast, there was a consistent positive complementarity effect. This supports the hypothesis that plant diversity influences primary production through local biological processes such as niche differentiation and facilitation. Similar conclusions have been obtained from long-term data in another large-scale biodiversity experiment at Cedar Creek, Minnesota (Tilman et al. 2001).

Thus, there is little doubt that plant species diversity does affect ecosystem processes such as primary production and nutrient retention in grassland ecosystems, even at the small spatial and temporal scales considered in recent experiments. What remains unclear, however, is how many species are involved in these effects. And what is still largely unexplored is whether similar effects of diversity also occur at other trophic levels, and in other ecosystems, such as forests, and freshwater and marine ecosystems.

32.2 Scaling Up in Time: Biodiversity As an Insurance Against Environmental Changes

Even when high diversity is not critical for maintaining ecosystem processes under constant or benign environmental conditions, it might nevertheless be important for maintaining them under changing conditions. The insurance hypothesis proposes that biodiversity provides an "insurance", or a buffer, against environmental fluctuations, because different species respond differently to these fluctuations, leading to functional compensations between species and hence more predictable aggregate community or ecosystem properties (McNaughton 1977; Yachi and Loreau 1999). A number of studies have recently provided theoretical foundations for this hypothesis (e.g., Doak et al. 1998; Yachi and Loreau 1999; Lehman and Tilman 2000). Several empirical studies have found decreased variability of ecosystem processes as diversity increases, despite sometimes increased variability of individual populations, in agreement with the insurance hypothesis (Fig. 32.4). The interpretation of these patterns, however, is complicated by the correlation of additional factors with species richness in these experiments, which does not fully preclude alternative interpretations (e.g., Huston 1997). Experiments in which both diversity and environmental fluctuations are controlled are needed to perform more rigorous tests of the insurance hypothesis.

The important message here is that time adds another dimension to the potential for complementarity among species and hence for biodiversity effects on ecosystem properties. Species that appear to be functionally redundant for an ecosystem process at a given time may no longer be redundant through time, because their variations in abundance or metabolism compensate each other. Society depends on the steady and predictable input of ecological services; biodiversity may provide greater stability in the production of these services, such as timber production, pollination levels, biomass production, and nutrient cycling. Functional compensations may be particularly important in a world that is changing in many ways at an unprecedented rate. Biodiversity loss may reduce the ability of natural and managed ecosystems to cope with other global changes. As an example, some recent experiments showed significant effects

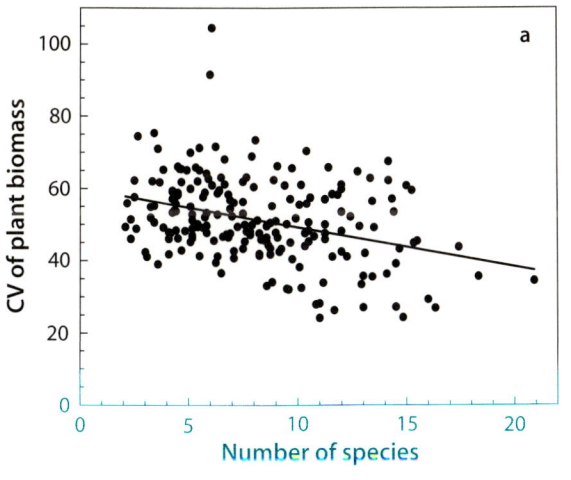

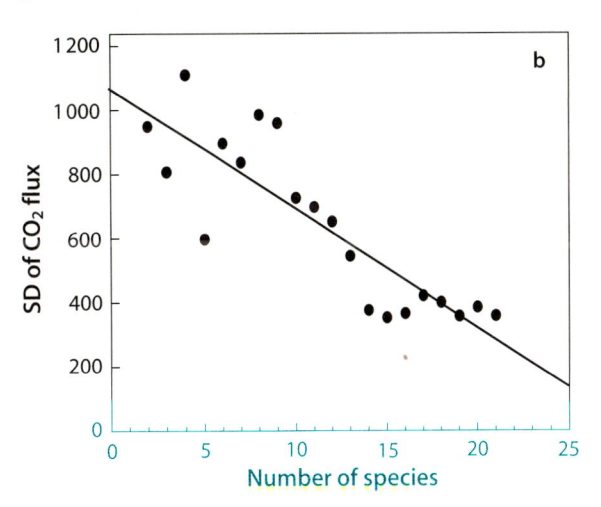

Fig. 32.4. Variability of ecosystem processes decreases as species richness increases. a Adjusted coefficient of temporal variation (*CV*) of annual total plant biomass (in g m^{-2}) over 11 years for plots in experimental and natural grasslands in Minnesota (reproduced from Tilman 1999a with permission, Ecological Society of America). b Standard deviation (*SD*) of CO_2 flux (in µl per 18 h) from microbial microcosms (reproduced with permission from Nature; McGrady-Steed et al. (1997) Nature 390:162–165, ©1997 MacMillan Magazines)

of plant diversity on the responses of grassland ecosystems to elevated atmospheric CO_2 concentration and nitrogen deposition (e.g., Reich et al. 2001). Thus, short-term experiments are very likely to underestimate the importance of biodiversity for the maintenance of ecosystem processes and services.

32.3 Scaling Up in Space: Biodiversity Effects at Landscape and Regional Scales

Small-scale experiments are equally likely to underestimate the functional significance of biodiversity. Many species that coexist locally occupy similar ecological niches at small scales because the niche differences that allow them to coexist are situated at larger scales, in their habitat differences from the scale of the landscape to that of the region. Just as diversity allows functional compensations between species through time, it allows functional compensations through space. Species replace each other along environmental gradients, because different species have different optimum abilities along these gradients. The larger the spatial scale, the greater the environmental heterogeneity, and the higher the biological diversity needed to take full advantage of these environmental differences. Species richness typically increases with surface area as a power function (Rosenzweig 1995). These species–area curves have been used to extrapolate the diversity that is needed at larger scales to perform the ecosystem processes that have been studied at small scales (Tilman 1999b).

There are good reasons to believe that these estimates are underestimates of the actual importance of bio-diversity at large scales. Habitat shifts and fragmentation following land-use and climate changes are transforming, and will increasingly transform landscapes into patch mosaics. Accordingly, spatially continuous communities are being broken up into "metacommunities", i.e., isolated local communities connected by dispersal fluxes. Since many species from stable natural ecosystems are poor dispersers, habitat shifts and fragmentation will increase recruitment limitation of the appropriate set of complementary species that best perform given ecosystem processes in each site, and thereby amplify the effects of biodiversity loss on ecosystem functioning. Another way of seeing this is through species-area curves. Isolation at both ecological or evolutionary time scales typically has the effect of increasing the exponent of the power function relating species richness to surface area (Rosenzweig 1995). This implies that an even higher diversity is needed at large scales to ensure a given level of diversity at small scales, or, equivalently, that biodiversity loss at regional scales increases the extent and impact of biodiversity loss at local scales.

32.4 Conclusions

Recent theoretical and experimental work provides clear indications that biodiversity loss can have profound impacts on the functioning of the Earth System and the maintenance of our life-support system. Biodiversity can both enhance some ecosystem processes, such as productivity and nutrient retention, and act as a biological insurance against potential disruptions caused by environmental changes. Therefore biodiversity can no longer be ignored in global change and environmental issues.

Despite the explosive growth of the biodiversity-ecosystem functioning area during the last few years, knowledge of the functional consequences of biodiversity loss is still limited. The main challenges that lie ahead of us are:

1. To extend current knowledge on plant-based processes in temperate grasslands to other organisms (animals and microorganisms), other trophic levels (herbivores, predators and decomposers) and other ecosystems (forest, tropical, fresh-water and marine ecosystems);
2. To understand impacts of biodiversity change at larger temporal and spatial scales in interaction with other environmental changes, in particular land-use change;
3. To extend current research beyond a basic science perspective and focus on impacts on the provision of ecosystem goods and services of relevance to human societies.

This science agenda is that of the joint project of DIVERSITAS and IGBP-GCTE on Biodiversity, global change and ecosystem functioning. DIVERSITAS, as the international programme on biodiversity science, further aims to integrate this research on the functional and societal impacts of biodiversity change into a broader picture that includes the causes and processes of biodiversity change as well as the ways by which biodiversity can be conserved and used in a sustainable manner.

References

Doak DF, Bigger D, Harding EK, Marvier MA, O'Malley RE, Thomson D (1998) The statistical inevitability of stability-diversity relationships in community ecology. Am Nat 151: 264–276

Hector A, Schmid B, Beierkuhnlein C, Caldeira MC, Diemer M, Dimitrakopoulos PG, Finn JA, Freitas H, Giller PS, Good J, Harris R, Högberg P, Huss-Danell K, Joshi J, Jumpponen A, Körner C, Leadley PW, Loreau M, Minns A, Mulder CPH, O'Donovan G, Otway SJ, Pereira JS, Prinz A, Read DJ, Scherer-Lorenzen M, Schulze E-D, Siamantziouras A-SD, Spehn EM, Terry AC, Troumbis AY, Woodward FI, Yachi S, Lawton JH (1999) Plant diversity and productivity experiments in European grasslands. Science 286:1123–1127

Hector A, Schmid B, Beierkuhnlein C, Caldeira MC, Diemer M, Dimitrakopoulos PG, Finn JA, Freitas H, Giller PS, Good J, Harris R, Högberg P, Huss-Danell K, Joshi J, Jump-ponen A, Körner C, Leadley PW, Loreau M, Minns A, Mulder CPH, O'Donovan G, Otway SJ, Pereira JS, Prinz A, Read DJ, Scherer-Lorenzen M, Schulze E-D, Siamantziouras A-SD, Spehn EM, Terry AC, Troumbis AY, Woodward FI, Yachi S, Lawton JH (2000) No consistent effect of plant diversity on productivity: Response. Science 289:1255a, *http://www.sciencemag.org/cgi/content/full/289/5483/1255a*

Hooper DU, Vitousek PM (1997) The effects of plant composition and diversity on ecosystem processes. Science 277: 1302-1305

Huston MA (1997) Hidden treatments in ecological experiments: re-evaluating the ecosystem function of biodiversity. Oecologia 110:449-460

Huston MA, Aarsen LW, Austin MP, Cade BS, Fridley JD, Garnier E, Grime JP, Hodgson J, Lauenroth WK, Thompson K, Vandermeer JH, Wardle DA (2000) No consistent effect of plant diversity on productivity. Science 289:1255a, *http://www.sciencemag.org/cgi/content/full/289/5483/1255a*

Lehman CL, Tilman D (2000) Biodiversity, stability, and productivity in competitive communities. Am Nat 156:534-552

Loreau M (1998) Biodiversity and ecosystem functioning: a mechanistic model. Proc. Natl Acad Sci USA 95:5632-5636

Loreau M (2000) Biodiversity and ecosystem functioning: recent theoretical advances. Oikos 91:3-17

Loreau M, Hector A (2001) Partitioning selection and complementarity in biodiversity experiments. Nature 412:72-76

Loreau M, Naeem S, Inchausti P, Bengtsson J, Grime JP, Hector A, Hooper DU, Huston MA, Raffaelli D, Schmid B, Tilman D, Wardle DA (2001) Biodiversity and ecosystem functioning: current knowledge and future challenges. Science 294: 804-808

McGrady-Steed J, Harris PM, Morin PJ (1997) Biodiversity regulates ecosystem predictability. Nature 390:162-165

McNaughton SJ (1977) Diversity and stability of ecological communities: a comment on the role of empiricism in ecology. Am Nat 111:515-525

Naeem S, Thompson LJ, Lawler SP, Lawton JH, Woodfin RM (1994) Declining biodiversity can alter the performance of ecosystems. Nature 368:734-737

Reich PB, Knops J, Tilman D, Craine J, Ellsworth D, Tjoelker M, Lee T, Wedin D, Naeem S, Bahauddin D, Hendrey G, Jose S, Wrage K, Goth J, Bengdton W (2001) Plant diversity enhances ecosystem responses to elevated CO_2 and nitrogen deposition. Nature 410:809-812

Rosenzweig ML (1995) Species diversity in space and time. Cambridge University Press, Cambridge

Tilman D (1999a) The ecological consequences of changes in biodiversity: a search for general principles. Ecology 80:1455-1474

Tilman D (1999b) Diversity and production in European grasslands. Science 286:1099-1100

Tilman D, Wedin D, Knops J (1996) Productivity and sustainability influenced by biodiversity in grassland ecosystems. Nature 379:718-720

Tilman D, Knops J, Wedin D, Reich P, Ritchie M, Siemann E (1997a) The influence of functional diversity and composition on ecosystem processes. Science 277:1300-1302

Tilman D, Lehman C, Thompson K (1997b) Plant diversity and ecosystem productivity: theoretical considerations. Proc Natl Acad Sci USA 94:1857-1861

Tilman D, Reich PB, Knops J, Wedin D, Mielke T, Lehman C (2001) Diversity and productivity in a long-term grassland experiment. Science 294:843-845

Yachi S, Loreau M (1999) Biodiversity and ecosystem productivity in a fluctuating environment: The insurance hypothesis. Proc Natl Acad Sci USA 96:1463-1468

Part IVc
Can Technology Spare the Planet?

Chapter 33

Maglevs and the Vision of St. Hubert – Or the Great Restoration of Nature: Why and How

Jesse H. Ausubel

33.1 Introduction

The emblems of my essay are maglevs speeding through tunnels below the surface of Earth and a crucifix glowing between the antlers of a stag, the vision of St. Hubert. Propelled by magnets, maglev trains levitate passengers with green mobility. Maglevs symbolise technology, while the fellowship of St. Hubert with other animals symbolises behaviour.

Better technology and behaviour can do much to spare and restore nature during the 21st century, even as more numerous humans prosper.

In this essay I explore the areas of human use for fishing, farming, logging, and cities. Offsetting the sprawl of cities, rising yields in farms and forests and changing tastes can spare wide expanses of land. Shifting from hunting seas to farming fish can similarly spare nature. I will conclude that cardinal resolutions to census marine life, lift crop yields, increase forest area, and tunnel for maglevs would firmly promote the Great Restoration of nature on land and in the sea. First, let me share the vision of St. Hubert.

33.2 The Vision of St. Hubert

In The Hague, about the year 1650, a 25 year-old Dutch artist, Paulus Potter, painted a multi-panelled picture that graphically expresses contemporary emotions about the environment (Walsh et al. 1994). Potter named his picture "The Life of the Hunter" (Fig. 33.1). The upper left panel establishes the message of the picture with reference to the legend of the vision of St. Hubert.[1] Around the year 700, Hubert, a Frankish courtier, hunted deep in the Ardennes forest on Good Friday, a Christian spring holy day. A stag appeared before Hubert with a crucifix glowing between its antlers, and a heavenly voice reproached him for hunting, particularly on Good Friday. Hubert's aim faltered, and he renounced his bow and arrow. He also renounced his riches and military honours and became a priest in Maastricht.

The upper middle panel, in contrast, shows a hunter with two hounds. Seven panels on the sides and bottom show the hunter and his servant hounds targeting other animals: rabbit, wolf, bull, lion, wild boar, bear, and mountain goat. The hunter's technologies include sword, bow and guns.

One panel on either side recognises consciousness, in fact, self-consciousness, in our fellow animals. In the middle on the right, a leopard marvels at its reflection in a mirror. On the lower left, apes play with their self-images in a shiny plate.

In the large central panels, Potter judges 17th century hunters. First, in the upper panel the man and his hounds come before a court of the animals they have hunted. In the lower central final panel, the animal jury celebrates uproariously, while the wolf, rabbit and monkey cooperate to hang the hunter's dogs as an elephant, goat and bear roast the hunter himself. Paulus Potter believed the stag's glowing cross converted St. Hubert to sustainability. The hunter remained unreconstructed.

With Paulus and Hubert, we can agree on the vision of a planet teeming with life, a Great Restoration of Nature. And most would agree we need ways to accommodate the billions more humans likely to arrive while simultaneously lifting humanity's standard of living. In the end, two means exist to achieve the Great Restoration. St. Hubert exemplifies one behavioural change. The hunter's primitive weapons hint at the second technology. What can we expect from each? First, some words about behaviour.

33.3 Our Triune Brain

In a fundamental 1990 book, *The Triune Brain in Evolution*, neuroscientist Paul MacLean explained that humans have three brains, each developed during a stage of evolution (MacLean 1990). The earliest, found in snakes, MacLean calls the reptilian brain (Fig. 33.2). In

---

[1] The upper right panel shows Diana and Acteon, from the *Metamorphosis* of the Roman poet Ovid. Acteon, a hunter, was walking in the forest one day after a successful hunt and intruded in a sacred grove where Diana, the virgin goddess, bathed in a pond. Suddenly, in view of Diana, Acteon became inflamed with love for her. He was changed into a deer, from the hunter to what he hunted. As such, he was killed by his own dogs. This panel was painted by a colleague of Potter.

Fig. 33.1. The Life of the Hunter by Paulus Potter. The painting hangs in the museum of the Hermitage, St. Petersburg

mammals, another brain appeared, the paleomammalian, bringing such new behaviour as care of the young and mutual grooming. In humans came the most recent evolutionary structure, the hugely expanded neocortex. This neomammalian brain brought language, visualisation, and symbolic skills. But conservative evolution did not replace the reptilian brain, it added. Thus, we share primal behaviour with other animals, including snakes. The reptilian brain controls courting mates, patrolling territory, dominating submissives, and flocking together. The reptilian brain makes most of the sensational news and will not retreat. Our brains and thus our basic instincts and behaviours have remained largely unchanged for a million years or more. They will not change on time scales considered for "sustainable development".

Of course, innovations may occur that control individual and social behaviour. Law and religion both try, though the snake brain keeps reasserting itself, on Wall Street, in the Balkans, and by clawing for Nobel prizes in Stockholm.

Pharmacology also tries for behavioural control, with increasing success. Having penetrated only perhaps 10% of their global market, sales of new "anti-depressants", mostly tinkering with serotonin in the brain, neared $10 billion in 2000. Drugs can surely make humans very happy, but without restoring nature.

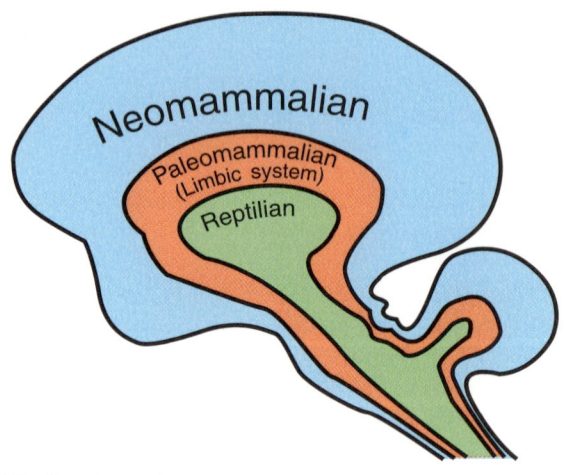

Fig. 33.2. Symbolic representation of the triune brain (after MacLean 1990)

Because I believe behavioural sanctions will be hard-pressed to control the eight or ten billion snake brains persisting in humanity, we should use our hugely expanded neocortex on technology that allows us to tread lightly on Earth. Since the beginning, *Homo faber* has been trying to make things better and to make better things. During the past two centuries, we have become more systematic and aggressive about it, through the diffusion of research and development and the institutions that perform them, including corporations and universities.

What can behaviour and technology do to spare and restore Nature during the 21st century? Let's consider the seas and then the land.

33.4 Sparing Sea Life

St. Hubert exemplifies behaviour to spare land's animals. Many thousands of years ago our ancestors sharpened sticks and began hunting. They probably extinguished a few species, such as woolly mammoths, and had they kept on hunting, they might have extinguished many more. Then without waiting on St. Hubert, our ancestors ten thousand years ago began sparing land animals in Nature by domesticating cows, pigs, goats, and sheep. By herding rather than hunting animals, humans began a technology to spare wild animals – on land.

In 2001 about 90 million t of fish are being taken wild from the sea and 30 from fish farms and ranches. Sadly, little reliable information quantifies the diversity, distribution, and abundance of life in the sea, but many anecdotes suggest large, degrading changes. In any case, the ancient sparing of land animals by farming shows us an effective way to spare the fish in the sea. We need to raise the share we farm and lower the share we catch. Other human activities, such as urbanisation of coastlines and tampering with the climate, disturb the seas, but today fishing matters most. Compare an ocean before and after heavy fishing.

Fish farming does not require invention. It has been around for a long time. For centuries, the Chinese have been doing very nicely raising herbivores, such as carp.

Following the Chinese example, one feeds crops grown on land by farmers to herbivorous fish in ponds. Much aquaculture of carp and tilapia in Southeast Asia and the Philippines and of catfish near the Gulf Coast of the USA takes this form. The fish grown in the ponds spare fish from the ocean. Like poultry, fish efficiently convert protein in feed to protein in meat. And because the fish do not have to stand, they convert calories in feed into meat even more efficiently than poultry. All the improvements such as breeding and disease control that have made poultry production more efficient can be and have been applied to aquaculture, improving the conversion of feed to meat and sparing wild fish.[2] With due care for effluents and pathogens, this model can multiply many times in tonnage.

A riskier and fascinating alternative, ocean farming, would actually lift life in the oceans (Ausubel 2000; Markels 1995). The oceans vary vastly in their present productivity. In parts of the ocean, crystal clear water enables a person to see 50 meters down. These are deserts. In a few garden areas, where one can see only a meter or so, life abounds. Water rich in iron, phosphorus, trace metals, silica, and nitrate makes these gardens dense with plants and animals. The experiments for marine sequestration of carbon demonstrate the extraordinary leverage of iron to make the oceans bloom.

Adding the right nutrients in the right places might lift fish yields by a factor of hundreds. Challenges abound, because the ocean moves and mixes both vertically and horizontally. Nevertheless, technically and economically promising proposals exist for farming on a large scale in the open ocean with fertilisation in deep water. One kg of buoyant fertiliser, mainly iron with some phosphate, could produce a few thousand tons of biomass.[3]

Improving the fishes' pasture of marine plants is the crucial first step to greater productivity. Zooplankton then graze on phytoplankton, and the food chain continues until the sea teems with diverse life. Fertilising 250 000 km^2 of barren tropical ocean, the size of the USA state of Colorado, in principle might produce a catch matching today's fish market of 100 million t. Colorado spreads less than 1/10th of 1% as wide as the world ocean.

The point is that today's depleting harvest of wild fishes and destruction of marine habitat to capture them need not continue. The 25% of seafood already raised by aquaculture signals the potential for Restoration (Fig. 33.3). Following the example of farmers who spare land and wildlife by raising yields on land, we can concentrate our fishing in highly productive, closed systems on land and in a few highly productive ocean farms. Humanity can act to restore the seas, and thus also preserve traditional fishing where communities value it. With smart aquaculture, we can multiply life in the oceans while feeding humanity and restoring nature. St. Hubert, of course, might improve the marine prospect by not eating fellow creatures from the sea.

33.5 Sparing Farmland

What about sparing nature on land? How much must our farming, logging, and cities take?

First, can we spare land for nature while producing our food (Waggoner and Ausubel 2000)? Yields per hec-

---

[2] In some fish ranching, notably most of today's ranching of salmon, the salmon effectively graze the oceans, as the razorback hogs of a primitive farmer would graze the oak woods. Such aquaculture consists of catching wild "junk" fish or their oil to feed to our herds, such as salmon in pens. We change the form of the fish, adding economic value, but do not address the fundamental question of the tons of stocks. A shift from this ocean ranching and grazing to true farming of parts of the ocean can spare others from the present, ongoing depletion.

[3] Along with its iron supplement, such an ocean farm would annually require about 4 million tons of nitrogen fertiliser, 1/20th of the synthetic fertilisers used by all land farms.

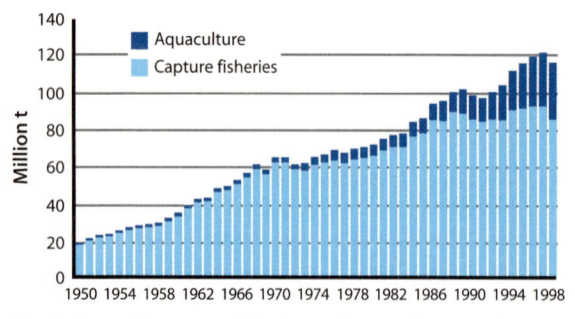

Fig. 33.3. World capture fisheries and aquaculture production. Note the rising amount and share of aquaculture (source: Food and Agriculture Organisation of the UN, The state of world fisheries and aquaculture 2000, Rome. http://www.fao.org/DOCREP/003/X8002E/X8002E00.htm)

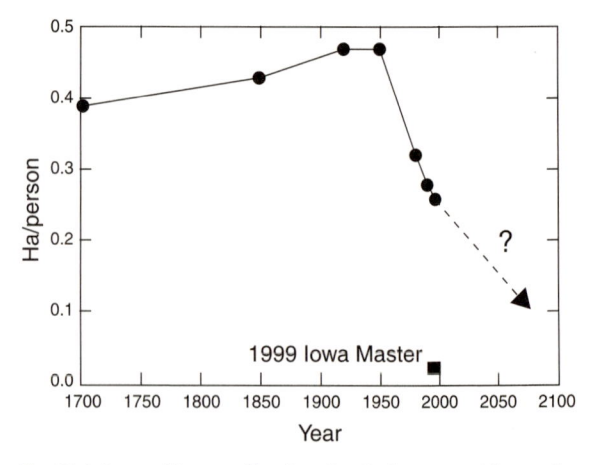

Fig. 33.4. Reversal in area of land used to feed a person. After gradually increasing for centuries, the worldwide area of cropland per person began dropping steeply in about 1950, when yields per hectare began to climb. The *square* shows the area needed by the Iowa Master Corn Grower of 1999 to supply one person a year's worth of calories. The *dashed line* shows how sustaining the lifting of average yields 2 percent per year extends the reversal (sources of data: Food and Agriculture Organisation of the United Nations, various Yearbooks; National Corn Growers Association, National Corngrowers Association Announces (1999) Corn Yield Contest Winners, Hot Off the Cob, St. Louis MO, 15 December 1999; Richards 1990

tare measure the productivity of land and the efficiency of land use. For centuries, land crops expanded faster than population, and cropland per person rose as people sought more proteins and calories. Fifty years ago, farmers stopped ploughing up nature (Fig. 33.4). During the past half-century, ratios of crops to land for the world's major grains-corn, rice, soybean, and wheat-have climbed fast on all six of the farm continents. Between 1972–1995, Chinese cereal yields rose 3.3% per year per hectare. Per hectare, the global Food Index of the Food and Agriculture Organisation of the UN, which reflects both quantity and quality of food, has risen 2.3% annually since 1960. In the USA in 1900, the protein or calories raised on one Iowa hectare fed four people for the year. In 2000, a hectare on the Iowa farm of master grower Mr. Francis Childs could feed eighty people for the year.

Since the middle of the 20th century, such productivity gains have stabilised global cropland, and allowed reductions of cropland in many nations, including China. Meanwhile, growth in the world's food supply has continued to outpace population, even in poor countries. A cluster of innovations including tractors, seeds, chemicals, and irrigation joined through timely information flows and better organised markets raised the yields to feed billions more without clearing new fields. We have decoupled food from acreage.

High-yield agriculture need not tarnish the land. Precision agriculture is the key. This approach to farming relies on technology and information to help the grower prescribe and deliver precise inputs of fertiliser, pesticides, seed, and water exactly where they are needed. We had two revolutions in agriculture in the 20th century. First, the tractors of mechanical engineers saved the oats that horses ate and multiplied the power of labour. Then chemical engineers and plant breeders made more productive plants. The present agricultural revolution comes from information engineers. What do the past and future agricultural revolutions mean for land?

To produce their present crop of wheat, Indian farmers would need to farm more than three times as much land today as they actually do, if their yields had remained at their 1966 level. Let me offer a second comparison: a USA city of 500 000 people in 2000 and a USA city of 500 000 people with the 2000 diet but the yields of 1920. Farming as Americans did 80 years ago while eating as Americans do now would require four times as much land for the city, about 450 000 hectares instead of 110 000.

What can we look forward to globally? The agricultural production frontier remains spacious. On the same area, the average world farmer grows only about 20 percent of the corn of the top Iowa farmer, and the average Iowa farmer lags more than 30 years behind the state-of-the-art of his most productive neighbour. On average, the *world* corn farmer has been making the greatest annual percentage improvement. If during the next 60 to 70 years, the world farmer reaches the average yield of today's USA corn grower, the ten billion people then likely to live on Earth will need only half of today's cropland. This will happen if farmers maintain on average the yearly 2% worldwide growth per hectare of the Food Index achieved since 1960, in other words, if dynamics, social learning, continues as usual. Even if the rate falls to 1%, an area the size of India, globally, could revert from agriculture to woodland or other uses. Averaging an improvement of 2% per year in the productivity and efficiency of natural resource use may be a useful operational definition of sustainability.

Importantly, as Hubert would note, a vegetarian diet of 3000 primary calories per day halves the difficulty or doubles the land spared. Hubert might also observe that eating from a salad bar is like taking a sport utility vehicle to a gasoline filling station. Living on crisp lettuce, which offers almost no protein or calories, demands many times the energy of a simple rice-and-beans vegan diet (Leach 1976). Hubert would wonder at the greenhouses of the Benelux countries glowing year-round day and night. I will trust more in the technical advance of farmers than in behavioural change by eaters. The snake brain is usually a gourmet and a gourmand.

Fortunately, lifting yields while minimising environmental fall out, farmers can effect the Great Restoration.

33.6 Sparing Forests

Farmers may no longer pose much threat to nature. What about lumberjacks? As with food, the area of land needed for wood is a multiple of yield and diet, or the intensity of use of wood products in the economy, as well as population and income. Let's focus on industrial wood – logs cut for lumber, plywood, and pulp for paper.

The wood "diet" required to nourish an economy is determined by the tastes and actions of consumers and by the efficiency with which millers transform virgin wood into useful products (Wernick et al. 1997). Changing tastes and technological advances are already lightening pressure on forests. Concrete, steel, and plastics have replaced much of the wood once used in railroad ties, house walls, and flooring. Demand for lumber has become sluggish, and in the last decade world consumption of boards and plywood has actually declined. Even the appetite for pulpwood, logs that end as sheets of paper and board, has levelled.

Meanwhile, more efficient lumber and paper milling is already carving more value from the trees we cut.[4] And recycling has helped close leaks in the paper cycle. In 1970, consumers recycled less than one-fifth of their paper; today, the world average is double that.

The wood products industry has learned to increase its revenue while moderating its consumption of trees. Demand for industrial wood, now about 1.5 billion $m^3 yr^{-1}$, has risen only 1% annually since 1960, while the world economy has multiplied at nearly four times that rate. If millers improve their efficiency, manufacturers would deliver higher value through the better engineering of wood products, and consumers would recycle and replace more; in 2050 virgin demand could be only about 2 billion m^3 and thus permit reduction in the area of forests cut for lumber and paper.

The permit, as with agriculture, comes from lifting yield. The cubic meters of wood grown per hectare of forest each year provide strong leverage for change. Historically, forestry has been a classic primary industry, as Hubert doubtless saw in the shrinking Ardennes. Like fishers and hunters, foresters have exhausted local resources and then moved on, returning only if trees regenerated on their own. Most of the world's forests still deliver wood this way, with an average annual yield of perhaps two cubic meters of wood per hectare. If yield remains at that rate, by 2050 lumberjacks will regularly saw nearly half the world's forests (Fig. 33.5). That is a dismal vision – a chainsaw every other hectare skinhead Earth.

Lifting yields, however, will spare more forests. Raising average yields 2% per year would lift growth over 5 m^3 ha^{-1} by 2050 and shrink production forests to just about 12% of all woodlands. Once again, high yields can afford a Great Restoration.

At likely planting rates, at least one billion cubic meters of wood – half the world's supply – could come from plantations by the year 2050. Semi-natural forests – for example, those that regenerate naturally but are thinned for higher yield – could supply most of the rest. Small-scale traditional "community forestry" could also deliver a small fraction of industrial wood. Such arrangements, in which forest dwellers, often indigenous peoples, earn revenue from commercial timber, can provide essential protection to woodlands and their inhabitants.

More than a fifth of the world's virgin wood is already produced from forests with yields above 7 m^3 ha^{-1}. Plantations in Brazil, Chile and New Zealand can sustain yearly growth of more than 20 m^3 ha^{-1} with pine trees. In Brazil eucalyptus – a hardwood good for some papers – delivers more than 40 m^3 ha^{-1}. In the Pacific Northwest and British Columbia, with plentiful rainfall, hybrid poplars deliver 50 m^3 ha^{-1}.

Environmentalists worry that industrial plantations will deplete nutrients and water in the soil and produce a vulnerable monoculture of trees where a rich diversity of species should prevail. Meanwhile, advocates for indigenous peoples, who have witnessed the harm caused by crude industrial logging of natural forests, warn that plantations will dislocate forest dwellers and upset local economies. Pressure from these groups helps explain why the best practices in plantation forestry now stress the protection of environmental quality and human rights. As with most innovations, achieving the promise of high-yield forestry will require feedback from a watchful public.

The main benefit of the new approach to forests will reside in the natural habitat spared by more efficient forestry. An industry that draws from planted forests rather than cutting from the wild will disturb only one-fifth or less of the area for the same volume of wood. Instead of logging half the world's forests, humanity can

---

[4] In the United States, for example, leftovers from lumber mills account for more than a third of the wood chips turned into pulp and paper; what is still left after that is burned for power.

Fig. 33.5.
Present and projected land use and land cover. Today's 2.4 billion ha used for crops and industrial forests spread on "Skinhead Earth" to 2.9 while in the "Great Restoration" they contract to 1.5 (reproduced with permission from Victor and Ausubel 2000)

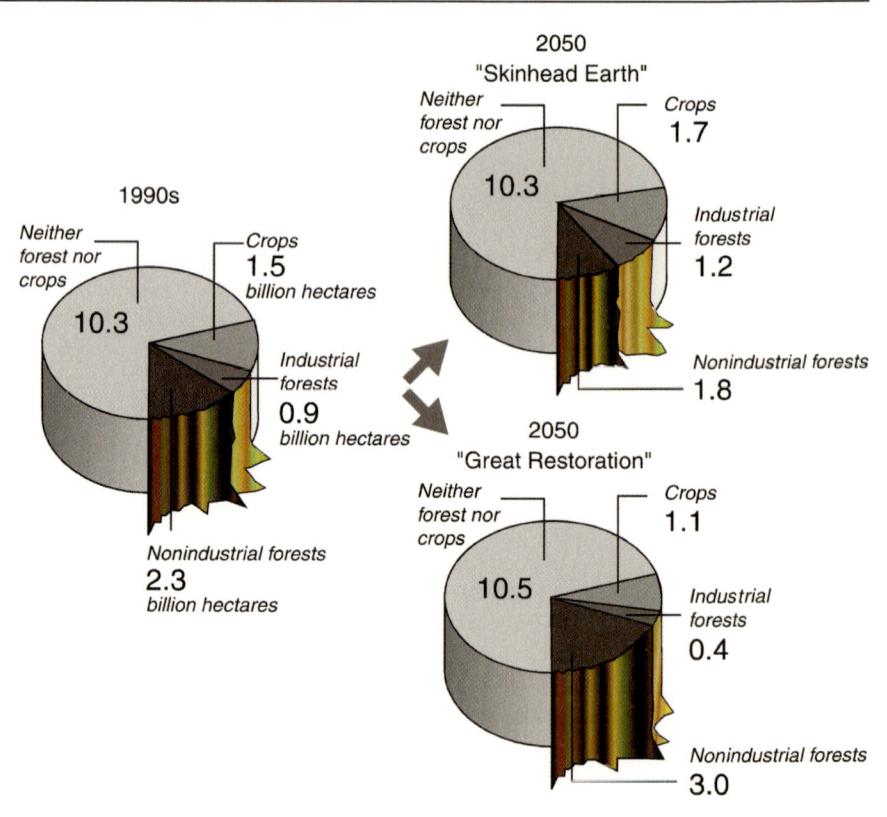

1990s
- Neither forest nor crops: 10.3
- Crops: 1.5 billion hectares
- Industrial forests: 0.9 billion hectares
- Nonindustrial forests: 2.3 billion hectares

2050 "Skinhead Earth"
- Neither forest nor crops: 10.3
- Crops: 1.7
- Industrial forests: 1.2
- Nonindustrial forests: 1.8

2050 "Great Restoration"
- Neither forest nor crops: 10.5
- Crops: 1.1
- Industrial forests: 0.4
- Nonindustrial forests: 3.0

leave almost 90% of them minimally disturbed. And nearly all new tree plantations are established on abandoned croplands, which are already abundant and accessible. Although the technology of forestry rather than the behaviour of hunters spared the forests and stags, Hubert would still be pleased.

33.7 Sparing Pavement

What then are the areas of land that may be built upon? One of the most basic human instincts, from the snake brain, is territorial. Territorial animals strive for territory. Maximising range means maximising access to resources. Most of human history is a bloody testimony to the instinct to maximise range. For humans, a large accessible territory means greater liberty in choosing the points of gravity of our lives: the home and the workplace.

Around 1800, new machines began transporting people faster and faster, gobbling up the kilometres and revolutionising territorial organisation (Ausubel et al. 1998). The highly successful machines are few – train, motor vehicle, and plane – and their diffusion slow. Each has taken from 50 to 100 years to saturate its niche. Each machine progressively stretches the distance travelled daily beyond the 5 km of mobility on foot. Collectively, their outcome is a steady increase in mobility. For example, in France, from 1800 to today, mobility has ex-

tended an average of more than 3% per year, doubling about every 25 years. Mobility is constrained by two invariant budgets, one for money and one for time. Humans always spend an average 12–15% of their income for travel. And the snake brain makes us visit our territory for about one hour each day, the travel time budget. Hubert doubtless averaged about one hour of walking per day.

The essence is that the transport system and the number of people basically determine covered land (Waggoner et al. 1996). Greater wealth enables people to buy higher speed, and when transit quickens, cities spread. Both average wealth and numbers will grow, so cities will take more land.

The USA is a country with a fast growing population, and expects about another 100 million people over the next century. Californians pave or build on about 600 m^2 each. At the California rate, the USA increase would consume 6 million ha, about the combined land area of the Netherlands and Belgium. Globally, if everyone new builds at the present California rate, 4 billion added to today's 6 billion people would cover about 240 million ha, midway in size between Mexico and Argentina.

Towering higher, urbanites could spare even more land for nature. In fact, migration from the country to the city formed the long prologue to the Great Restoration. Still, cities will take from nature.

But, to compensate, we can move much of our transit underground, so we need not further tar the land-

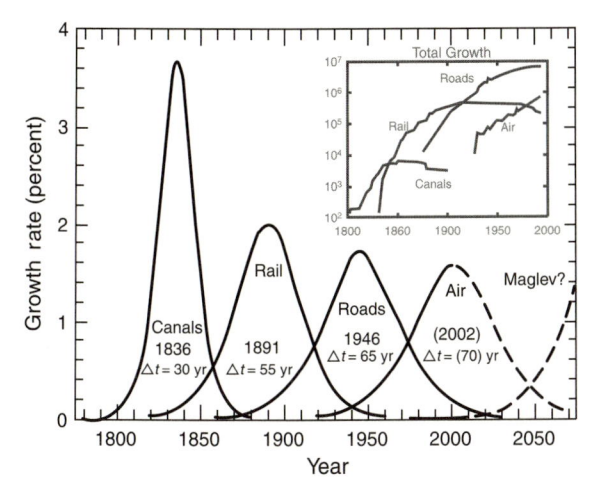

Fig. 33.6. Smoothed historic rates of growth (*solid lines*) of the major components of the US transport infrastructure and conjectures (*dashed lines*) based on constant dynamics. Rhythm evokes a new entrant now, maglevs. The inset shows the actual growth, which eventually became negative for canals and rail as routes were closed. Delta t is the time for the system to grow from 10% to 90% of its extent (source: Toward Green Mobility, Ausubel et al. 1998)

scape. The magnetically levitated train, or maglev, a container without wings, without motors, without combustibles aboard, suspended and propelled by magnetic fields generated in a sort of guard rail, nears readiness (Fig. 33.6). A route from the airport of Shanghai to the city centre will soon open. If one puts the maglev underground in a low-pressure or vacuum tube, as the Swiss think of doing with their Swissmetro, then we would have the equivalent of a plane that flies at high altitude with few limitations on speed. The Swiss maglev plan links all Swiss cities in 10 minutes (www.swissmetro.com).

Maglevs in low-pressure tubes can be ten times as energy efficient as present transport systems. In fact, they need consume almost no net energy. Had Hubert crossed the USA in 1850 to San Francisco from St. Louis on the Overland Stage, he would have exhausted 2 700 fresh horses.

Future human settlements could grow around a maglev station with an area of about 1 km^2 and 100 000 inhabitants, be largely pedestrian, and via the maglev form part of a network of city services within walking distance. The quarters could be surrounded by green land. In fact, cities please people, especially those that have grown naturally without suffering the sadism of architects and urban planners.

Technology already holds green mobility in store for us. Naturally, maglevs want 100 years to diffuse, like the train, auto or plane. With maglevs, together with personal vehicles and aeroplanes operating on hydrogen, Hubert could range hundreds of kilometres daily for his ministry, fulfilling the urges of his reptilian brain, while leaving the land and air pristine.

33.8 Cardinal Resolutions

How can the Great Restoration of Nature I envision be accomplished? Hubert became only a Bishop, but in his honour, I propose we promote four cardinal resolutions, one each for fish, farms, forests, and transport.

Resolution one: The stakeholders in the oceans, including the scientific community, shall conduct a worldwide Census of Marine Life between now and the year 2010. Some of us already are trying (Ausubel 2001). The purpose of the Census is to assess and explain the diversity, distribution, and abundance of marine life. This Census can mark the start of the Great Restoration for marine life, helping us move from uncertain anecdotes to reliable quantities. The Census of Marine Life can provide the impetus and foundation for a vast expansion of marine protected areas and wiser management of life in the sea.

Resolution two: The many partners in the farming enterprise shall continue to lift yields per hectare by 2% per year throughout the 21st century. Science and technology can double and redouble yields and thus spare hundreds of millions of hectares for nature. We should also be mindful that our diets, that is, behaviour, can affect land needed for farming by a factor of two.

Resolution three: Foresters, millers, and consumers shall work together to increase global forest area by 10%, about 300 million ha, by 2050. Furthermore, we will concentrate logging on about 10% of forest land. Behaviour can moderate demand for wood products, and foresters can make trees that speedily meet that demand, minimising the forest we disturb. Curiously, neither the diplomacy nor science about carbon and greenhouse warming has yet offered a visionary global target or timetable for land use (Victor and Ausubel 2000).

Resolution four: The major cities of the world shall start digging tunnels for maglevs. While cities will sprawl, our transport need not pave paradise or pollute the air. Although our snake brains and the instinct to travel will still determine travel behaviour, maglevs can zoom underground, sparing green landscape.

Clearly, to realise our vision we shall need both maglevs and the vision of St. Hubert. Simply promoting the gentle values of St. Hubert is not enough. Soon after he painted his masterpiece, Paulus Potter died of tuberculosis and was buried in Amsterdam on 7 January 1654 at the age of 29. In fact, Potter suffered poor engineering. Observe in The Life of the Hunter that the branch of the tree from which the dogs hang does not bend.

Because we are already more than six billion and heading for ten billion in the new century, we already have a Faustian bargain with technology. Having come this far with technology, we have no road back. If Indian wheat farmers allow yields to fall to the level of 1960, to sustain the present harvest they would need to clear nearly 50 million ha, about the area of Madhya Pradesh or Spain.

So, we must engage the elements of human society that impel us toward fish farms, landless agriculture, productive timber, and green mobility. And we must not be fooled into thinking that the talk of politicians and diplomats will achieve our goals. The maglev engineers and farmers and foresters are the authentic movers, aided by science. Still, a helpful step is to lock the vision of the Great Restoration in our minds and make our cardinal resolutions for fish, farms, forests, and transport. In the 21st century, we have both the glowing vision of St. Hubert and the technology exemplified by maglevs to realise the Great Restoration of Nature.

Acknowledgements: Georgia Healey, Cesare Marchetti, Perrin Meyer, David Victor, Iddo Wernick, Paul Waggoner, and especially Diana Wolff-Albers for introducing me to Paulus Potter.

References

Ausubel JH (2000) The great reversal: Nature's chance to restore land and sea. Technology in Society 22(3):289-302

Ausubel JH (2001) The census of marine life: Progress and prospects. Fisheries 26 (7):33-36

Ausubel JH, Marchetti C, Meyer PS (1998) Toward green mobility: The evolution of transport. European Review 6(2):143-162

Leach G (1976) Energy and food production. IPC Science and Technology Press, Guildford UK

MacLean PD (1990) The triune brain in evolution: Role in paleocerebral functions. Plenum, New York

Markels M Jr. (1995) Method of improving production of seafood. US Patent 5,433,173, July 18, 1995, Washington DC

Richards (1990) Land transformations. In: Turner II BL, Clark WC, Kates RW, Richards JF, Mathews JT, Meyer WB (eds) The Earth as transformed by human action. Cambridge University, Cambridge, UK

Victor DG, Ausubel JH (2000) Restoring the forests. Foreign Affairs 79(6):127-144

Waggoner PE, Ausubel JH (2000) How much will feeding more and wealthier people encroach on nature? Population and Development Review 27(2):239-257

Waggoner PE, Ausubel JH, Wernick IK (1996) Lightening the tread of population on the land. American Examples, Population and Development Review 22(3):531-545

Walsh A, Buijsen E, Broos B (1994) Paulus Potter: Schilderijen, tekeningen en etsen, Waanders, Zwolle

Wernick IK, Waggoner PE, Ausubel JH (1997) Searching for leverage to conserve forests: The industrial ecology of wood products in the U.S. Journal of Industrial Ecology 1(3):125-145

Chapter 34

Industrial Transformation:
Exploring System Change in Production and Consumption

Pier Vellinga

There are three critical issues when discussing research regarding global environmental change and industrial transformation. The first is the issue of human choice to be based on a scientific assessment of biophysical processes that operate on a time scale of 50 to 500 years or more. This implies an unprecedented reliance on science and our capabilities to predict the impact of human activity on life support systems. The notion of human choice also implies that it is important to engage social sciences and focus more than before on the questions regarding the interaction between natural and social systems.

The second critical issue is exploring future development trajectories. Limiting climate change and reversing the trend of increasing loss of biodiversity requires a major transformation of the ways societal needs are met in the fields of energy, transport, food, and water. New ways of production and consumption are to be explored through research and experiments. This includes research on the incentives that shape the interaction between production and consumption (e.g., institutional frameworks regarding property, liability and fiscal systems).

The third critical issue is international cooperation in creating effective incentives and institutions that would support a transformation towards a significantly smaller impact on global life support systems. This includes research on the relevant actors and the dynamics of their interaction. Governments including the UN system, international corporations and civil society should be considered as the major actors.

34.1 Human Choice on Issues Involving a Time Scale of Decades to Centuries

Many governments and international corporations promote a further opening up of global markets as a way to enhance development and income levels worldwide. However, the present rate of globalisation raises concern about growing disparities in income levels at national and international levels and a further degradation of the quality of life support systems such as biological diversity and the global climate. Illustrating why global environmental issues are particularly challenging the relation between environment and development requires three different geographic scales: local, regional and global, each with its specific environmental problems.

Figure 34.1 reflects empirical evidence that people tend to solve their *local* environmental problems as income levels go up, such that growing income levels can be combined with improvement of the local environmental quality. Many cities/countries in the industrialised part of the world have gone through this curve, while many cities in the developing countries are in the upward part of the curve. The reasoning behind the curve is that as income levels go up and as local environmental/health problems become manifest, there is a driver and there are financial means to introduce technologies and regulations (incentives and institutions) that reduce pollution and protect the health of the population. Two critical factors for success can be identified. One: people take measures based on health impact observations. Two: the costs and benefits play out at the same (local/national) level.

A similar curve can be developed for environmental problems that are manifest at a *regional* level (Fig. 34.2), such as acidification and water quantity/quality issues

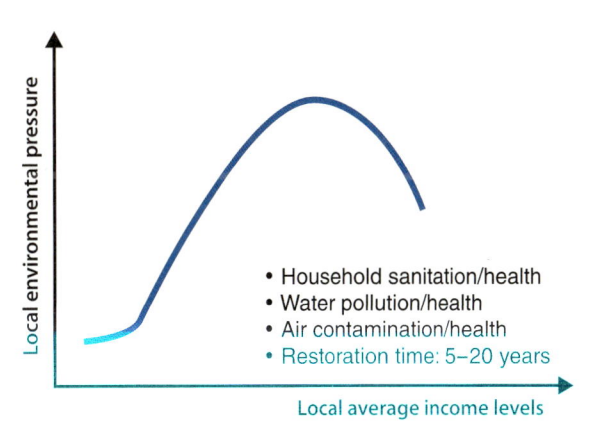

- Household sanitation/health
- Water pollution/health
- Air contamination/health
- Restoration time: 5–20 years

Fig. 34.1. Local average income levels and environmental quality (reproduced with permission from Panayotou 1997, © Cambridge University Press)

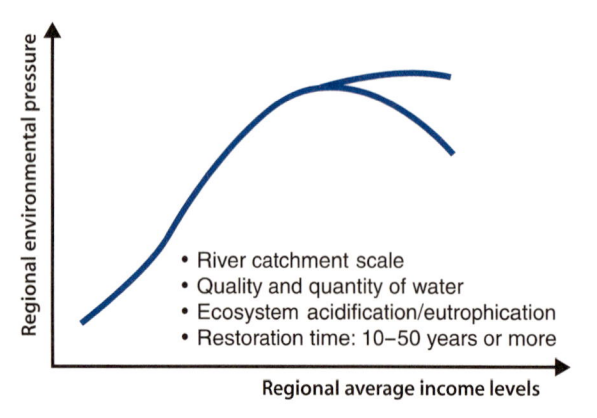

Fig. 34.2. Regional average income levels and environmental quality

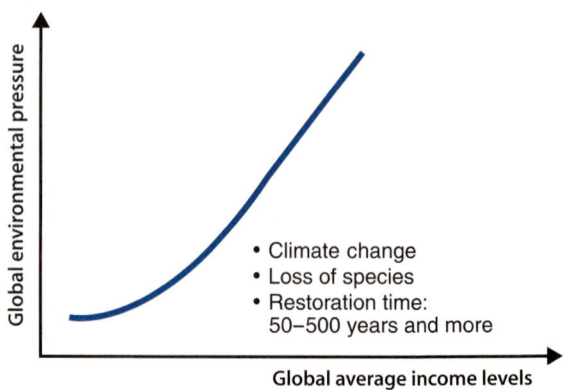

Fig. 34.3. Global average income levels and environmental quality

at the scale of river catchments. There is less evidence that people successfully address these problems as income levels go up. One reason is that upstream and upwind industrial and agricultural activities benefit from polluting and overuse of environmental resources such as water and air while downstream, and downwind people and nations experience the negative impacts. Another reason for continued environmental degradation as income levels go up is the time delay between the act of polluting and the effect of pollution downstream. There are some examples of regions and environmental problems where the curve has been pulled downward, but this is not a general empirical finding. For most regions of the world the jury is still out.

Such a curve can also be developed for *global* environmental problems such as climate change and loss of species and habitats (Fig. 34.3). Empirical data illustrate that there is no income-related levelling off point when we look at the relation between income and emissions of greenhouse gases. Income levels correlate with energy use, and present day energy use is coupled with CO_2 emissions. Similarly, the space we use for our activities (housing, transport and recreation) grows linearly with income projections going up; this is at the expense of natural habitats. A most critical feature of global environmental change is the time scale of biophysical response. Climate responds to changes in the concentration of greenhouse gases at a time scale in the order of decades to centuries or more. Loss of species including their ecosystem is considered irreversible at human time scales (although some genetic modification experts may challenge this statement).

Given the observations described above, it is clear that global environmental change poses an unprecedented challenge for society. It requires a pro-active approach: we have to act before the effects of our actions have become visible. How can research help in clarifying the issues at stake? For sure this will require a better understanding of the interaction between social and natural systems. This implies a broader engagement of social sciences: what makes people and institutions move, what the barriers and the opportunities are?

34.2 Transformation: Why, How Does It Work and What Are the Options?

Global environmental change will test, in an unprecedented way, the capacity of the human species to manage their activities in a pro-active manner. Guidance for the future may be found in analysing trends in societal response to environmental problems over the last 50 years. Figure 34.4 presents a number of stages of societal response to environmental problems. In most of the OECD countries, environmental policies were initiated in the period between 1960 and 1970. The first set of policies can be characterised as predominantly reactive: policies driven by visible negative effects such as massive fish killing in polluted rivers and health problems related to air pollution and chemical waste. The response can be characterised as "end of the pipe", implemented by technical specialists. An important philosophy was keeping the cost down. In response to the oil crises in the 1970s, policies were complemented with ideas about efficiency gains in the production process. This required the involvement of managers. Optimisation of resource use became the major driving philosophy.

From, say, 1990 onward, new approaches can be recognised where environmental concerns are transformed into opportunities for developing and selling new products. The driving philosophy for such strategies is acceleration: developing new markets based on environmental performance and green image. A major question is: will the approach focused on green products be powerful enough to limit global environmental change to acceptable proportions? The answer is: probably not. Green products are only marketable when they fit in the larger physical and institutional framework. A hydrogen car needs an infrastructure with hydrogen service stations, and renewable energies can only compete in a market where the price of fossil fuel use reflects all environmental cost.

A more far-reaching approach will be required when we want to combine growing income levels with a sig-

Fig. 34.4.
Development stages in corporate and societal response (adapted with permission from Vellinga and Herb 1999, and Winsemius and Guntram 1992)

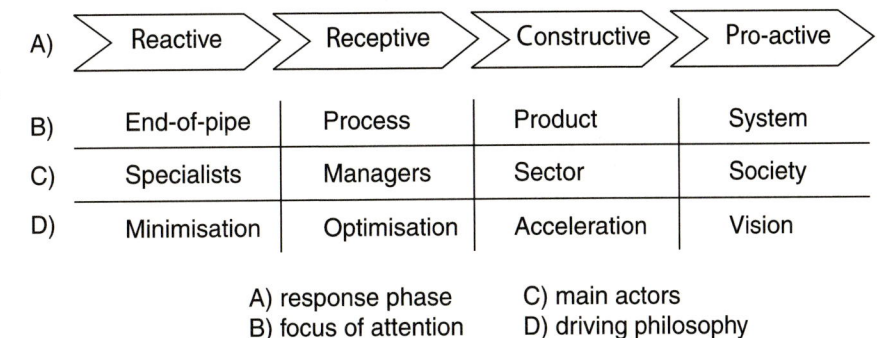

A)	Reactive	Receptive	Constructive	Pro-active
B)	End-of-pipe	Process	Product	System
C)	Specialists	Managers	Sector	Society
D)	Minimisation	Optimisation	Acceleration	Vision

A) response phase   C) main actors
B) focus of attention   D) driving philosophy

nificant reduction of the impact of human activities on global life support systems. Such an approach will have to focus on systems and system change. A system is defined as a chain of production, distribution, consumption and disposal activities including the incentives that shapes this system (i.e., property, liability and fiscal laws and regulations). Given the complexity of such chains and given the need for a pro-active approach, such system changes will require the involvement of society as a whole and an inspiring vision to mobilise all participants (see Fig. 34.4). It is clear that such visions are likely to compete with one another. This can slow down change, but is not necessarily counter productive, as competition is a driving force in itself. The implications for research are overwhelming. The need to engage society as a whole requires a more participatory process in defining research priorities. Moreover, doing research should include stakeholder analysis and to some extent stakeholder participation. It should be clear that I do not want to compromise the objectivity and the independence of research. To safeguard these values, new and transparent procedures will have to be developed for assessing and reviewing the results of research. In fact, we are presently witnessing the development of such procedures in some of the global assessment procedures like the Intergovernmental Panel on Climate Change and the Ozone Panel.

34.3 From Green Products to System Transformation

Moving from end of the pipe and efficiency measures towards green products and systems innovation reflects a societal development from reactive to pro-active environmental policies. When the incentive system such as fiscal systems, property and liability laws reflect the (environmental and societal) cost of production and consumption, green products provide a market opportunity, and a more fundamental change can occur.

There are plenty of technologies available or within reach for clean production, such as zero emission power plants and zero emission cars and ecological and/or low input farming systems. Still, they do not easily enter the market for two reasons: (*i*) they may not fit in the present system of energy, transport and food production and consumption system: the present system is a "lock-in" with high status quo interests including education, research, infrastructure and not in the least commercial interests; and (*ii*) the incentive structure through its historical development favours the present system of sharing private and public costs (in any fiscal, property, international trade and liability system of rules and regulations some costs are internalised in the price of production and consumption while other costs are borne by the public (e.g., health costs), or future generations (e.g., climate change) or other species (e.g., loss of biodiversity)).

In fact system change is a combination of technological and societal change. The systems at hand in global environmental change (energy, transport, food, and water) are difficult to change as these systems are global in scope through trading and they are usually deeply embedded in the local and national economies and cultural systems.

Therefore transformation to more sustainable systems is only partially a matter of technology. Economic, social-cultural and institutional change plays an equally important role. Moreover transformation can only be successful when societal change and technological change are mutually reinforcing at different levels, as illustrated in Fig. 34.5: this includes the micro scale (niche), the meso scale (regimes) and the macro scale (landscapes) (see Kemp et al. 2000; Geels and Kemp 2000; Rotmans et al. 2000). Industrial Transformation is usually initiated as the result of a local (or national) innovation, serving as a "niche market" (technological and/or institutional). When this innovation fits in a regime change that occurs (maybe for completely other reasons) at a regional or continental scale, the innovation is reinforced. When finally internationally socio-cultural changes are occurring that favour the new way of doing things, then the (system) innovation can be absorbed at the global level.

This illustrates how important it is to include the human dimensions in global change research. Many promising ideas have failed as society had its reasons not to adopt or lock out certain technologies.

Fig. 34.5.
Industrial transformation occurs through mutually reinforcing technological and societal changes at micro, meso and macro scales (reproduced with permission from Geels 2002)

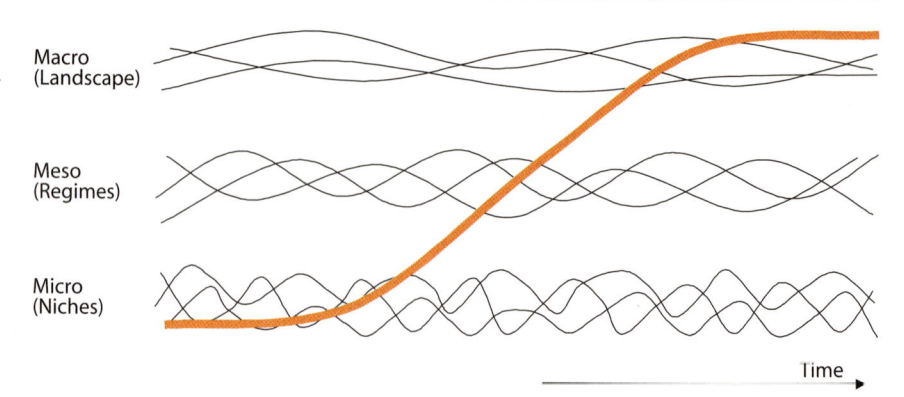

34.4 Five Major Foci for Transformation Research

The most relevant topics for global change industrial transformation research are (Fig. 34.6):

1. Energy and material flows, in view of the global climate and development nexus;
2. Food cycle (food production, processing, transport, consumption, and waste) in view of its major impact on the environment, its development challenges, its international interdependencies, and the complexity of its connections with energy, climate change and biodiversity issues;
3. Cities, focusing on water and transport, as a major challenge for development, the quality of life and the global environment;
4. Information and communication as a major driving force in societal transformation and a crosscutting theme deeply embedded in all "systems";
5. Governance and transformation processes in view of analysing and understanding the driving forces for changing the way society relates to its environment.

Industrial transformation research should build on the foundations of a range of social science disciplines including economics, sociology, psychology, human ecology, anthropology, political science, geography, and history, as well as on the foundations of natural sciences such as physics, chemistry, biology, and technological sciences.

Examples of priority research projects include:

A. Energy and material flows
   1. The technical, institutional and economic feasibility of a transport system based on bio-fuels;
   2. Economic theories regarding a transformation of the energy system;
   3. A hydrogen economy with CO_2 underground storage.
B. Food
   - Environmental benefits, social desirability and technical feasibility of a (partial) replacement of animal protein by plant protein;
   - Feasibility and implications of a switch to organic farming and the role of developing countries;
   - Feasibility and implications of sustainable food consumption and production systems at a regional level.
C. Cities
   1. Comparative studies of cities in terms of their effects on the water and carbon cycles;
   2. What the options are for decoupling urban activities from the carbon and water cycles?;
   3. Cities as complex systems – biocomplexity analysis.
D. Information and communication
   1. How information and communications technology will influence society and lifestyle and through this alter the way of environmental resource use?;
   2. Information and communication system organisers and their role in global environmental change
E. Governance and transformation processes
   1. Analysis of trends in corporate strategies to develop new (green) products for low budget developing country consumers;
   2. Analysis of socially responsible investment trends as a mechanism for transformation of capital markets towards sustainability;
   3. Co-evolution of society-nature interactions at various stages of development.

34.5 International Cooperation

A critical issue in the management of human use of natural resources is cooperation at the global scale. Section 34.1 indicates that human choice and time scale issues complicate international cooperation. Section 34.2 indicated that transformation of meeting needs and preferences in the field of energy, food and transport requires the engagement of society as a whole, and it requires new ways of science-society interaction. Different cultures will deal with this differently. Section 34.2 also mentions that system change may well be initiated

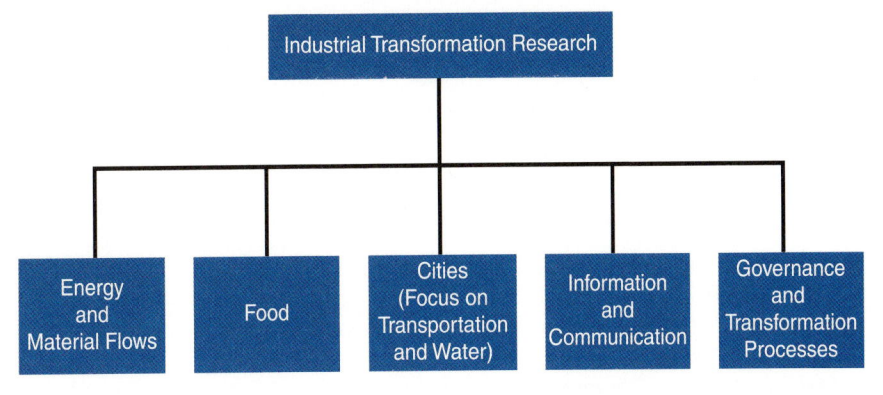

Fig. 34.6.
Research foci and framework for industrial transformation research (see IHDP Science Plan – Vellinga and Herb 1999)

```
                    Industrial Transformation Research
         ┌──────────┬──────────┬──────────┬──────────┐
      Energy       Food      Cities    Information  Governance
        and                 (Focus on      and          and
     Material              Transportation Communication Transformation
      Flows                 and Water)                  Processes
```

at the local level, but to be successful it will have to proliferate at a global scale.

All these factors complicate international cooperation. Still there are precedents indicating that such barriers can be overcome. The necessary ingredients are: (a) increasing scientific understanding of global environmental change; (b) increasing visibility of early impacts; (c) market penetration of more environmentally friendly ways of meeting human needs and preferences. A certain mix of such ingredients can make the system move as we have seen in the case of ozone depletion and CFC phase out.

The question is who will take the lead in organising the decision making about the various trade-offs involved. In the historic model of addressing environmental issues we can recognise a leading role for governments and the UN system. However, this is not necessarily so for all environmental issues. For example in climate change policies we observe a limited capacity of governments and the UN system in coming to grips with global energy and investment policies. Some international corporations take a more advanced position than most of the governments.

In general, the latter part of the 20th and the early years of the 21st century are characterised by a global wave of liberalisation and privatisation and a parallel withdrawal of government in many fields relevant to environmental resource use such as energy, agricultural policies and water. This could make one wonder about the prospects for managing the human use of environmental resources.

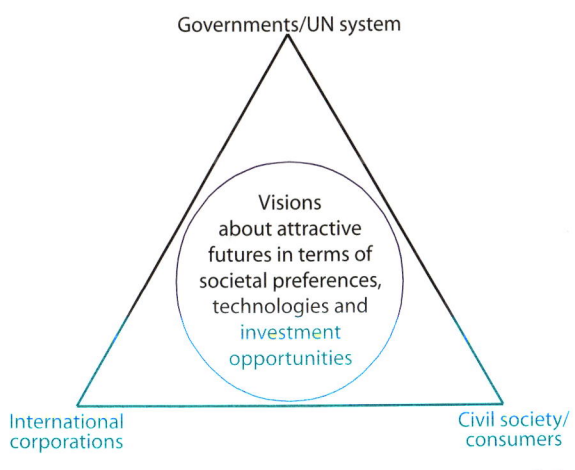

However, new players are entering the field. Non-Governmental Organisations have become more powerful over the last few decades. Where they primarily addressed governments and the UN system in the 1980s and 1990s, NGOs are now paying equal attention to international corporations, in particular in the energy, food and financial sectors. Simultaneously, corporations have started to develop strategies for dealing with societal concerns. The notion of Triple P Bottom Line Performance is an example of how they integrate sustainable development concerns in their strategies (triple P stands for profit, planet and people).

In the understanding that there are good reasons for society to come to grips with global environmental change, we may expect new models of cooperation emerging between the three major actors: civil society, international corporations and government (including the UN system). This tri-sectoral arrangement is illustrated in Fig. 34.7. This tri-sectoral arrangement is particularly relevant for research on global environmental change and industrial transformation. As a consequence, the research scene is likely to become more complex; however, there are also new opportunities for doing research. The bottom line for research is transparency in financing, in peer review and in assessment procedures.

34.6 In Summary

The triple challenge for industrial transformation research can be summarised in three questions:

1. How to focus human choice on processes that operate on a time scale of decades to centuries?
2. Which future development trajectories can be identified that both meet human needs and preferences

Fig. 34.7. System change requires engagement of society as a whole

(Triangle diagram with vertices: Governments/UN system, International corporations, Civil society/consumers; center: Visions about attractive futures in terms of societal preferences, technologies and investment opportunities)

regarding energy, transport, food, and water; and have a significantly smaller impact on global environmental resources?
3. Which mechanisms for effective international cooperation can be mobilised?

Regarding technologies and technology research:

1. To solve global environmental problems "end-of-pipe" and efficiency approaches will not be sufficient;
2. Green products may help but only in the context of systems change,
3. System change requires mutually reinforcing societal, institutional and technological changes;
4. While there are many technologies available, the human dimension is critical for their adoption;
5. The time scale and momentum of global environmental change implies that actions have to be based on scientific projections.

Research focused on the interaction between natural and social systems as promoted by the International Human Dimensions Programme (IHDP) and in particular its Industrial Transformation project plays an important role in exploring development trajectories that have a significantly smaller impact on all global life supporting systems. It is considered important that global change research programmes pay attention to the critical issues summarised above.

References

Geels F, Kemp R (2000) Transities vanuit socio-technisch perspectief. Achtergronddocument bij hoofdstuk 1 van het rapport Transities en transitiemanagement (Rotmans et al. 2000), Maastricht, Nederland

Geels FW (2002) Technological transitions as evolutionary configuration processes: A multi-level perspective and a case-study, Research Policy, forthcoming (November 2002)

Industrial Transformation Project of the IHDP http://www.vu.nl/ivm/research/ihdp-it

International Human Dimensions Programme on Global Environmental Change (IHDP) http://www.uni-bonn.de/IHDP/

IPPC Report (2001) in print

IUCN (2000) 2000 IUCN Red List of Threatened Species. IUCN Species Survival Commission, International Union for Conservation of Nature and Natural Resources, Gland

Kemp R, Rip A, Schot J (2000) Constructing transition paths through the management of niches. In: Garud R, Karnoe P (eds) Path creation and dependance. Lawrence Erlbaum Associates Publ, forthcoming

National Research Council (1999) Our common journey, a transition toward sustainability. National Academy Press, Washington DC

Panayotou T (1997) Demystifying the environmental Kuznets curve: turning a black box into a policy tool. Environment and Development Economics 2:465–484

Rotmans J, Kemp R, van Asselt M, Geels F, Verbong G, Molendijk K (2000) Transities and transitiemanagement. De casus van een emissiearme energievoorziening. International Centre for Integrative Studies (ICIS B.V.), Maastricht, Nederland

Vellinga P, Herb N (eds) (1999) Industrial Transformation Science Plan. International Human Dimensions Programme, IHDP Report No. 12, http://www.uni-bonn.de/ihdp/ITSciencePlan/

Winsemius P, Guntram U (1992) Responding to the environmental challenge. Business Horizons, Vol. 35, No. 2, Indiana University Graduate School of Business, March–April 1992, pp 12–20

Chapter 35

Will Technology Spare the Planet?

Will Steffen

In terms of global change and the Earth System, there is no doubt that technology is a two-edged sword. Bad technology and poorly applied or uncontrolled technology contribute strongly to significant environmental degradation. On the other hand, the two presentations in the session 'Can Technology Spare the Planet?' offer some hope that technology can be turned into a strong ally in the battle to slow or reverse many of the negative global environmental trends currently threatening Earth's life support system.

The presentations offered complementary views on the role of technology in addressing global change. Jesse Ausubel used examples of cutting-edge technologies to highlight their potential to increase production while simultaneously reducing environmental impacts. Pier Vellinga analysed the nature of industrial systems and noted the significant challenges that must be met in converting promising technologies into widely-used industrial systems. In summary, the main points were:

- Increases in productivity in land systems (food and wood) promise significant release of land from production systems to return to more natural systems. Similar increases in productivity may also be achievable in managed ocean production systems;
- If achieved globally, these trends (now apparent regionally in North America and Europe) could reverse many of the *global* trends that threaten Earth's environment;
- Basic industrial systems (energy, transport, food and water) are increasingly global in scale but embedded in national and regional economies and cultures. They are difficult to change;
- A proactive approach to global environmental change is required. A business-as-usual (end-of-pipe), reactive approach will not work.

To realise the potential of technology to spare the planet from further pollution, degradation and global change, two system-level challenges must be met if technology can succeed in taking the pressure off the environment at the global scale.

First, new technologies must be adopted in human-environment systems that are vastly different in different parts of the world. Technologies are embedded in a large array of cultures, institutional systems, economies, political structures, and ecological and climatic regimes. Different combinations of these present unique challenges. Most daunting of all, however, is the growing divide between rich and poor countries, and between rich and poor sectors within particular countries.

Modern technologies require, at a minimum, stable political systems, a well-educated population, adequate and reliable infrastructure, and an ability to work in digitally-based information systems. The last requirement, in particular, presents a major hurdle to many regions of the world in adopting environmentally friendly technologies developed in the high-tech societies of North America and Europe. Bridging this *digital divide* (Kates et al. 2001) and closing the gap between the wealthy and the poor in many other ways, are the most important obstacles in finding global-scale solutions to global change.

The second major challenge is related to the dynamics of the Earth System itself, which provide insights into the application of technology – where, when and how technology is applied matters!

Where. Some regions of the world are much more important than others in controlling the planetary machinery. Figure 35.1 shows one representation of so-called switch and choke points of the Earth System (Schellnhuber, this volume). These are particular places or regions where changes in critical processes may have a large impact on the functioning of the Earth System as a whole, well beyond the particular place or region itself. Based on our current level of understanding, the Earth System may generally be much more sensitive to perturbations in the Tropics and in the high latitudes than in the Temperate Zone.

For example, maintaining land in or returning it to natural (less intensively managed) ecosystems may be much more important in Amazonia than in North America. Figure 35.2 shows schematically the results of model simulations, suggesting that deforesting the Amazon Basin will change atmospheric circulation patterns in the Western Hemisphere and, through teleconnections, affect climate in other parts of the world. Other model-based projections indicate that deforestation of the Amazon will change rainfall patterns in both central Africa and Southeast Asia (Graf et al. 2001). Thus, the widespread deforestation of the European and North American temperate zones in historical

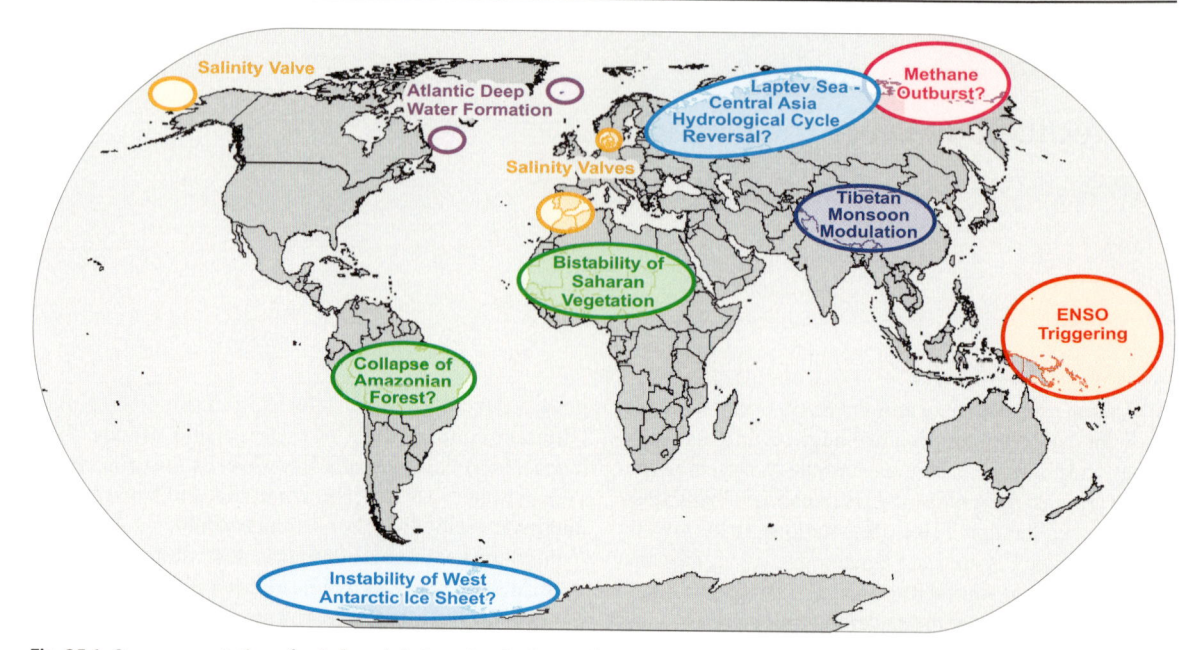

Fig. 35.1. One representation of switch and choke points in the Earth System – those regions where a change in biogeochemical or physical subsystem can cause significant changes to the functioning of the Earth System as a whole (from Schellnhuber, this volume)

Fig. 35.2.
Deforestation of the Amazon Basin could act as a switch point in the Earth System. It could lead to fundamental shifts in the atmospheric circulation of the entire Western Hemisphere, with significant consequences for atmospheric circulation and climate of the Earth as a whole (figure courtesy of R. Avissar, unpublished)

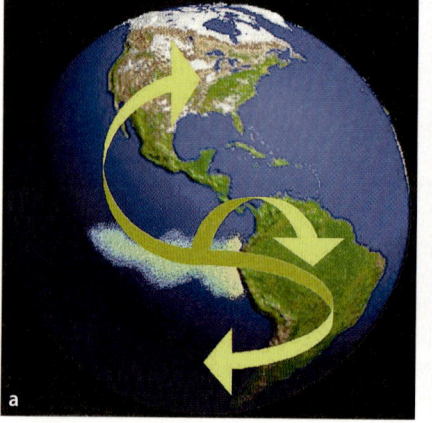

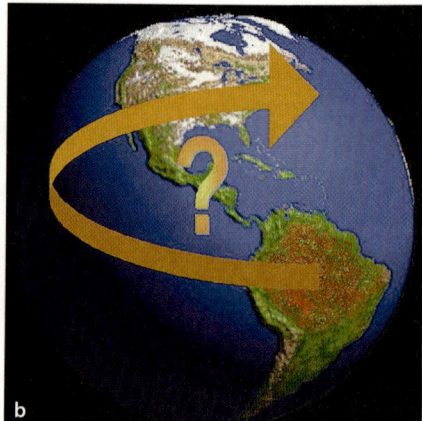

times and the current trend toward reforestation as a result of improved farming technologies may have had little or no effect on the dynamics of the Earth System as a whole, whereas widespread deforestation of Amazonia may have profound implications. This suggests even more strongly that the cultural, economic and institutional constraints that may prevent high technology farming and forestry systems from being applied in the countries of the Amazon Basin must be overcome to lessen the pressure on the forests of that region.

When. Terrestrial and marine systems currently provide a free service to humanity by absorbing nearly half of the anthropogenic CO_2 emitted to the atmosphere and thus slowing the rate of climate change. However, the terrestrial sink is expected to level off around 2050 and according to some model projections, diminish in strength during the second half of the century (Fig. 35.3). One projection suggests that the terrestrial biosphere could even become a net source of carbon to the atmosphere by the end of the century. Similarly, oceanic carbon sinks are expected to decrease in effectiveness during the second half of the century, although not as dramatically as the terrestrial sinks (Fig. 35.4). Overall, a much higher fraction of anthropogenically emitted CO_2 is likely to remain in the atmosphere during the second half of the century than now, creating the risk of strong positive feedbacks between a warming climate and weakening carbon sinks.

Thus, the next 50 years present a fleeting opportunity to replace fossil fuel-based energy systems while avoiding a much higher risk of a strong surge in climate change that may be exceptionally hard to mitigate and significantly more complex and costly to adapt to. Given the long lead times in transforming entire industrial and energy systems, especially at the global scale, the time scales associated with the carbon cycle dynamics of a perturbed

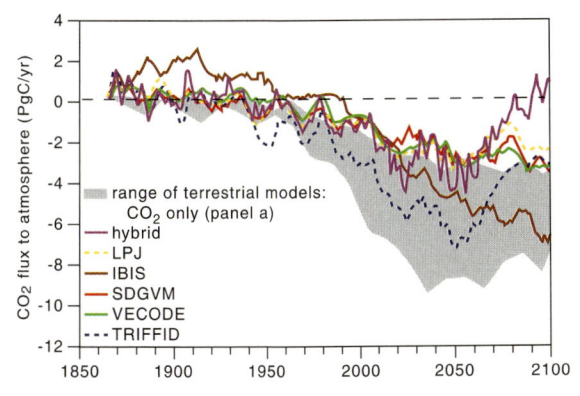

Fig. 35.3. Projections of uptake of atmospheric carbon by terrestrial ecosystems to the year 2100. The projections are based on simulations of Dynamic Global Vegetation Models run with IPCC projections of CO_2 increase and transient predictions of climatic change (Source: Cramer et al. 2001)

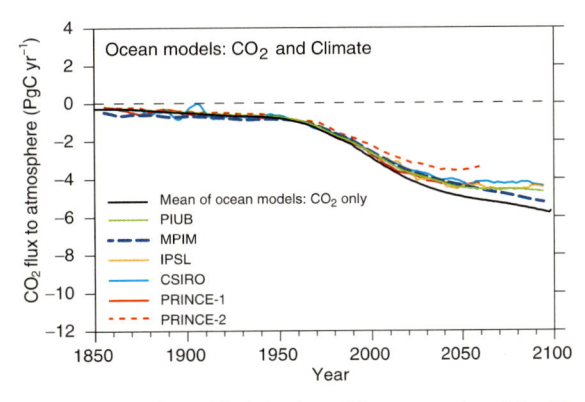

Fig. 35.4. Projections of the behaviour of the ocean carbon sink with increasing atmospheric CO_2 concentration and changing climate (reproduced with permission from Prentice et al. 2001, © IPCC 2001)

Earth System signal a strong sense of urgency in dealing with the energy problem. There simply is no time to lose in beginning the transformation to non-fossil fuel based energy systems. The transformation must begin now.

How. There are two fundamentally different ways in which technology can help to solve environmental problems. First, clean technologies attack the environmental problem at its source and prevent it from occurring in the first place. Second, technological solutions can be applied at the point of the environmental impacts themselves, to lessen the environmental damage after it occurs. For example, the impacts of oil spills can be avoided by improving ship safety technology or rerouting cruise tracks, or alternatively the impacts can be ameliorated through improving technologies for treating the spills once they occur. Clearly, a proactive approach to prevent the problem from occurring in the first place is preferable to attempting to deal with the problem once it has occurred in reactive mode with so-called end-of-pipe technologies.

For an environmental problem as complex, extensive and long-lasting as global change, there is no option as to which technological approach to use. It is mandatory that the problem be solved proactively at its point of origin. The only appropriate use of technology is to take the pressure off the planetary environment at its source, as described in both Jesse Ausubel's and in Pier Vellinga's papers.

In global change, there is no rational approach to dealing with the problem once it has occurred; at the scale of the Earth System, end-of-pipe technologies are not an option. Geo-engineering solutions, in which major Earth System functions are deliberately modified to counteract an aspect of global change, are sometimes proposed. An example might be the injection of aerosols high in the atmosphere to cool the climate in opposition to the effect of greenhouse gases. Such geo-engineering approaches are exceptionally dangerous. They are based on simple cause-effect logic and ignore the fact that global change is a complex, interactive phenomenon acting on a single interlinked planetary system. Unintended consequences of geo-engineering approaches are highly likely, are very difficult to predict, and could well lead to problems as severe as those they were intended to solve. Geo-engineering solutions should never be confused with the appropriate use of technology to reduce the pressure of human activities on the Earth System.

In summary, technology can indeed spare the planet. However, to ensure that a Great Transformation is effective in meeting both legitimate human needs for a better life and in protecting the Earth's environment on which all life depends, the following guidelines must be met:

1. The transformation must be accessible to all nations and cultures, not just to the developed world;
2. The transformation must be pursued with the spatial and temporal dynamics of the Earth System providing the ultimate guardrails.

References

Cramer W, Bondeau A, Woodward FI, Prentice IC, Betts RA, Brovkin V, Cox PM, Fisher V, Foley JA, Friend AD, Kucharik C, Lomas MR, Ramankutty N, Sitch S, Smith B, White A, Young-Molling C (2001) Global response of terrestrial ecosystem structure and function to CO_2 and climate change: results from six dynamic global vegetation models. Global Change Biology 7:357–374

Graf H-F, Nober J, Rosenfeld D (2001) Sensitivity of global climate to the detrimental impact of smoke on rain clouds. Max Planck Institute for Meteorology, Hamburg, Germany. Report No. 316

Kates RW, Clark WC, Corell R, Hall JM, Jaeger CC, Lowe I, McCarthy JJ, Schellnhuber HJ, Bolin B, Dickson NM, Faucheux S, Gallopin GC, Grubler A, Huntley B, Jager J, Jodha NS, Kasperson RE, Mabogunje A, Matson P, Mooney H, Moore B, O'Riordan T, Svedin U (2001) Environment and development – Sustainability science. Science 292 (5517):641–642

Prentice IC, Farquhar GD, Fasham MJR, Goulden ML, Heimann M, Jaramillo VJ, Kheshgi HS, Le Quéré C, Scholes RJ Wallace DWR (2001) The carbon cycle and atmospheric carbon dioxide. In: Houghton JT, Ding Y, Griggs DJ, Noguer M, van der Linden PJ, Dai X, Maskell K, Johnson CA (eds) Climate change 2001: The scientific basis. Contribution of Working Group I to the Third Assessment Report of the Intergovernmental Panel on Climate Change. Cambridge University Press, Cambridge New York, 881 pp

Part IVd
Towards Global Sustainability

Chapter 36

Challenges and Road Blocks for Local and Global Sustainability

Julia Carabias Lillo

After 30 years from Stockholm, 13 from Our Common Future and 9 from UNCED, we have to recognise that very important advances have been made to achieve sustainable development. However, neither the multilateral principles, commitments and agreements adopted nor the local national efforts have been enough to curb deterioration and impoverishment trends, much less revert them.

36.1 Advances

Among the advances we can mention the following:

- Sustainable development as a concept has been endorsed in most countries and has gradually brought closer the different perspectives of development: the economic, the social and the environmental;
- It has been accepted that sustainability entails common, but differentiated responsibilities that demand from countries their best possible effort, nevertheless proportional to the specific economic, institutional and cultural capacities of every country;
- Better monitoring, diagnosis and understanding of the environmental processes has been produced;
- Most countries have, in some way or an other, a programmatic platform and institutional capacity to manage the environment;
- Conscience and participation are increasing.

Unfortunately, trends in forest degradation, biodiversity extinction, soil erosion, overexploitation of important fishing resources, water pollution and depletion, dangerous unprocessed wastes, air pollution in large cities, and severe distortions in the ecological planning of the territory regarding our regional variety in resources still persist and in some cases have even increased.

It is necessary to recognise advances and identify the powerful obstacles that inhibit a more agile progress, starting with the insufficient magnitude of international cooperation, the dispersion of the present agendas and their lack of coordination. There are gaps that have to be addressed with new instruments and commitments that connect the efforts in time, space, scale, and perspective.

36.2 International Instruments: Urgency of Goals and Synergies

The current international context shows important advances with the creation of binding, multilateral environmental instruments (Framework Convention on Climate Change, Convention on Biological Diversity, Convention to Combat Desertification, Forestry Panel, Responsible Fisheries Code, etc.). However, at the same time it shows enormous limitations for the following main reasons:

- No quantitative targets exist to ensure that carrying capacity is not exceeded (except for the Montreal Protocol);
- No synergies among conventions or with established programs and agencies exist;
- Biases are in some of the thematic agendas where the main orientation is lost;
- Lack of compliance with commitments by developed countries further undermines their effectiveness.

Negotiations regarding different aspects of sustainable development have kept the polarisation between developed and developing countries. Among the first are concentrated the binding commitments. There is a trend to elude or postpone compliance of some of them. Among the second, a growing tension is perceived between the intensity of transformation that is required by sustainable development and the limited capacity to assume it.

Let's have a closer look at the two main conventions: Climate Change and Biological Diversity.

36.3 The Biodiversity Convention

Eight years ago the CBD took on three main objectives to address the critical problems of genetic, species and ecosystem loss. These were:

- Conservation, in particular *in situ* conservation;
- Sustainable use of all components of biodiversity;
- Equitable sharing of benefits of use of genetic resources.

During this time, progress has been made on technology discussions, access to genetic resources, financing, the clearing house mechanism, and a protocol on biosafety has been signed by 97 countries but ratified only by three.

But, unfortunately, these collective efforts have been completely insufficient and there are multiple indicators that show that after eight years of existence of the CBD, there is no change in the global trends of rapid loss of the various components of biodiversity: genes, species and ecosystems.

The Parties to the Convention have walked away from the main objective: *in situ* conservation. Governments are not assuming the structural risk the planet is being put in, when the loss of vegetation cover will lead to the loss of fundamental environmental services, altering the hydrological cycles, permanence of soils, carbon fixation, pollutant absorption, and oxygen production. Even further, the convention makes no reference to those environmental services.

Those countries that have already lost much of their biodiversity and their natural resources may not see this as their own problem; but they have been very actively involved in the access to genetic material. This can be explained probably because it is one of the few subjects for which clear economical value has been identified.

It is urgent to do an in-depth revision in the context of the broad discussion leading to the Conference of Rio plus 10 in 2002.

Important advances would occur by including

- *In situ* conservation as the main thrust of the CBD;
- The adoption of quantitative targets and potential commitments to arrest biodiversity and ecosystem loss;
- Performance indicators to measure progress;
- Adequate mechanisms to achieve equitable benefits sharing from sustainable use of all biodiversity;
- Giving the right value to the sustainable use of all biodiversity, including species and ecosystems, and not only to genetic material, as as it is now limited;
- Instruments and mechanisms that build positive synergies among the fora that address various aspects of biodiversity like CITES, Ramsar, UNESCO, the Turtle Convention, the Convention on Migratory Species and the UN Framework Convention on Climate Change, the Convention to Combat Desertification and the Panel on Forests;
- Instruments that allocate economic values for the goods and services generated by natural ecosystems with the purpose of both incorporating those values in the national accounting and recognising them as global services.

It is clear that the uneven distribution of natural wealth among countries means different efforts and different responsibilities. In contrast to what happens with respect to climate change, the effort cannot only be made by individual countries, i.e., those that have biodiversity. It is a must to make a collective, international effort.

36.4 The Climate Change Convention

The Framework Convention on Climate Change and the Kyoto Protocol are, together with the Montreal Protocol the only two instruments that contain quantitative goals, and that is a very important advance.

However, in this case, and in contrast to the Montreal Protocol, these goals are not related to the carrying capacity of the atmosphere as noted in the Convention's objective, which is to avoid dangerous anthropogenic interference with the climate system.

Unfortunately, the goals respond instead to political and economic considerations, making the effective implementation of the protocol very much uncertain. Annex I and Annex B countries treat the problem as an industrial and economic issue, and some important discussions seem to concentrate more on issues of international competitiveness than on mitigation of climate change. The problem is an environmental one with obvious economic repercussions, but ultimately it is environmental and its solutions must achieve the goal of reverting the trends of temperature increase in the atmosphere. It will be a big mistake to find solutions that will be economically suitable but that do not solve the environmental problem.

If the two major components of the problem are the consumption of fossil fuels as well as emissions from changes in vegetation cover, both elements should be part of the solution as well.

That is why the resistance to incorporating the forestry sinks as a mitigating instrument for non-annex countries in the framework of a global effort is not congruent with the importance of the impact of the change in the vegetation coverage. Also, from a wider perspective than just climate change, forestry sinks have local and regional benefits in terms of recuperation, restoration, maintenance, and sustainable use of vegetative cover, as well as in the protection of biodiversity, forests, soils, water, wildlife, and diversification of the rural economy. All of this represents synergies with the objectives of other international conventions.

It is obvious that the inclusion of the the forest sink must be considered seriously in order to avoid biases in achieving the goals. For example, in the Clean Development Mechanism, there must be a limited and controlled inclusion of projects related to land use, subject to strict social and environmental regulations. If we are able to guarantee these conditions we will have the possibility of enhancing the territorial scope of the global climate change regime and to reduce direct and indirect pressures on native forests, particularly tropical ones.

It is necessary to make the decisions that will allow the operation of the flexibility mechanisms provided by Kyoto Protocol, the development of an efficient and strict compliance system, and a comprehensive cooperation package that fulfils the legitimate demands of developing countries, which are the most vulnerable to climate change.

However, these discussions have less to do with the global problem than with national and local interests. The same is true for full access to the flexible mechanisms.

If the Kyoto Protocol does not enter into force by the year 2002, a decade after Río Summit, it might then be too late to meet the targets agreed upon as a first step, which is quite modest in any case.

36.5 Synergies

Each of these international instruments has its own objectives to solve one crucial environmental problem. Unfortunately, the links and interactions between the problems and, moreover, the possible solutions are not yet explored, and the advantages of the synergies that the different instruments could provide are lost.

The Convention to Combat Desertification has experienced serious delays in its development, in spite of the fact that it may be the widest-ranging and most integrating of the conventions. The Panel on Forests and the CDB have very clear synergetic potential with the CCD and with the use of carbon sinks. All of them have potential synergy with the FCCC.

Establishing synergies among the conventions would allow for significant advances in the resolution of these problems. But the governments of the Earth seem to lack the capacity to consider after anything other than short-term sectoral interests. As a result, multilateral solutions are negotiated in terms of the positions adopted by nations and groups of nations, determined by their geopolitical and geo-economic interests and as a function of the interests and possibilities of their businesses and pressure groups. Frequently, conventions end up being utilised more as adjustment mechanisms to settle economic and political accounts among nations than effective fora for constructive negotiations on the strategic issues about the environment.

In sum, there are no set targets, substantive issues are not addressed and synergies are nonexistent. Governmental systems appear to be inefficient and inappropriate mechanisms for assuming the management of long-term global problems, that is, the intergenerational dimension of sustainable development.

36.6 New Inputs

To limit the human impact on the biosphere to a level that is within the carrying capacity and to make this compatible with alleviating poverty we need to:

- Fully integrate social and economic considerations with environmental ones, which also implies different mechanisms to achieve this integration at the national level;
- Define strategies, priorities, quantitative and qualitative specific goals, timetables, objective indicators of compliance, economic instruments such as ecological footprints, green accounting systems and follow-up and assessment mechanisms;
- Enhance and integrate the presently dispersed financial resources to support national and regional agendas and to define a new generation of cooperation mechanisms;
- Expand the participation spaces for independent organisations;
- Increase government's and social capacities to build consensus between them to achieve sustainability both at the national and global level;
- Favour the development of environmental regulations, especially in standards and preventative actions, in order to induce changes in patterns of production and consumption;
- Combine in a synergistic way the protection and conservation of the environment and natural resources with a sustainable and more diversified use;
- Display alternatives of natural resources use that favour equity and alleviate poverty;
- Articulate an active participation in international meetings and agreements with definitions of internal policies and priorities;
- Implement innovations for a decentralised and efficient public policy;
- Link the knowledge generated in academia with public policy and the productive sector;
- Promote economic incentives to open the path for clean and sustainable technologies, in green and brown production.

Seventeen years after the Brundtland report, and with an eye on the proximity of Rio plus 10, temporary restrictions in the economies of countries cannot be a reason to postpone the commitments to sustainable development. There is no more time to lose.

Chapter 37
Research Systems for a Transition Toward Sustainability

William C. Clark

Sustainability concerns have occupied a place on the global agenda since at least the Brundtland Commission's 1987 report "Our Common Future" (World Commission on Environment and Development 1987; Clark 1986). The prominence of that place has been rising, however. UN Secretary General Kofi Annan reflected a growing consensus when he wrote in his Millennium Report to the General Assembly that "Freedom from want, freedom from fear, and the freedom of future generations to sustain their lives on this planet" are the three grand challenges facing the international community at the dawn of the 21st century (Annan 2000). Sustainability has become a "high table" issue in international affairs.

Science and technology are increasingly recognised to be central to both the origins of Secretary General Annan's three challenges, and to the prospects for successfully dealing with them (United Nations Development Programme 2001; World Bank 1998; Sachs 2000). But there is a great imbalance in the resources and attention devoted to harnessing science and technology in the service of these three transcendent goals. Efforts to achieve "freedom from fear" are supported by a mature, well-funded, problem-driven R&D system based in the world's military establishments. Efforts to achieve "freedom from want" have created and been supported by several effective R&D systems, for example those engaged in international agricultural research and in certain global disease campaigns. In contrast, efforts to achieve sustainability are relatively new, because in the words of the Secretary General, the "founders of the UN could not imagine that we would be capable of threatening the very foundations for our existence" (Annan 2000). As a result, efforts to harness science and technology for sustainability have largely had to draw on R&D systems built for other purposes – begging monitoring data from the world's military establishment, piggybacking on the already over-extended international agricultural research system, and borrowing insights gained from basic research programs on global environmental change. With a few important but relatively small and under-funded exceptions, efforts to "sustain the lives of future generations on this planet" still lack dedicated, problem-driven R&D systems of anything like the scale or maturity of those devoted to security and development.

The World Summit on Sustainable Development, scheduled for August/September of 2002 in Johannesburg, South Africa, represents the best opportunity in a decade to construct a global R&D system tailored to the particular needs and magnitude of the sustainability challenge. Seizing that opportunity will require a strategic approach that transcends the interests of individual nations, policy initiatives and research programs. Fortunately, important elements of the foundation for such a strategy have been laid out over the last several years through a rapidly expanding discourse on the relationships among science, technology and sustainability.

Many of the earliest and most thoughtful contributions to this discourse have come from the developing world through the work of individual scholars and of institutions such as the Third World Network of Scientific Organizations (TWNSO), the Commission on Science and Technology for Sustainable Development in the South (COMSATS), the Society for Research and Initiatives for Sustainable Technologies and Institutions (SRISTI), and the South Center[1]. European thinking of the late 1990s is exemplified in Schellnhuber and Wenzel's *Earth Systems analysis: Integrating Science for Sustainability* and the European Union's *Fifth Framework Programme* (Schellnhuber and Wenzel 1998; European Commission 1998). A synthesis of US views from the same period is given in the National Research Council's *Our common journey: A Transition Toward Sustainability* (United States National Research Council, Board on Sustainable Development 1999). Initial efforts to capture an international cross-section of perspectives include the special issue on *Sustainability Science* published by the International Journal of Sustainable Development in 1999, and the World Academies of Science report on a *Transition to Sustainability in the 21st Century*

---

[1] Third World Network of Scientific Organizations (TWNSO), http://www.ictp.trieste.it/~twas/TWNSO.html; Commission on Science and Technology for Sustainable Development in the South (COMSATS), http://www.comsats.org.pk/index.html; Society for Research and Initiatives for Sustainable Technologies and Institutions (SRISTI), http://www.sristi.org/; Policy statements by the International Foundation for Science (IFS), http://www.ifs.se/index.htm; International Science Programme (ISP), http://www.isp.uu.se/Home.htm.

(Funtowicz and O'Connor 1999)². In addition, international environmental assessments are increasingly reaching out to connect with sustainability issues, as are research planning efforts for global environmental change programmes at both national and international levels (Watson et al. 1998; IPCC Plenary Seventeenth Session 2001; Millennium Ecosystem Assessment 2002). A number of academies of science have also recently addressed the links between sustainability and global change (Rocha-Miranda 2000; African Academy of Sciences 1999; but see also German Advisory Council on Global Change (WGBU) 1997)³. Most recently, the ad-hoc International Initiative on Science and Technology for Sustainability has been sponsoring a series of workshops around the world to help regionalise the discourse on core questions, research strategies, action priorities and institutional needs for mobilising knowledge in the service of sustainable development (Kates 2001).

This widening discourse on science, technology and sustainability has revealed profound differences in perspectives and priorities (Fig. 37.1) between rich and poor people, northern and southern regions and public and private sectors (Kates 2001). But it has also demonstrated broad agreement on a number of characteristics that effective R&D systems for sustainability might be expected to exhibit.

First and foremost, effective R&D systems for promoting sustainability will need to be structured so that they are driven by the most pressing problems of sustainable development as defined by stakeholders in those problems. This will almost certainly result in a much different agenda than would be obtained by continuing to allow priorities to reflect primarily the most interesting problems in science and technology as defined by stakeholders in research and innovation. But while the specific character of those "most pressing problems" of sustainability will need to be assessed on a regional and even local basis, a general consensus is emerging that they involve discovering and inventing ways of simultaneously meeting human needs with special attention to the reduction of hunger and poverty while protecting the Earth's essential life support systems and biodiversity.

There is also general agreement that to accomplish these goals, R&D systems for sustainability will have to be extraordinarily integrative, encompassing the communities engaged in promoting not only environmental conservation, but also human health and economic development. They will need to entrain formal expertise from the public and private sectors, the natural and social sciences, and engineering. Perhaps most challenging, they will need to find ways of identifying, utilising and honouring the vast resources of informal expertise derived from practical experience in grappling with particular sustainability problems in particular social and ecological settings (Gupta 1999).

As implied above, much of the knowledge needed for advancing sustainability goals involves making sense of how multiple environmental stresses, social institutions and ecological conditions interact in particular places. This means that R&D systems for sustainability will need to give special emphasis to integration at intermediate or regional scales (National Research Council, Committee on Global Change 2001). From this base, they will need to be structured to facilitate "vertical" connections between the best research anywhere in the world and practical experience in particular field situations. At the same time, they will need to foster "horizontal" connections among regional research and application centres that might learn from one another (Knowledge Network of Grassroots Green Innovators 2002).

Finally, effective R&D systems for sustainability will need to bridge the artificial but pernicious divide between "basic" and "applied" research (Branscomb et al. 2001). Progress on some of the most urgent problems of sustainability will almost certainly require fundamental improvements in our understanding of nature-society interactions: sustainability science needs to be fundamental research. But on other issues, the requirement is less for new knowledge than for learning how to apply what is already known in an experimental, problem-solving mode: sustainability science needs to be learn-

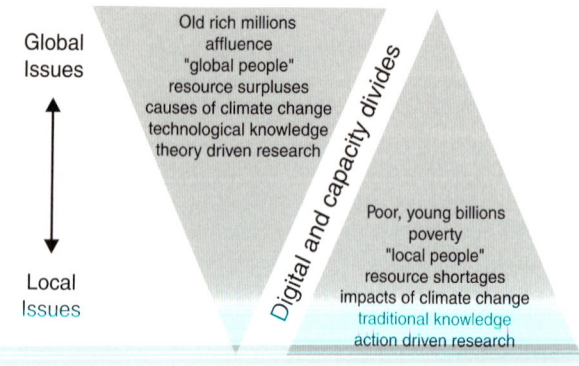

Fig. 37.1. The challenge to achieve a sustainable future in the context of a divided world. The socioeconomic, environmental and knowledge dichotomies are exacerbated by the deepening digital divide (reproduced with permission from Kates et al. 2001, © 2001 American Association for the Advancement of Science)

---

² World's Scientific Academies' Transition to Sustainability in the 21ˢᵗ Century (Tokyo Summit of May 2000), http://interacademies.net/intracad/tokyo2000.nsf/all/sustainabilitystatement

³ The Global Environmental Change Programmes of the International Council of Science (ICSU) have made "global sustainability" a central point of their research planning for the coming years (see IGBP 2001).

ing-by-doing. More generally, promoting sustainability needs integrated *knowledge systems* that connect what have too often been the "island empires" of research, monitoring, assessment and operational decision support.

With these widely shared criteria for effective sustainability R&D systems in mind, an initial set of goals for the World Summit to pursue with regard to science and technology might include the following:

- Secure continued support for the core disciplinary and integrative R&D programs on which sustainability science and technology must build;
- Launch focused, action initiatives in priority problem areas (e.g., sustainable urban growth, carbon management) where we know enough to complement learning-by-studying with learning-by-doing;
- Initiate focused R&D efforts on fundamental scientific questions (e.g., determinants of the vulnerability of nature-society systems) arising from attempts to resolve priority problems of sustainability;
- Increase the world's capacity for regionally-based, problem-driven, integrated R&D in support of a sustainability transition.

Effective R&D systems to mobilise science and technology for sustainable development should not be impossible to design and implement. Some relatively successful international programmes exhibiting many of the characteristics outlined here have already been developed to address problems ranging from increasing agricultural productivity to combating human disease, to protecting the Earth's ozone layer. Likewise, there already exist efforts such as START's Southeast Asia Regional Center that have made a good beginning in implementing integrated, problem-driven, place-based research and applications programs in support of sustainability (Southeast Asia START Regional Center n.d.). To date, however, these successes reflect idiosyncratic, if invaluable, exceptions rather than general rules.

Needed over the period leading up to the World Summit is a systematic and critical effort to learn from both successes *and* failures of the past lessons that have the most to offer the design of effective R&D systems for promoting a transition toward sustainability. Such learning will in turn require a determination to move beyond the advocacy of existing programs that have been built for other (often excellent) reasons, toward a critical dialogue about the science and technology strategies most needed to support sustainable development per se. Above all, it will demand a unified campaign by the scientific, engineering and development communities to build the political support needed to implement – at a scale worthy of the challenges before us – an R&D system for sustainability.

Acknowledgements

This paper is based on a collaborative study of "Research, Assessment and Decision Support Systems for Sustainability" conducted under grants from the US National Science Foundation, the National Oceanic and Atmospheric Administration, and the Packard Foundation. I have drawn particularly heavily on ideas of David Cash, Calestous Juma and Nancy Dickson. Portions of this paper were published as an editorial in the October, 2001 issue of *Environment* magazine.

References

African Academy of Sciences (1999) Tunis Declaration: Millennial perspective on science, technology and development in Africa and its possible directions for the twenty-first Century. Fifth General Conference of the African Academy of Sciences, Hammamet, Tunisia, 23–27 April 1999, http://www.unesco.org/general/eng/programmes/science/wcs/meetings/afr_hammamet_99.htm

Annan K (2000) We, the peoples: The role of the United Nations in the 21st Century. United Nations, New York, http://www.un.org/millennium/sg/report/full.htm

Branscomb L, Holton G, Sonnert G (2001) Science for society: Cutting-edge basic research in the service of public objectives. http://www.cspo.org/products/reports/scienceforsociety.pdf

Clark C (1986) Sustainable development of the biosphere: Themes for a research program. In: Clark WC, Munn RE (eds) Sustainable development of the biosphere. Cambridge Univ. Press, Cambridge, pp 5–48

Commission on Science and Technology for Sustainable Development in the South (COMSATS) (2002) http://www.comsats.org.pk

European Commission (1998) Fifth framework programme: Putting research at the service of the citizen. http://www.cordis.lu/fp5/src/over.htm

Funtowicz S, O'Connor M (eds) (1999) Science for sustainable development. Special issue of *International Journal of Sustainable Development* 2:3

German Advisory Council on Global Change (WGBU) (1997) World in Transition: The Research Challenge, Annual Report 1996, Springer-Verlag, Heidelberg Berlin, http://www.wbgu.de/wbgu_publications.html

Gupta A (1999) Science, sustainability and social purpose: barriers to effective articulation, dialogue and utilization of formal and informal science in public policy. Int J Sustainable Development 2(3):368–371

IGBP (2001) Global change and the Earth system: A planet under pressure. IGBP Science Series No. 4., Paris, ICSU, and http://www.igbp.kva.se/

International Foundation for Science (IFS) (2002) Policy statements. http://www.ifs.se/index.htm

International Science Programme (ISP) (2002) http://www.isp.uu.se/Home.htm

IPCC Plenary Seventeenth Session (2001) Special report on climate change and sustainable development. IPCC Plenary Seventeenth Session, Nairobi, April 2001, http://www.ipcc.ch/meet/p17.pdf

Kates RW, Clark WC, Corell R, Hall JM, Jaeger CC, Lowe I, McCarthy JJ, Schellnhuber HJ, Bolin B, Dickson NM, Faucheux S, Gallopin GC, Gruebler A, Huntley B, Jäger J, Jodha NS, Kasperson RE, Mabogunje A, Matson P, Mooney H, Moore B III, O'Riordan T, Svedin U (2001) Sustainability science. Science 292:641–642, and http://sustainabilityscience.org

Knowledge Network of Grassroots Green Innovators (2002) *http://www.sristi.org/Nissat.htm*.

Millennium Ecosystem Assessment (2002) *http://www.millenniumassessment.org/en/index.htm*

National Research Council, Committee on Global Change (2001) The science of regional and global change: Putting knowledge to work. National Academy Press, Washington, *http://books.nap.edu/catalog/10048.html*

Rocha-Miranda CE (ed) (2000) Transition to global sustainability: The contributions of Brazilian science. Academia Brasiliera de Ciências, Rio de Janeiro, 2000, *http://www.abc.org.br/eventos/trabsim99_en.htm*

Sachs JD (2000) A new map of the world. The Economist 355: 81–83

Schellnhuber HJ, Wenzel V (eds) (1998) Earth system analysis: Integrating science for sustainability. Springer-Verlag, Heidelberg Berlin

Society for Research and Initiatives for Sustainable Technologies and Institutions (SRISTI) (2002) *http://www.sristi.org/*

Southeast Asia START Regional Center (2002) Program, *http://www.start.or.th*

Third World Network of Scientific Organizations (TWNSO) (2002) *http://www.ictp.trieste.it/~twas/TWNSO.html*

United Nations Development Program (2001) Making new technologies work for human development: The human development report (2001). Oxford Univ. Press, Oxford

United States National Research Council, Board on Sustainable Development (1999) Our common journey: A transition toward sustainability. National Academy Press, Washington, DC, *http://www.nap.edu/catalog/9690.html*

Watson R, Dixon JA, Hamburg SP, Janetos AC, Moss RH (1998) Protecting our planet, securing our future. UN Environment Programme, Nairobi, 1998, *http://www-esd.worldbank.org/planet/*

World Bank (1998).Knowledge for development: The world development report for 1998/9. Oxford Univ. Press, Oxford

World Commission on Environment and Development (1987) Our common future. Oxford University Press, Oxford

World's Scientific Academies (2002) Transition to Sustainability in the 21st Century (Tokyo Summit of May 2000). *http://interacademies.net/intracad/tokyo2000.nsf/all/home*

Chapter 38

Summary: Towards Global Sustainability

Jill Jäger

Both of the papers in this section give us some good news and some bad news.

Julia Carabias points out that the concept of sustainable development has been endorsed in many countries, and understanding of environmental processes has increased, as demonstrated strongly throughout this volume.

She also shows, however, that the pressure on the environment has increased and asks what the obstacles to progress are. She notes weaknesses in the international instruments, which have failed to halt global trends of rapid loss of biodiversity and have not dealt effectively with the issue of climate change.

Julia Carabias calls for the exploitation of synergies among the conventions and for a number of steps, including

- Integration of socioeconomic and environmental considerations;
- Strategies, priorities and goals;
- Support for national and regional agendas;
- Expanded participation.

She also calls for a stronger link between science and public policy, knowledge and action – a point that is also picked up by Bill Clark.

Bill Clark illustrates the challenges of developing research systems for the transition to sustainability – research systems that can help to remove some of the roadblocks identified by Julia Carabias.

He asks how we can design research systems to support the needed work – and argues that the existing research agendas must be extended – to be strongly linked to the development agenda, integrative across natural, social and engineering sciences, and echoing Julia Carabias, across the worlds of knowledge and action.

The science agenda for sustainability is challenging – not all parts of it are new – but what is new is the call for the simultaneous inclusion of environmental and development sciences: integrative, place-based, problem-solving approaches that are responding to urgent policy needs.

This agenda challenges our current funding structures and research systems. There are, however, some examples of effective systems on which we could build – to create an R&D system in response to the sustainability challenge.

Where do the GEC Programmes that organised the Global Change Open Science Conference and this volume fit in this dialogue? Clearly, the programmes will continue to contribute to a strengthening of the link between knowledge and action – not a priority in the first phase of GEC research but now, with a synthesis of scientific advances over the last 15 or so years, a restructuring of the research agenda and a much stronger contribution from both the human dimensions and biodiversity research communities, the GEC programmes will surely play an important role.

The GEC programmes should also be major contributors to the sustainability science research systems. In particular, the joint projects on food systems, carbon and water provide useful models of integrative research closely linked to the policy agenda.

I close my summary by reading two short paragraphs from the report "Global change and the Earth System: A Planet under Pressure" released at the Open Science Conference.

> The present international global environmental change programmes represent a first step forwards towards such a global system (for sustainability science), but they are subject to the vagaries of uncorrelated and insecure sources of funding and must rely on voluntary contributions of research from the scientific community. On the other hand, traditional branches of so-called big science, like particle physics or space exploration, enjoy continued international support through well-funded mega-projects. Ironically, the science instrumental for the preservation of the global life support system does not.
>
> Any global system of sustainability science will fail, however, if it does not stimulate and achieve the active participation of the developing world, i.e., of those regions where the capacity to undertake sustainability science is the weakest but where the impacts of global change will unfold in the most dramatic ways. Without this participation, the systems-order problems of the planet cannot be solved. The challenge of achieving a truly global system of science is substantial; the challenge of bridging the divide between developed and developing countries across a broader range of human endeavours is much more daunting. Both challenges must be met to achieve a sustainable future for planet Earth.

Part IVe

Closing Session

Chapter 39

Closing Address

Michael Zammit Cutajar

It is not easy to speak at the end of a four-day scientific conference with little input from all that has gone before. Nor is it easy to look forward two days to a political conference that is shrouded in political uncertainty. I have accepted to do this for two reasons. First, a friend invited me: Larry Kohler, the Executive Director of ICSU; and, second, I knew that I would be speaking under the chairmanship of a well-known colleague, Professor Bert Bolin. When I first started to work on climate change over ten years ago, Bert presented me with the human face of a scientific community which I held in awe while at the same time being ignorant about it. On account of his wise guidance, I am now more familiar with the way that community thinks and it it has my great respect.

My remarks today will address two topics: the interaction between science and policy, as exemplified in the area of climate change, and the prospects for the negotiations on climate change that resume in Bonn on Monday. In both cases, you will hear the personal views of an observer of the scientific and negotiating processes.

39.1 Science and Policy

The development, integration and diffusion of the science of climate change is one of the success stories of the last decade. For this we have to thank the IPCC, the foresight of its founding fathers, UNEP (Dr. Mostafa Tolba) and WMO (Professor Godwin Obasi), the leadership of its successive Chairmen, Bert Bolin and Bob Watson, and the scientific bodies – such as those organising this conference – that have stimulated the basic work.

It is worth recalling that *climate science has proved remarkably robust*. Before coming here, my attention was drawn to a rather dusty document, dated 1985 – the report of the second Villach Conference on the role of GHGs in climate variations. Statements from that seminal meeting – e.g., on the rise in concentrations of GHGs and their impact on temperature and sea level – read remarkably well today, after three IPCC assessments.

The difference, of course, is that Villach was a small meeting; the IPCC is a global network. Through the IPCC, the patient work of scientific peer review and the careful formulation of assessments have had a tremendous impact on political and public awareness of the problem of climate change and the need to respond to it. Each IPCC assessment has underpinned a phase of negotiation. *Science has driven the politics.*

The IPCC is writing the bible of climate science. Dissent from its central message is dwindling. While more chapters remain to be written, this bible gives the climate negotiators a basis for precautionary action, as envisaged by the Convention. Even at this time of political uncertainty about the path forward, this body of scientific information and assessment gives us hope that all governments will honour their commitment to respond to the serious nature of the climate change phenomenon.

Looking back over my contact with this scientific process, I conclude that if the science is to continue guiding the politics, it is essential to *keep the politics out of science*. This risk of political contamination arises in different ways. It occurs, for example, in the negotiation of the policy makers' summaries from the IPCC, which are the documents that most nonscientists read. The demography and the dynamics of such negotiation are not unlike those of the political negotiations under the Convention. Coherence is thus assured, of course, but the dividing line between science and politics is blurred. The preservation of scientific integrity demands the full engagement of the leaders of any scientific process that touches upon the political domain.

Another form of contamination arises when nonscientific lobbies peddle "pseudo-scientific" opinions with the aim of destabilising the work of serious scientists. I do not mean to say that any dissent from mainstream opinion is to be dismissed as "pseudo-science". I have in mind pronouncements that are advanced in the cause of misinformation, in the context of campaigns to derail a political process by casting doubts on the credibility of the scientific assessment that underlies it. We have seen some of this around the climate change issue. And the press is quick to give credence to controversy around a topic that is in political contention.

Nevertheless, an editorial in *Science* magazine recently stated, with reference to climate change, "Consensus as strong as the one that has developed around this topic is rare in science." Even in the USA, where the controversy is greatest, the recent report to the President by the National Academy of Sciences seems to have scotched the view that the current state of science is insufficient to justify action.

This leads me to a third point: *the importance of communication of scientific messages*. The audience out there receives these messages as sound bites and interprets them in a political context. Scientists need to be good communicators, with access to the arts of communication, if they are to get their messages across and beyond their own community.

I realise that this can run counter to scientific discipline. I have discovered that, despite the popular image of the eccentric or 'mad' scientist, scientists are in fact excessively prudent people. They move towards certainty by disproving hypotheses. They are sceptics, if not born then bred. Qualifiers such as "likely" or "highly probable" hedge their reports. They are adept at digging deep into a specialised subject whose link to a broader context is not well-understood by the profane observer.

When scientists deal with a subject as controversial as climate change – and I am sure that this same observation could be applied in other fields – they need to be on guard against the risks of misinterpretation of their conclusions. They need to devote resources and time to presenting those conclusions in ways that can be understood by journalists, economists, policy makers, their parents and their children.

When I think of misinterpretation, I can see it going in different directions. For example, a leading climate scientist has gone on record to suggest that neither the role of black soot nor the potential of reductions in methane emissions have received sufficient attention – by scientists in the first case and by policy makers in the second. Up went the clamour: "Stop the clock. We cannot move until we've sorted this one out." Now the scientist, Dr. James Hansen, was not saying that carbon dioxide was no longer relevant. He was pointing to other factors that may have been underestimated. Nevertheless, the journalists smelled controversy, and the dissenters jumped on this opportunity to try to slow down action on the political front.

Equally problematic, in a way, is the well-meaning propensity of politicians to say, "Climate change is happening. The proof: the latest storm or flood, wherever it happens to be: in Mozambique, Orissa, Central America, or England." Politicians love certainty, even where it doesn't exist, and this puts advocates of responsible action on climate change in the uncomfortable position of having to dampen political enthusiasm for our own cause. "No, no," we say. "This particular storm cannot be pinned down to climate change. All we can affirm is that climate change will increase the frequency of such events in the future."

All this is to say that the effective communication of the results of scientific assessment and research is an extremely important part of your responsibilities as socially aware, politically aware scientists.

My last comment on this borderline between science and policy concerns the need for *targeted scientific guidance to policy makers* on where it is that they should be heading. Putting myself in the shoes of a busy politician with little time to digest complex briefs, I can summarise the message I get on climate change as follows: "Emissions of GHGs caused by human activity are accumulating in the atmosphere and causing global temperatures to rise. The effects of this are generally bad. Therefore, governments must guide economic actors and consumers to actions that limit and reduce emissions at the least cost. But I have no idea – based on science – by how much emissions should be reduced and when. Some of my constituents push for quick and radical action and take a sanguine view of economic impacts. Others are concerned about economic costs and are convinced that a technological fix is just around the corner. Moreover, looking to the next election, I have little basis for judging how this global phenomenon is going to influence my particular electorate."

I allude here to two big unknowns, which have a major influence on the readiness of politicians to act. Taking the latter first: *we do not know enough about regional impacts of climate change*. People react better to changes in their own surroundings than to abstract global averages. The IPCC has addressed this issue in its Third Assessment Report. More needs to be done to take the science to a level where it could satisfy the political demand for more focussed impact assessment.

The other bigger issue is that while we are confident that we should be emitting less or less intensively, we do not know what the level is at which the accumulation of emissions should stop. *What is the safe equilibrium level of atmospheric concentrations?* The objective of the Convention is to avoid dangerous human interference with the climate system, but nobody has yet defined what is dangerous and what is safe. Policy makers, unable to make such judgements, turn to the scientists for answers, as a patient turns to a medical doctor for advice on limits. A doctor prescribes the limits and the medication. A "climate doctor" says: "Well, it depends on the risks you are prepared to bear."

We are in something of a trap here. I believe that the political process on climate change would be greatly assisted by agreement on a target for atmospheric concentrations, at least an intermediate target. This would give a sense of where the whole international community should be heading and a basis for apportioning responsibility for getting there.

How do we get out of this trap? I am not sure. I guess that politicians could lay out a set of possible targets, scientists could tell them the implications of each one, and policy makers could then choose.

39.2 Prospects for the Climate Change Negotiations

Having gotten so close to the current debate, let me move into the second phase of my statement, which is about the negotiations that resume next week.

The President of the Conference of the Parties, Jan Pronk, through a representative, gave you his views at the beginning of this conference on the politics of the current negotiations and the political arithmetic which makes it necessary for the EU, the Russian Federation and Japan to be pulling on the same rope if the Kyoto Protocol is to move ahead without the initial participation of the USA. I will not go over that political analysis again but will try to put it in a broader context.

First, the imminent Bonn Conference is not just about the Kyoto Protocol. It takes place under the sign of the Buenos Aires Plan of Action, which also seeks to drive action under the Convention. Specifically, agreements are sought on enhanced financial and technological support, including capacity building, for the efforts of developing countries to integrate climate change in their development programmes and policies, to launch adaptation strategies and to start contributing to the global objective of maintaining concentrations at a safe level, while also pursuing their immediate priorities of economic growth and poverty eradication. A long-term climate strategy must involve emission limitation by developing countries, and they need encouragement and incentives to move their development strategies in that direction.

Second, a Conference of the Parties is an occasion for the near-universal membership to take stock of the political level of where matters stand and to map out ways of maintaining political momentum towards the objective of the Convention. This has become especially important on account of the announcement by the United States administration of its opposition to the Kyoto Protocol, which has induced a crisis in the climate change community. The community must not allow itself to be blocked by this crisis. It must overcome it and move ahead together, in the unifying framework of the Convention. The Bonn Conference is to be followed within precisely three months by the seventh session in Marrakech, Morocco. One important political output of Bonn would be a sense of purpose for COP 7 and a sense of direction beyond that, on a road that passes through the Johannesburg Summit on Sustainable Development in 2002.

Nevertheless, you will rightly observe, that it is the Kyoto Protocol that will be in the news. Here too, I would like to invite you to a broader perspective.

In my opinion, the most important aspect of the Kyoto Protocol is that it establishes a system – what some have called an "architecture" – for limiting GHG emissions: legally-binding targets, supported by rigorous performance indicators, a range of flexible options – including market mechanisms and the controlled use of sinks – for achieving these targets at lowest economic cost, and incentives for emission-saving investments in countries that are not bound by targets. In this perspective, the Protocol is important as a market-based economic instrument for achieving an environmental goal over the next two or three decades.

For the last four years, negotiators have sought to complete the first operational rules for this instrument. It would be a waste of a costly investment if this rulebook were to be left unfinished. I believe that the rulebook has a value that is independent of a political accord on the Protocol itself. For a start, the rules will tie up the loose ends left by the heroic negotiation of the Protocol in 1997. Heroism is not inherently tidy, and the rules will make it clear what the consequences of ratification are. Secondly, even if the Kyoto Protocol were not to enter into force in its present form, I am prepared to bet that its mechanisms and procedures will be found useful in any alternative multilateral instrument that might be developed. Logically, this work could be completed by the end of this month in Bonn.

The other key feature of the Protocol is that it sets emission targets for an initial five-year period and promises further commitment periods to follow. The US declaration that its target would cause it serious economic harm has heightened the political debate on these targets and on efforts to soften their economic impact through the details of the rulebook, for example through the rules on sinks. Unfortunately, the economic costs of climate change itself and the economic opportunities of technological innovation are not given the same weight in these political calculations.

Nevertheless, the numbers and dates for the first commitment period are neither magic nor scientific. When our successors look back on such targets in 50 years, their importance will be neither the percentages nor the dates – but the fact that they triggered movement in the right direction by governments, corporations and consumers. These initial targets are just the first touches on the tiller of the global economy. The longer it takes to start the required moves, the more they may cost. However, if a shift in the target dates were to be the cost of a comprehensive political deal now, a long-range view would suggest that this might be a cost worth bearing.

Such a deal would still leave outstanding the engagement of developing countries – especially populous and industrialising countries – in emission limitation commitments. This is the focus of the other US objection to the Protocol. On this, I will express two personal opinions: that it would be unfair to expect developing countries to take on the same type of quantitative target as developed countries, and that it would be unrealistic to imagine that they might take on any limitation commitment whatsoever before developed countries demonstrate that they are making serious progress in reducing their emissions.

The way ahead on this front could involve "demonstrable progress" by developed countries, recognition of significant emission limitations being achieved by developing countries without commitments, design work on suitable qualitative targets, and a realistic time frame – which could be set once it is clear that all the major industrial countries are committed to taking the lead. And, to repeat, I believe that an intermediate target for atmospheric concentrations would give this "common but differentiated" effort a sense of direction and of equity.

Note: This address preceded the Bonn Agreements of July 2001 and the Marrakech Accords of November 2001, by which the Conference of the Parties to the UNFCCC determined the operational rules for the Kyoto Protocol. The USA did not join in these agreements.

Chapter 40
The Amsterdam Declaration on Global Change

The scientific communities of four international global change research programmes – the International Geosphere-Biosphere Programme (IGBP), the International Human Dimensions Programme on Global Environmental Change (IHDP), the World Climate Research Programme (WCRP), and the international biodiversity programme DIVERSITAS – recognise that in addition to the threat of significant climate change, there is growing concern over the ever-increasing human modification of other aspects of the global environment and the consequent implications for human well-being. Basic goods and services supplied by the planetary life support system, such as food, water, clean air, and an environment conducive to human health are being affected increasingly by global change.

Research carried out over the past decade under the auspices of the four programmes to address these concerns has shown that:

- *The Earth System behaves as a single, self-regulating system comprised of physical, chemical, biological and human components.* The interactions and feedbacks between the component parts are complex and exhibit multi-scale temporal and spatial variability. The understanding of the natural dynamics of the Earth System has advanced greatly in recent years and provides a sound basis for evaluating the effects and consequences of human-driven change.
- *Human activities are significantly influencing Earth's environment in many ways in addition to greenhouse gas emissions and climate change.* Anthropogenic changes to Earth's land surface, oceans, coasts and atmosphere and to biological diversity, the water cycle and biogeochemical cycles are clearly identifiable beyond natural variability. They are equal to some of the great forces of nature in their extent and impact. Many are accelerating. Global change is real and is happening *now*.
- *Global change cannot be understood in terms of a simple cause-effect paradigm.* Human-driven changes cause multiple effects that cascade through the Earth System in complex ways. These effects interact with each other and with local- and regional-scale changes in multidimensional patterns that are difficult to understand and even more difficult to predict. Surprises abound.
- *Earth System dynamics are characterised by critical thresholds and abrupt changes. Human activities could inadvertently trigger such changes with severe consequences for Earth's environment and inhabitants.* The Earth System has operated in different states over the last half million years, with abrupt transitions (a decade or less) sometimes occurring between them. Human activities have the potential to switch the Earth System to alternative modes of operation that may prove irreversible and less hospitable to humans and other life. The probability of a human-driven abrupt change in Earth's environment has yet to be quantified but is not negligible.
- *In terms of some key environmental parameters, the Earth System has moved well outside the range of the natural variability exhibited over the last half million years at least.* The *nature* of changes now occurring *simultaneously* in the Earth System, their *magnitudes* and *rates of change* are unprecedented. The Earth is currently operating in a non-analogue state.

On this basis, the international global change programmes urge governments, public and private institutions and people of the world to agree that:

- *An ethical framework for global stewardship and strategies for Earth System management are urgently needed.* The accelerating human transformation of the Earth's environment is not sustainable. Therefore, the business-as-usual way of dealing with the Earth System is *not* an option. It has to be replaced – as soon as possible – by deliberate strategies of good management that sustain the Earth's environment while meeting social and economic development objectives.
- *A new system of global environmental science is required.* This is beginning to evolve from complementary approaches of the international global change research programmes and needs strengthening and further development. It will draw strongly on the existing and expanding disciplinary base of global

change science, integrate across disciplines, environment and development issues and the natural and social sciences, collaborate across national boundaries on the basis of shared and secure infrastructure, intensify efforts to enable the full involvement of developing country scientists, and employ the complementary strengths of nations and regions to build an efficient international system of global environmental science.

The global change programmes are committed to working closely with other sectors of society and across all nations and cultures to meet the challenge of a changing Earth. New partnerships are forming among university, industrial and governmental research institutions. Dialogues are increasing between the scientific community and policy makers at a number of levels. Action is required to formalise, consolidate and strengthen the initiatives being developed. The common goal must be to develop the essential knowledge base needed to respond effectively and quickly to the great challenge of global change.

Berrien Moore III   Chair, IGBP
Arild Underdal   Chair, IHDP
Peter Lemke   Chair, WCRP
Michel Loreau   Co-Chair, DIVERSITAS

Challenges of a Changing Earth:
Global Change Open Science Conference
Amsterdam, The Netherlands
13 July 2001

Index

A

acid
   –, nucleic 83
   –, rain, North East Asian 50
acidification 183
advection 77, 98, 119
aerosol, sulphate a. 45, 48, 95, 146
AEZ, resources database 33
Africa 23, 24, 27, 31–34, 39, 42, 94, 107, 118, 119, 123–128, 189, 197
African
   –, tropics, precipitation 127
   –, Wet Period 126
Agenda 21 51
agriculture 21, 23, 32–34, 37–42, 48, 57, 62, 73, 97, 101, 103, 138, 140, 178–182
   –, recession 39, 40
air pollution 32, 48–50, 54, 87, 89, 91, 101, 184, 193
   –, Asia 49
Alaska 27, 68, 98, 114
albedo 64, 125, 129, 134, 135, 152
alpine glacier, melting 108
aluminium 34
Amazon
   –, Basin 32, 124, 137, 189, 190
      –, mean outflow 137
   –, River 137
Amazonia 22, 40, 137–140, 189
ammonia 93–95
Amu Darya River 21
analysis
   –, eigenfunction 158
   –, socioeconomic 33
   –, syndrome 153, 158
anchovy 28
animal, pelagic 77
Annan, Kofi 197
Antarctic
   –, food web 28
   –, Polar Frontal Zone 163
Antarctica 45
Anthropocene 45
anthroposphere 151, 157
approach, system 157
archive, palaeoclimatic 109
Arctic 45, 113–115, 127
   –, biota 115
   –, hydrology 114
   –, sea ice 127
   –, tundra 113
aridification 126, 136
   –, North Africa 126
ASEAN
   –, cooperation 51
   –, Ministerial Meeting on Haze, Singapore 52
   –, region 49
   –, Strategic Plan of Action 51

Asia 24, 27, 31–34, 49–54, 69, 71, 94, 98–104, 118, 124, 129–136, 189, 199
   –, air pollution 49
Asian
   –, Development Bank 52, 54
   –, monsoon 124, 129, 135
      –, region 133, 135
      –, system 130
assessment method 24
Aswan Dam 97
Atlantic
   –, Ocean 39, 78, 119
   –, thermohaline circulation 118, 147
atmosphere
   –, chemical composition 45, 148
   –, oxidation efficiency 139
atmospheric
   –, absorption 85
   –, carbon dioxide 21, 60, 83, 84, 94, 125, 127, 128, 162, 172, 191
   –, chemistry 47, 109, 148, 149
      –, role of tropics and subtropics 47
   –, deposition processes 93
   –, inputs to the oceans 93
   –, nitrogen deposition 94
   –, regimes 89, 90
   –, residence time 47
   –, transport 52, 59
Australia 34, 118, 165

B

bacteria, marine 94
Bangladesh 31
biodiversity 32–34, 49, 63, 64, 69, 73, 80, 103, 114, 129, 140, 156, 161, 162, 165–167, 169–172, 183, 185, 193, 194, 198, 201, 207
   –, convention 161, 193
   –, extinction 193
   –, loss 183, 185, 201
   –, marine systems 161
biological pump 58, 77, 84
biomass, burning 48, 138, 139
biosphere 21, 62, 78, 81, 82, 85, 123–125, 149, 151, 157, 167, 190, 195
   –, terrestrial 62, 81, 190
biota
   –, mass extinction 21
   –, restructured global 21
Black Sea 99
Bonn Conference 205
boreal forest 22
boundary, continental 77
Bowen ratio 136
Box model 98
Brazil 34, 41, 42, 118, 119, 137, 138, 180
bromoform 94
Brunei 49
Buenos Aires Plan of Action 205
business 50, 54, 61, 65–68, 70, 71, 149, 156, 189, 207

C

Cairo 97
Canada 34, 37, 51, 53, 108, 113, 119
capital 31, 186
capture fishery 27, 178
carbon
  –, capture 68
  –, credit 63
  –, cycle 58, 61, 77, 81–85, 103, 124, 138, 139, 161, 186, 191
    –, Amazonia 138
    –, dynamics 81
    –, global 58, 77, 81, 84, 103
    –, history of 83
    –, linkage to nitrogen cycle 82
    –, oceanic 77
  –, dioxide
    –, atmospheric 21, 60, 83, 84, 94, 125, 127, 128, 162, 172, 191
    –, upper bound of atmospheric 84
    –, concentration 34, 45, 58, 64
    –, emissions 138
    –, fertilisation 60, 61
  –, dissolved
    –, inorganic 77, 79, 80
    –, organic 77
    –, pool 77
  –, management 70
  –, monoxide 47
  –, oceanic reservoir 77
  –, particulate
    –, pool 77
  –, pool 64
    –, conservation of 64
    –, land 81
  –, sequestration 64, 68
    –, calculation 24
  –, sink 34, 81, 190, 195
    –, land 81
    –, missing 21
  –, trading 61
  –, transfer from land to ocean 81
carbonyl sulphide 94, 95
cave deposits (speleothems) 109
Census of Marine Life 181
cereal production 34–38
challenge, intellectual 151
chemical cycle 93
chemistry, cold 45
Chernobyl 51
Chile 180
China 22–24, 32, 34, 37, 40, 50, 53, 54, 67, 98, 108, 129–136, 178
  –, Atmospheric Pollution Control Law 54
chlorofluorocarbons (CFCs) 45, 47, 87, 89
circulation, thermohaline 107, 113, 118, 119, 147
Clean Development Mechanism (CDM) 61, 63, 64, 70, 194
clean technology 52, 54
clear cutting 21
climate
  –, change
    –, impact 35–38
    –, impact on cereal production 37
    –, scenario 28
  –, data set 33
  –, model 28, 35, 45, 123, 127, 145, 146, 148
  –, predictability 117
  –, regime 91
  –, regulation 93, 95
  –, system 45, 59, 84, 107, 113, 117, 118, 120, 123, 124, 131, 134, 145, 149, 153, 194, 204
  –, variability 77, 34, 80, 109, 111, 117–120, 145
    –, decadal 118
    –, interannual 117
    –, seasonal 117
  –, warming 45, 60

Climate Change Convention 194
cloud 85, 139, 162
  –, condensation nuclei (CCN) 139, 140
  –, feedback 149
  –, formation 139
coastal
  –, current 28
  –, ecosystem 103
  –, plain 101
  –, water 95
  –, zone 101, 102, 104, 113
Colombia 34, 42
Colorado River 21
Columbia River 98
combustion of fossil fuel 45, 57, 58
commerce, maritime 27
communication 31, 32, 120, 186, 204
community
  –, assembly 170
  –, forestry 180
concept of equity 41
Conference of the Parties to the UN Framework Convention on Climate Change (COP6) 60
Congo 34
convergence 129
cooperation, international 54, 183, 187, 188
Cooperative Programme for the Monitoring and Evaluation of the Long-Range Transmission of Air Pollutants in Europe (EMEP) 89, 90
coral 103, 104, 109, 162
  –, banded 109
  –, reef 57, 101
crop 32–34, 169, 178, 180
  –, modelling 33
  –, production 33, 34
  –, productivity 33
  –, yield 175
cropland 22, 23, 180
cropping
  –, multiple 34, 35
  –, strategy 22
cryosphere 113, 123, 157
  –, definition 113
cyanobacteria 161, 162
cycle
  –, biogeochemical 77, 207
  –, carbon 82
  –, chemical 93
  –, glacial 84
  –, hydrological 132, 169
  –, interacting cycles 95
  –, linkage between carbon nitrogen 82
  –, nitrogen 82
  –, nutrient 73, 138, 171
  –, oceanic carbon 77
  –, physical 93

D

Daihai Lake 136
dam
  –, construction 39
  –, effects on deltas and estuaries 97
  –, project
    –, benefits 41
    –, equity performance 41, 42
Danube River 99
data
  –, satellite 24
  –, vegetation cover 134, 135
decomposition 83, 162, 166
deforestation 21–26, 45, 47–51, 61–64, 81, 108, 124, 129, 130, 137–140, 189
  –, cryptic 22

-, model 24
-, tropical 21–25, 47, 51
   -, model 24
degradation pattern, dynamical 153
denudation, terrestrial 97
deposition
   -, acid 50, 52
   -, atmospheric processes 93
   -, nitrogen 60, 94, 95
desertification 23, 32, 123
detection method 24
deterministic process 169
developing country 32–38, 48, 57, 61, 63, 65, 69, 70, 91, 183, 186, 193, 195, 201, 205, 206
development
   -, sustainable 31, 38, 53, 64, 73, 140, 151, 156, 176, 187, 193, 195, 198–201
   -, trajectory 183, 187, 188
Diama Dam 39
diffusion 57, 77, 176, 180, 203
dimethylsulphide 94, 95, 162
discharge, standard 88
disease 21, 32, 42, 57
dissolved
   -, carbon pool 77
   -, inorganic
      -, carbon 77–80
      -, nitrogen 103
   -, organic carbon 77
   -, substrate 78
diversity, genetic 32
DMS 94, 95, 162
   -, oxidation 95
Dnieper, river 99
Don, river 99
drought 34, 39, 50, 107, 110, 118, 119
   -, El Niño Southern Oscillation (ENSO) 24
   -, episodic 107
   -, reconstruction 110
dry season 50

E

Earth Models of Intermediate Complexity (EMICS) 158
Earth Summit 51
Earth System
   -, complexity 152
   -, indeterminacy 152
   -, interactions 158
   -, nonlinearity 151
   -, processes 158
   -, science 104
   -, switch and choke points 153, 189
East Asia, land cover and use change 133
East China Sea 98
East Kalimantan 49
Eastern Canada 113
Eastern Europe 22
ECHAM4 model 35
ecological system 57
ecology, insect 108
ecosystem
   -, marine 27, 77, 171
      -, investigation 77
   -, terrestrial 58–60, 73, 82, 84, 133
      -, carbon uptake 59, 60
      -, investigation 77
eigenfunction analysis 158
El Niño Southern Oscillation (ENSO) 24, 28, 49, 50, 52, 60, 61, 79, 110, 117, 118, 120, 124, 147, 152
   -, drought 24
electricity 40–42, 68
   -, access to 40–42
embayment 101

Emiliania huxleyi 163
emissions
   -, carbon dioxide
      -, Amazonia 138
   -, particulate matter 48
   -, reduction 57, 66
   -, rights trading 61
   -, sulphur dioxide 48
   -, trading system 66, 70
energy 40, 45, 50, 54, 57, 63–68, 71, 77, 79, 89, 91, 111–113, 130–139, 162, 167, 179–190
   -, balance 113
   -, demand 57, 65, 67, 71
   -, efficiency 54, 66, 67
   -, flow 186
   -, inefficient production 50
   -, primary e. need 50
   -, renewable 68
   -, use
      -, inefficient 50
      -, per unit of GDP 67
environmental
   -, factor 77
   -, fluctuation 171
   -, policy 53, 184
   -, problem 87, 113, 183, 184, 188, 191
   -, local 183
equilibrium climax vegetation 133
equipment standard 89
equity performance of dam project 41, 42
erosion 32, 97, 193
eruption, volcanic 81, 109, 147
estuary 97, 101, 103
euphotic zone 77
Europe 22, 34, 48, 49, 52, 67, 71, 89, 145–147, 169, 170, 189
European
   -, grasslands 169
   -, Round Table 70
   -, Union 31, 51, 68, 197, 205
evaporation 28, 115, 129–131, 161
evapotranspiration 129–132, 138
event, extreme 34, 102
evolution tree 155
extinction of species 32
extreme event 34, 102

F

factor, environmental 77
farming
   -, smart f. 22
   -, system 185
farmland
   -, sparing 177
   -, transitioning 22
feedback loop 152
fertilisation
   -, carbon dioxide 61
   -, nitrogen 60, 61, 83
   -, of plants 81
fertiliser 32, 103, 177, 178
field experiment 48
fire 49–53, 60, 108, 139, 140, 166
   -, Indonesia 49
fish
   -, farming 177
   -, population 27
fishery 27, 28, 57, 101, 178
   -, capture f. 27, 178
   -, management strategy 29
   -, Peruvian anchovetta f. 28
fishing 27, 29, 39, 42, 94, 98, 103, 175, 177, 193
   -, activity 39
   -, pressure 27

fixation, nitrogen 73, 79
flaring 66, 67
flood
    -, artificial 42
    -, control 34
fluctuation, environmental 171
food 27, 28, 31–34, 37, 38, 41, 49, 64, 101, 104, 162, 165, 166, 169, 177–179, 183–189, 201, 207
    -, index 178, 179
    -, production system 32
    -, security 27, 32, 33
foraminifera 162
forest 21–24, 32–34, 47, 49, 50, 60, 61, 64, 81, 104, 108, 135, 138, 139, 148, 165, 171, 175, 179–182, 190, 194, 195
    -, access 24
    -, age structure change 60
    -, boreal 22
    -, degradation 193
    -, ecosystem 32–34
        -, cultivation potential 34
    -, fire, prevention and mitigation 52
    -, fragmentation 24, 139, 140
    -, protection 33
    -, regrowth 81
    -, sparing 179
    -, temperate 22
forestry 39, 57, 61, 97, 140, 179, 180, 190, 194
    -, community 180
fossil fuel 45, 57, 58, 63, 89, 194
    -, combustion 45, 57, 58, 138
framework
    -, political 54
    -, social 54
Framework Convention on Climate Change 58, 87, 193, 194
France 37, 42, 180
fresh water 40, 73, 97–99, 113, 118, 120, 137, 171
function, key f. 88

G

Ganges 22, 32
    -, Basin 22
gas
    -, dissolved 78
    -, halogen 94
    -, natural 54, 65, 67, 68, 71
general circulation model (GCM) 111, 145, 146, 148
genetic
    -, diversity 32
    -, revolution 31
Gephyrocapsa 163
Germany 37
glacial cycle 84
glacier 113
    -, melting 108
global
    -, carbon cycle 58, 77, 81, 84, 103
    -, ocean primary production 77
    -, temperature change 110
    -, warming 34, 35, 38, 49, 65, 113, 117, 119, 147, 153, 162
        -, impact in high latitudes 114
Global Climate Observing System (GCOS) 109, 158
Global Monitoring for Environment and Security (GMES) 158
Global Ocean Observing System (GOOS) 109
Global Palaeoclimate Observing System (GPOS) 158
Global Terrestrial Observing System (GTOS) 110
grassland 23, 60, 63, 165, 169, 171, 172
green
    -, product 184, 185
    -, revolution 31
greenhouse gas 35, 45, 48, 58, 60, 61, 66, 70, 77, 80, 84, 89, 91, 103, 107, 109, 110, 133, 135, 138, 145–147, 162, 191
    -, emissions 45, 57, 61, 70, 156, 207
Greenland 113, 119, 158

Gross Domestic Product (GDP) 37, 61, 67, 101
groundwater
    -, extraction 108
        -, subsidence 108
    -, recharge 39
Guatemala 41
Gulf
    -, of Alaska 98
    -, of Bohai 108
    -, Stream 28, 146

H

habitat, loss 27
halogen gas 94
halogenated organic compound 47
Harare 41
harvesting-regeneration cycle 61
Hawaii Ocean Time-series programme (HOT) 79, 80
haze 49, 51–53, 55
health, human 22, 32, 57, 198, 207
heat
    -, latent 113, 129–131
        -, flux 131
    -, sensible h. flux 131
    -, surface flux 145
herding 39, 177
Holland 65, 108
Holocene optimum 126
Homo sapiens 21, 45
hot spot
    -, land-use change 123
    -, vegetation 125
Huang Ho River 21
human
    -, health 22, 32, 57, 198, 207
    -, settlement 57, 181
    -, society 27, 107, 169, 172
humidity 136, 145–148
hunger 31, 38, 198
hybrid poplar 180
hydrocarbons 47, 71
hydrogen sulphide 99
hydrological cycle 132, 169
hydrosphere 123, 151, 157
hydroxyl
    -, concentration 47
    -, radical 47, 139

I

ice 21, 28, 83–85, 94, 107–115, 127, 147, 148, 152, 158
    -, body 85
    -, cap 109, 110, 113
    -, core 107, 109, 111
    -, lake i. 113
    -, river i. 113
    -, Vostok core 107
indeterminacy 152
India 23, 31, 32, 34, 37, 40, 48, 110, 179
Indian
    -, Ocean 48, 94, 120, 129
    -, Subcontinent 129
Indonesia 21, 34, 41, 49–53, 103, 118
    -, fire 49
industrial
    -, age 83
    -, era 84
    -, production 22
    -, transformation 185–188
    -, research 186
industrialisation 50, 111, 133, 140
industry organisation 70
information revolution 31

infrared climate warming forcing 45
insect ecology 108
integrated
 –, assessment
  –, approach 152
  –, model 155
 –, land science 21, 25
Integrated Global Observing Strategy (IGOS) 158
interaction, biogeophysical 125
international
 –, carbon trading 61
 –, cooperation 54, 183, 187, 188
 –, laws on transboundary pollution 51
International Biodiversity Programme DIVERSITAS 207
International Climate Change Partnership 70
International Geosphere-Biosphere Programme (IGBP) 23, 25, 77, 85, 95, 153, 172, 207, 208
International Human Dimensions Programme (IHDP) 23, 25, 87, 187, 188, 207, 208
International Satellite Land Surface Climatology Project (ISLSCP) 134
International Tropical Timber Organization 52
intertropical convergence zone 48
iron 79, 94, 95, 177
irrigation 21, 22, 32–35, 40–42, 178
 –, access to 40
isoprene 47

J

Japan 37, 40, 42, 50, 54, 134, 205
Johannesburg Summit on Sustainable Development 205
Joint Global Ocean Flux Study (JGOFS) 77–80

K

Kalimantan 49, 52
Kappa statistic 25
Kariba Dam 42
Kedung Ombo Dam 41
key
 –, function 88
 –, nutrient 81, 93
 –, phytoplankton growth 93
knowledge 21, 31, 37, 45, 52, 54, 68, 69, 73, 79, 87–89, 94, 101, 138, 140, 152–159, 161, 172, 195, 198, 201, 208
 –, share 52
Korean Peninsula 50
krill 28
Kyoto Protocol 37, 59–64, 87, 89, 91, 194, 195, 205

L

lake
 –, ice 113
 –, sediment 109
land
 –, agricultural 22, 32
 –, arid 23
 –, carbon sink 81
 –, change 21
  –, detection 24
  –, modelling 24
 –, clearing 50
 –, cover, change 123, 133
 –, cultivated 22
 –, cultivation potential 34
 –, evaluation method 33
 –, resource 32, 33, 41
 –, surface transformation 73
 –, system, productivity 189
 –, use 23, 33, 114, 115, 124, 133, 138–140, 165, 178, 180, 181, 194
  –, change 21, 25, 123, 133
Land Use and Cover Change Project (LUCC) 104

Land-Ocean Interactions in the Coastal Zone Project (LOICZ) 98, 104
Large Scale Biosphere-Atmosphere Experiment (LBA) 138–140
latent heat 113, 129–131
Latin America 24
leaf area index 134, 135
Lena River 131
life, marine 101, 175, 181
Lisutu 41
lithosphere 151, 157
Little Ice Age 111
livestock 39, 83
logging 22, 50, 138–140, 175, 177, 180, 181
land-use, land-use change and forestry (LULUCF) 61, 63, 64
 –, sequestered carbon 64
lumber 179

M

MacLean, Paul 175
maglev 175, 181, 182
Maldives 48
management, adaptive 156
Manantali Dam 39, 42
mangrove 28, 101–104
Maranhão State 41
marine
 –, bacteria 94
 –, ecosystem 27, 77, 171
 –, life 101, 175, 181
 –, species 28
 –, system, production characteristics 27
maritime commerce 27
marsh 28
material flow 186
matter, particulate 48, 77, 78, 97
Mauna Loa 79
mechanism, behavioural 88, 89, 91
mega-biodiversity country 49
megacity 23, 107
megafauna 21
Meiyu frontal zone 130
mercury pollution 138
metacommunity 172
methane 47
 –, atmospheric concentration 45
 –, chemistry 70
 –, reaction with OH 47
methyl
 –, bromide 94
 –, iodide 94, 95
Mexico 34, 53, 180
mining 97, 138, 140, 152
mitigation, biological 61, 64
mobility 27, 69, 70, 175, 180–182
model
 –, atmosphere-vegetation 125, 126
 –, behavioural m. 25
 –, ECHAM4 35
 –, general circulation model (GCM) 111, 145, 146, 148
 –, integrated assessment 155
 –, qualitative 155
 –, structural m. 25
 –, tropical deforestation 24
modelling 21–24, 33, 52, 59, 62, 117, 118, 120, 123, 133, 145, 151
 –, land change 24
monitoring technique 52
monsoon
 –, Asian 129 ff
 –, North African 126
 –, summer m. 126, 135, 136
Montreal Protocol 87, 90, 91, 193, 194
 –, Multilateral Fund 90, 91
multiple cropping 34, 35
municipal law 53

N

natural
- , gas 54, 65, 67, 68, 71
- , sulphur emissions 45
- , vegetation cover 136, 137

navigation 40
New Zealand 52, 180
NGOs 51–54, 187
niche 169
Nile River 32, 97
nitrogen 25, 50, 60, 68, 73, 79–84, 90, 93–95, 98, 102, 103, 162, 166, 177
- , atmospheric deposition 94
- , biological fixation 83, 95
- , cycle 81, 82
  - , linkage to carbon cycle 82
- , deposition 60, 94, 95
- , dioxide emissions 50
- , dissolved inorganic 103
- , fertilisation 60, 61, 83
- , fixation 73, 79, 81, 83, 94

North African monsoon 126
North American Free Trade Agreement (NAFTA) 53
North Atlantic 78, 94, 107, 113, 118, 120, 127, 146, 147
- , Oscillation (NAO) 118, 119, 147

North East Asian Cooperation 53
North Korea 54
North Pacific 78, 80
North Sea 68, 94
Northern Hemisphere 22, 94, 126, 127
nucleic acid 83
nutrient 77, 79, 81, 93, 97, 98, 101, 140, 177, 180
- , cycling 73, 138, 171
- , key n. 81, 93

Nyaminyami 41

O

observation, importance 45
ocean
- , atmospheric inputs 93
- , carbon uptake 58
- , current 28, 85, 145, 152
- , deep circulation 107
- , emissions from 94
- , global primary production 77
- , physical structure 27
- , sediment 109
- , thermohaline circulation 113
- , warming 107

oceanography 79
oil palm production 50
oil spill 51, 191
Opuntia 165
organic
- , compound, halogenated 47
- , matter 77–80, 103

organisation, industry 70
Organisation pour la Mise en Valeur du Fleuve Sénégal (OMVS) 39
overexploitation 193
ozone 22, 45–47, 87–91, 95, 187, 199
- , -depleting substances 89
- , atmospheric concentration 45
- , hole 45, 47
- , tropospheric 22, 45

P

Pacific Decadal Oscillation 27, 118, 120
- , Index 27
Pakistan 40
palaeoclimatic archive 107, 109
palaeodata 107
Pará State 41

particulate
- , carbon, pool 77
- , matter 48, 77, 78, 97

pastureland 23
pavement sparing 180
Pearl River 23
pelagic
- , animal 77
- , species 28

permafrost 113–115, 131, 132
- , zone 115, 131

peroxy radical 47
pest 21, 32
pesticide 32
Pew Centre 70
Philippines 49, 103
phosphate 79, 177
phospholipids 83
phosphorus 81–84, 93, 98, 162, 177
- , cycle 81

photosynthesis 28, 34, 77, 83, 94, 161, 162
- , rate 83

physical
- , cycle 93
- , structure of the oceans 27

phytoplankton 28, 58, 77, 80, 93, 94, 98, 103, 161, 162, 177
- , cell 28

pink salmon 27
plant
- , benthic 103
- , fertilisation 81

Pleistocene 81, 107
plywood 50, 179
- , industry 50

Poland 111
polar bear 115
policy
- , economic 53
- , environmental 53, 184

pollution 23, 27, 32, 47–54, 87–91, 101, 103, 138, 183, 184, 189, 193
- , international laws 51
- , mercury 138
- , transboundary 49–54, 87, 91

Pollution Standard Index (PSI) 49
Pongolapoort Dam 42
population
- , Brazilian Amazonia 138
- , displacement 34
- , growth 32, 65, 101, 111
- , world p. 32, 47, 112, 129

potential vegetation 133–135
Potter, Paulus 175, 176, 181, 182
poverty 31, 32, 64, 138, 140, 195, 198, 205
- , reduction 31, 32

precipitation 27, 28, 45, 50, 102, 107, 110, 113, 114, 117–119, 127, 129–133, 136, 145–148, 165
- , acid 45, 50
- , African tropics 127
- , pattern 27, 28, 107, 118
- , Sudan 127

prediction limitations 45
primary
- , energy need 50
- , production 27, 77, 93, 98, 169–171

problem, environmental 87, 113, 183, 184, 188, 191
process
- , biological 77–80, 94, 148
- , deterministic 169
- , interact 73
- , sampling p. 170
- , stochastic 170

product, green 184, 185
production
- , cereal p. 34–38

-, clean 185
-, intensified 22
-, primary 27, 77, 93, 98, 169–171
proteins 83, 178
pump
   -, biological 58, 77, 84
   -, solubility p. 58
Punjab 40

Q

qualitative model 155

R

R&D system 197–199
radiative
   -, cooling 48
   -, forcing 48, 84, 95, 147
   -, heating 145, 147
rain, acid 49–55, 88–90
redundancy 166, 167
regime, atmospheric 89, 90
Regional Haze Action Plan 52
Regional Integrated Environmental Model System (RIEMS) 134
remote, climatic effects 124
renewable energy 68
reservoir 40, 70, 98
residence time, atmospheric 47
resource management
   -, agro-halio-pastoral 39
   -, water 32, 118
revolution
   -, genetic 31
   -, green 31
Rio plus 10 195
Rioni 99
river
   -, ice 113
   -, regulate 97
   -, runoff 28, 103
rotational cultivation 24
rRNA 83
Rubisco 83
runoff 28, 103, 113, 114, 123, 136
Russia 34, 115
Russian Federation 61, 205

S

safe water 40
Sahara 123, 125–128
Sahel Syndrome 153, 155
Sakarya 99
salinity 28, 32, 79, 113, 145
   -, change 28
   -, surface water 113
   -, tolerance level 28
sampling process 170
Sandoz Spill 51
Sarawak 49
sardine 28
satellite data 24
Scotland 66, 68
sea
   -, grass 101
      -, bed 28
      -, species 101
   -, ice 28, 113–115, 127
   -, level rise 28, 107, 113
   -, life, sparing 177
   -, salt 93, 95
      -, aerosol 95
   -, surface temperature 117

-, variation 110
sea floor 77, 162
seal 28, 115
sea water
   -, density 28
   -, stratification 28
SEC International Policy Dialogue 50
sediment 84, 97, 101, 103, 109, 110
   -, lake 109
   -, ocean 109
Senegal River 39, 40, 42
   -, valley 40, 42
sensible heat flux 131
sensitive steering system 90
Seran River 41
settlement, human 57, 181
shallow water 101, 103
Shanghai 108, 181
shellfish population 27
shelve 98, 99, 101
Siakobvu 41
Siberia 22, 114, 129–132
   -, forests 22
silica 81, 162, 177
silicate retention 99
Sindh province 40
sink
   -, permanence 63
   -, terrestrial 59, 60
      -, location 59
      -, origin 60
smart farming 22
snow 113
social
   -, framework 54
   -, practice 88, 89
society, human 27, 107, 169, 172
soft law 51
soil
   -, biology 73
   -, erosion 193
   -, fertility 32, 169
   -, moisture 123, 136
   -, structure 73
   -, toxicity 34
solar
   -, irradiation 48
   -, power 69
   -, radiation 48, 84, 85, 139
      -, ultraviolet 47
solidarity mechanism 42
solubility pump 58
South Africa 42, 118, 197
South Atlantic Convergence Zone (SACZ) 139
Southeast Asia 24, 101–104, 124, 189, 199
Southern Indian Ocean 94
Southern Ocean 28, 94, 120, 163
Spain 40, 182
species
   -, abundant 165
   -, distribution pattern 28
   -, extinction 32
   -, marine 28
   -, number 166
   -, pelagic 28
   -, richness 172
speleothems 109
St. Hubert 175, 177, 181, 182
steering system 88–91
stewardship, deterministic 156
stochastic process 170
subsidence 108
   -, groundwater extraction 108
substrate, dissolved 78

subtropics, importance in atmospheric chemistry 47
Sudan, precipitation 127
sulphate
– , aerosol 45, 48, 95, 146
– , particle 95
sulphur
– , dioxide emissions 48, 50
– , natural emissions 45
Sumatra 49, 52
summer monsoon 126, 135, 136
Sunda Shelf 21
surface
– , parameter 134, 135
– , pressure 145
– , roughness 129, 134, 135
– , temperature 57, 109, 110, 113, 117–119, 126, 146, 159
Surface Ocean Lower Atmosphere Study (SOLAS) 95, 96
sustainability science 21, 22, 25, 198–201
sustainable development 31, 38, 53, 64, 73, 140, 151, 156, 176, 187, 193–201
Swissmetro 181
syndrome analysis 153, 158
system
– , approach 157
– , ecological 57
– , farming 185
– , land, productivity 189
– , marine, production characteristics 27

T

Taiga 131, 132
Tarbela Dam 40, 42
technology 22, 31–33, 52, 54, 57, 65, 67, 69, 90, 175–178, 180, 182, 185, 186, 188–191, 194, 197–199
– , clean 52, 54, 191
– , share 52
temperate forest 22
temperature
– , average regional 107
– , global mean surface, prediction 159
– , global change 109, 110
– , sea surface 110, 117
– , surface 57, 109, 110, 113, 117–119, 126, 146, 159
terrestrial
– , biosphere 62, 81, 190
– , denudation 97
– , ecosystem 58–60, 73, 82, 84, 133
– , sink 59, 60
– , location 59
– , origin 60
– , system, carbon uptake 58
Texas 66
Thailand 49, 129, 130
thermohaline circulation 107, 113, 118, 119, 147
Third World Network of Scientific Organizations (TWNSO) 197
Three Gorges Dam 97, 98
Tianjin 108
Tibetan Plateau 129
timber production 50
Tocantins River 40
total vegetation fractional coverage 134
trajectory, development t. 183, 187, 188
transfer, biogeochemical 85
transport 21, 50–52, 58, 59, 68–71, 94, 97, 129, 136, 146, 148, 163, 180–189
– , atmospheric 52, 59
– , fuel 69, 71
trap, sediment 97
tree ring 109
Trichodesmium 91
triple P bottom line performance 187
tropical
– , Atlantic dipole mode 118
– , deforestation 21–25, 47, 51
– , estuaries 103

Tropical Ocean Global Atmosphere Programme 117
Tropics, importance in atmospheric chemistry 47
Tucuruí Dam 41

U

UN Framework Convention on Climate Change (UNFCCC) 58–60, 87, 90
United Kingdom 37
upwelling 28, 98, 118
– , system 28
urbanisation 23, 50, 133, 177
– , of coastlines 177
US Clean Air Act 51
USA 31, 34, 37, 40, 51–53, 59, 60, 66, 87, 97, 110, 119, 177–181, 197, 204, 205

V

vegetation
– , cover, data 134, 135
– , natural cover 136, 137
– , potential v. 133–135
volcanic eruption 81, 109, 147
Vostok ice core 8, 9, 107, 152

W

walrus 115
warming, global 34, 35, 38, 49, 65, 113, 117, 119, 147, 153, 162
– , impact in high latitudes 114
waste 193
water
– , access to 39
– , conservation 32, 33
– , cycle 81
– , cycling system 129
– , deep w., metabolic processes 77
– , pollution 32, 193
– , quality 183
– , quantity 183
– , resources 33, 39, 42, 57, 63, 64, 73, 137
– , management 32, 118
– , safe w. 40
– , salinity of surface w. 113
– , shallow 101, 103
– , vapour transport 129
watershed management 34
West Africa 39
wetland 33, 115, 133, 138
wind 28, 117, 140, 145, 146
– , field 27, 28
– , power 65, 68
World Bank 54, 197
World Business Council for Sustainable Development (WBCSD) 70
World Climate Research Programme (WCRP) 117, 207, 208
World Ocean Circulation Experiment (WOCE) 117
world population 32, 47, 112, 129
World Summit on Sustainable Development 197

Y

Yangtze
– , River 98, 108
– , valley 108
– , watershed 108
Yellow River 133

Z

zero emission
– , car 185
– , power plant 185
Zimbabwe 42, 118
zooplankton 80